Migration and Gender in the Developed World

The subject of migration has traditionally been analysed through the lens of economic factors. The importance of adopting a gender-sensitive perspective to academic work is now generally appreciated however, and *Migration and Gender in the Developed World* contains chapters from a diverse range of leading contributors to apply such a perspective to the study of migration in the countries of the developed world.

Each chapter demonstrates how migration is highly gendered, with the experiences of women and men often varying markedly in different migration situations. The volume covers a wide range of migration issues including: dual-career households, regional migration patterns, emigration from Ireland and Hong Kong, elderly migration, the migration decision-making process, and the costs and benefits attached to migration and draws out the importance of gender issues in each area. Approaching the subject from a variety of academic traditions including Geography, Sociology and Social Policy, the volume combines both quantitative analysis of factual data and qualitative analysis of interview material to demonstrate the importance of studying migration through gender-sensitive eyes.

Paul Boyle is a lecturer in Geography at the University of Leeds.
Keith Halfacree is a lecturer in Geography at the University of Wales, Swansea.

Routledge Research in Population and Migration
Series Editors Paul Boyle and Mike Parnwell

1 **Migration and Gender in the Developed World**
Edited by Paul Boyle and Keith Halfacree

Migration and Gender in the Developed World

Paul Boyle and Keith Halfacree

London and New York

First published 1999 by Routledge
11 New Fetter Lane, London EC4P 4EE

Simultaneously published in the USA and Canada
by Routledge
29 West 35th Street, New York, NY 10001

Typeset in Garamond 3 by Stephen Wright-Bouvier
of the Rainwater Consultancy, Faringdon, Oxfordshire.
Printed and bound in Great Britain by
Biddles Ltd, Guildford and King's Lynn

British Library Cataloguing in Publication Data
A catalogue record for this book is available
from the British Library

Library of Congress Cataloging-in-Publication Data

Migration and gender in the developed world / edited by Paul Boyle and
Keith Halfacree.
p. cm. — (Routledge research in population & migration)
Includes bibliographical references and index.
ISBN 0-415-17144-x
1. Population geography. 2. Feminist geography. 3. Sex role.
I. Boyle, P. J. II. Halfacree, Keith. III. Series.
HB1951.M515 1999
304'.6-dc21 98-44333
CIP

ISBN 0-415-17144-x

Contents

Tables and figures

Tables

Figures

1 Introduction

Gender and migration in developed countries

Keith Halfacree and Paul Boyle

Introduction

In a short paper presented to the British limited life Migration Research Network a few years ago, one of the current authors (Halfacree 1993) argued that the interface between migration and gender remained 'a neglected area of research'. This was in spite of the substantial leaps made in both feminist scholarship and the study of human migration over the previous couple of decades. Moreover, this neglected status of geography and gender was seen to be especially acute with respect to work undertaken in so-called developed nations, since much more recognition has been paid to gender-specific migration patterns and processes in the developing world (see Chant 1992; Lawson 1998).

Five years on, it behoves us to consider the current state of play. On the one hand, it is clear that much progress has been made. This is evidenced in the key role given to gender in recent migration publications (for example, Boyle, Halfacree and Robinson 1998; King, Connell and White 1995) and in the high profile of gender issues in the grants from which the research for many of the chapters in this collection are taken. None the less, as will also become clear from the contributions given here, there is still much to do before we have satisfactorily answered the questions raised within these chapters. Gender and migration in developed countries is a rich area of ongoing research.

There are two main aims of this introductory chapter. In the latter half we provide a flavour of each chapter's subject, approach and principal findings, set in the context of the overall ordering of the book. Initially, though, we describe the current state of play in research on the gendering of human migration, contextualizing this work in the context of a mapping of the evolution of feminist scholarship within geography.

Gender recognition in migration research

Within migration research in general, an economic focus on labour migration has been and in many respects remains predominant from studies within the neo-classical tradition to those more rooted in radical interpretations. This concern to understand the factors responsible for either encouraging or discouraging the movement of economically active people between or within countries is apparent from various migration overviews (cf. Boyle, Halfacree and Robinson 1998; Clark 1986; Greenwood 1985; Massey 1990; Molho 1986). Clearly, there are good reasons for this bias, notably the central role played by the (capitalist) economy within all of our lives and, more specifically, the questions raised by such indicators of the imperfections within the labour market as persistent geographical inequalities in the distribution of unemployment. As we shall see below, this economic bias is equally apparent in studies of gender and migration. However, from such recognition we can begin to appreciate some of the reasons for the masculinism apparent within migration research. For example, the implied institutional nexus of the 'traditional' nuclear family in which 'rational economic man' (*sic*) is embedded allows little space for migration to be interpreted in ways other than as a (male) response to employment factors. Indeed, the public/private dichotomy (Bondi 1992) of husband, sphere of production and waged labour versus wife, sphere of reproduction and domestic labour has been a powerful organizing principle behind migration research, whether this is stated explicitly or, as more often, implicitly, as has been the case within almost all areas of social science (Rose 1993). Such a dualism is now, of course, increasingly being undermined both at the ideological level from within feminism and at the more material level, reflected most simply in the increasing engagement of women with waged labour. Such is the sense of change and evolution within which studies of the interface between gender and migration must be located.

Recognizing women in the data: a first indication of the importance of gender?

> Females are more migratory than males within the Kingdom of their birth, but males more frequently venture beyond.
>
> (Ravenstein 1885: 199)

Perhaps the first stage in recognizing the potential significance of gender to our understanding of human migration was the acknowledgement of differences of migration behaviour by sex within migration data sets. Thus, data sets obtained from censuses, migration records, specialist surveys, and so on tend these days to be routinely disaggregated by sex as well as by other key categories, notably class and age. Moreover, such a breakdown is not just concerned with questions of representativeness but forms a part of their description and analysis.

Two recent studies of British migration demonstrate this recognition. Using data from the 1987 Labour Force Survey, Owen and Green (1992) found males to be slightly more mobile than females, especially within regions as compared to between regions of the country. Halfacree, Flowerdew and Johnson (1992) obtained a similar finding from a slightly more detailed analysis of data from a specially commissioned survey. Men showed slightly higher migration rates over a one-year time period than women, a greater association than women with moves between towns within the same county, and a lesser association than women with moves within the same town.

Studies such as these can be used to modify generalizations such as Ravenstein's 'law'. However, the overwhelming impression stated in both of the studies cited was that the category 'sex' was a poor discriminant of migration behaviour, especially when compared to other characteristics such as class, age, housing tenure, and so on. Thus, taking the implications of this finding, gender[1] seems a relatively unimportant variable in the migration equation. However, is this a correct impression to take away? On the one hand, it may be, since there are good reasons for us to expect to find similarities in the migration behaviour of men and women. In particular, many people move as part of a family unit, typically with a male and female partner. On the other hand, however, the apparent unimportance shown by gender in these studies is misleading. First, there may be distinctive gender-specific migration patterns for certain population subgroups, such as single young adults (Boyle and Halfacree 1995), which are not evident when aggregate groupings of males and females are considered. Second, and developing the last point, the seeming unimportance of gender is often based on measurable, quantifiable outcomes alone – the 'end of the line' – neglecting the processes which go into constituting the migration decision, which may not have empirically gendered results but which themselves may be highly gendered. Thus, we turn to a more social perspective and to the migration decision-making process, notably as it affects dual-career households.

Dual-career households: negotiation and compromise

> To date . . . a rather disturbing feature of most of this work [on labour migration] has been the implicit assumption that household moves are made within the context of a single career.
>
> (Snaith 1990: 155)

As one would expect, recent trends within the character of the labour force have stimulated much of the ongoing interest in the gendering of labour migration. In particular, much work has focused on the dual-career household, defined as a household containing two principal adults in which both partners are engaged in occupations with a distinctive and progressive career path based on 'commitment' to the job. They thus differ, to a greater or lesser extent, from dual-income households (Bonney 1988), where both partners are in paid employment but where either or both are in jobs with relatively indistinct career paths. Dual-career households, in short, tend to comprise 'professional couples'.

A number of inter-linked factors are responsible for the ongoing rise in the number of dual-career households or, more specifically, the rise in the number of women in professional and managerial jobs (Bonney 1988; Green 1995, 1997). These include:

- demographic restructuring: including, low and/or declining female fertility rates and less time away from paid work for childrearing;
- economic restructuring: including, increased participation by women in waged labour, the growth of service class occupations at the expense of the traditionally very male-dominated blue-collar sector, and the decline of 'organizational careers' with their employer-driven migrations; in short, the feminization of the labour market (Green 1994; McDowell 1991) or the rise of a less patriarchal 'post-Fordist' division of labour (Esping-Anderson 1993);
- socio-cultural restructuring: including the rise of more 'egalitarian' households (Kiernan 1992), where both partners are career-minded and share domestic tasks.

Consequently, women are increasingly developing careers and, where partnered by someone (usually a man) similarly occupationally orientated, we have the growth of the dual-career household.

In spite of recent interest shown in these dual-career households in western Europe (for example, Brun and Fagnani 1994; Camstra 1994; Domsch and Kruger-Basener 1993) and a longer pedigree of research in North America (for example, Holmstrom 1973; Hunt and Hunt 1977; Hertz 1987; Rapoport and Rapoport 1976), Green (1995; also Champion 1992) identifies a lack of a strong geographical focus within such work. The migration decision-making processes within these households have been poorly detailed, although they are the subject of ongoing work by Green and others (for example, Dudleston et al. 1995; Green 1997).

Decision making in dual-career households has been defined by Hertz (1987) as involving three perspectives: those of 'her career', 'his career' and 'their career' (since the relationship between the two people within a household must not be neglected). Fundamentally, with respect to migration, this challenges the predominant model of migration decision making within couples, which tends to prioritize the demands of the husband's career. The latter gave rise to the notion of the woman moving as a 'trailing wife' (Bonney and Love 1991; Bielby and Bielby 1992; Bruegel 1996; Clark 1986) – 'married' to her husband's job (Finch 1983) – as a 'tied migrant', a 'constrained migrant', or a 'secondary migrant' (the very number of these terms suggesting how prevalent the idea is). Unsurprisingly perhaps, it has been shown that such migration typically has negative consequences for married women's employment (for example, Long 1974; Morrison and Lichter 1988; Rives and West 1992; Shihadeh 1991). Thus, on marriage, it was observed that there was a tendency for wives' careers to become 'secondary' (for example, Dex 1987; Finch 1983; Kanter 1977).

Of course, awareness of the uneven geography of employment opportunities (discussed below) and the gendered character of this geography means that migration for a husband's career could be beneficial for the wife's career also. This has been implied in a number of studies (for example, Bonney and Love 1991; Cooke and Bailey 1996). Notwithstanding such complicating factors, instead of the trailing wife set-up within a dual-career household, we have a much more egalitarian decision-making scenario. This is because the benefits that may accrue to one partner's career through a migration must now be weighed up against the damage that this might do to the other career; we might expect it to be 'a question of compromise' (Green 1997).

The constraints of adopting a dual-career structure have migration implications other than just being linked to reduced levels of household migration (Abercrombie and Urry 1983).[2] Many of these implications develop and sharpen lessons learnt and questions raised in research on other labour migration situations. For example, the question of the ability of different places to satisfy the career demands of both partners has demonstrated the need to acknowledge how the geography of opportunity can shape migration even for single people. Within Britain, both the urban environment and the South East region of England seem especially conducive in this respect (Green 1995). Particularly significant here have been studies which highlight how certain areas of a country are conducive to career development,[3] such as the work by Fielding on the South East of England as an 'escalator region'. In a series of papers (for example, Fielding 1989, 1992a), Fielding has developed the idea that the South East region acts as a 'machine for upward social mobility' (Savage et al. 1992: 182), attracting young, highly qualified individuals at the start of their careers and then transporting them rapidly up the escalator of occupational success. The power of this machine can be explained largely through reference to the occupational geography of the region (see also Hanson and Pratt 1995; Ward and Dale 1992). Moreover, as Fielding went on to emphasize in his work with Susan Halford (Fielding and Halford 1993; also Boyle and Halfacree 1995), the escalator had interesting gendered aspects. In particular, migration flows into the region were especially advantageous to women, with there being an 'excess' of single women in London (Hall, Ogden and Hill 1998). Attention must then be given to the ability of London to provide these opportunities (see also Lelièvre and Bonvalet 1994 on Paris) – from the perspective of both the labour market supply of jobs and the environmental qualities of the area which may make it more supportive to women's everyday lives, at least in respect of them developing a career (Duncan 1991).

Regarding labour market factors, the geography of opportunity (Galster and Killen 1995) must also be recognized in the context of the types of jobs pursued by men and women, since many occupations remain strongly sex-typed (Bradley 1989; Reskin and Hartmann 1986; Reskin and Roos 1990). Thus, whilst many 'female' part-time jobs tend to be quite geographically ubiquitous, being especially concentrated in the service sector, their full-time occupations often show a much stronger geography (Ward and Dale 1992).

None the less, Green (1997) also suggests that one of the reasons why wives in dual-career households compromise more with respect to residential location is because their jobs are typically less geographically constrained.

As regards an area's other qualities, the gentrification of many inner-city neighbourhoods has been given a strong gender-sensitive interpretation (for example, Bondi 1991; Butler and Hamnett 1994; Rose 1989; Warde 1991), notably in respect of the supposedly more 'tolerant' ambience (Rose 1989: 131) that gentrified neighbourhoods provide for women. However, this process, too, is open to discussion, as Green (1997) found a strong attraction for accessible rural locations amongst dual-career households in the English Midlands.

The dual-career arrangement is also likely to necessitate the renegotiation of domestic responsibilities within the household (England 1991), since we must not

> ignore the household roles husbands and wives occupy, the gender-role beliefs they subscribe to regarding these roles, and the effect of these beliefs on both the process and outcome of couples' decision making.
>
> (Bielby and Bielby 1992: 1,245)

This draws attention to the need to investigate the sex roles adopted by each spouse, since Finch's (1983) idea of a wife being 'married' to her husband's job meant much more than just her willingness to move to benefit his career. Thus, issues such as the division of labour within the home must be considered (Pinch and Storey 1992). In a dual-career arrangement, any assumption of a 'tolerance of domestication' (Van Den Berghe 1970) on behalf of the wife must be severely challenged, given her role as a breadwinner (Bird and Bird 1985). Thus, we have classifications of sex roles, such as the 'supportive husband', willing to move for his wife's career (Houseknecht and Macke 1981; Rank 1982), the household with 'androgynous' gender identities (Hiller and Philliber 1982), or the spectrum of households between the 'unified' and the 'conflicting' (Bruegel 1996).

Recognizing patriarchy: migration and power

> Existing economic and social theories seem most apposite to the 'traditional' male breadwinner / female homemaker or dual earner households with a 'conventional' division of labour, where the man's career / job would be expected to take precedence over the woman's job (if she had one).
>
> (Green 1995: 33)

The work on dual-career households has drawn attention to the need to focus on the household as a decision-making unit in which questions of negotiation and role adoption feature strongly (Pahl and Pahl 1971). A 'supply-side'

(Smith 1997) analysis of the household as acting to encourage or constrain women's employment is required. This takes us into a more explicit consideration of the power relations within the household and a recognition of patriarchy, defined here as 'a system of social structures, and practices in which men dominate, oppress and exploit women' (Walby 1989: 214). Paramount here are relations of patriarchy within the workplace, supported by those within the domestic sphere. Developing this emphasis, we can analyse the way in which migration becomes gendered from what has been termed, respectively, 'external' and 'internal' perspectives (Halfacree 1995).

External perspectives focus on the sex-typing of waged labour (mentioned earlier) and the presence of sex discrimination both within the workplace and within society more generally. For example, Walby (1986) has demonstrated how 'patriarchy at work' has been increasing within British society, where segregatory practices have come to predominate over cruder practices of exclusion. She recognizes (Walby 1990) a rise of 'public patriarchy', or the collective exploitation of women within the workplace. Likewise, Acker (1990) talks of the 'gendered substructure' of workplace organizations.

Internal perspectives, by contrast, focus on the household as the prime site of patriarchal structuring. They no longer assume that the sex roles adopted by each spouse can be explained simply through the cold logic of the 'allocative efficiency' (Ermisch 1993) of tasks within the household. Human capital approaches to migration research (for example, Mincer 1978; Sandell 1977) saw any gendering of migration as being the unintended consequence of the rational (income-maximizing) household-level decision making: 'investment increasing the productivity of human resources, an investment which has costs and which also renders returns' (Sjaastad 1962: 83). Thus, for example, a wife might be a tied migrant because the net benefits to the household of moving for her husband's work outweighed the net gains of either staying put or moving for her work (see, for example, Long 1974). In contrast, the critique of this approach argues that the sex roles adopted within the decision-making processes, plus their effects in terms of migration, may be regarded as being highly patriarchal; it is *from within this context* that any 'rational' gendering of labour migration is formulated.

An example of the way external and internal perspectives come together is given by Nelson's (1986) study of employers in San Francisco. Some of them located to the city's suburbs to recruit white, college educated, married women who were 'family orientated', thus helping to perpetuate patriarchal relations both in the waged workplace and in the home. A similar gender engineering of employment opportunities has been shown by the comprehensive study of Worcester, Massachusetts, by Hanson and Pratt (1995). This study focused on the sustained sex-typing of occupations within this city. Denying that this segregation could be explained solely through reference to gendered variations in human capital assets, Hanson and Pratt told some 'stories of containment', whereby households, employers, residential location, and so on, all fed into creating strongly place-related landscapes of employment.

Recognition of patriarchy helps us to understand why, even within supposedly dual-career households, migration is still led much less by the demands of the woman's career than by those of the man (Green 1995). However, it also speaks of the need to look beyond just employment considerations in explaining the detailed playing out of a migration, with a broader gendering of migration being fundamentally important.

Beyond separate spheres: the biographical structuration of migration

> . . . sex inequality in the home and workplace serves to attenuate the link between geographic and social mobility, which in turn perpetuates existing sex inequalities.
>
> (Morrison and Lichter 1988: 171)

The 'internal' versus 'external' dualism may be useful in organizing our analysis of the gendering of migration (for example, Halfacree 1995) but it does tend to oversimplify the pervasiveness of the patriarchal system – Walby (1989) sees the latter reflected in a mode of production, capitalist waged work, the state, male violence, sexuality, and culture – and the reproduction of gender inequality within everyday life. This broadening of the field of interest can be considered from the perspectives of a biographical approach to the study of migration and, hence, to its structuration within everyday life.

A 'biographical approach' (Halfacree and Boyle 1993; Courgeau and Lelièvre 1992; Green 1997) moves away from the commonplace assumption that migration is stress-induced, stimulated almost purely by particular events and circumstances. Instead, in line with Shotter's (1984) emphasis on social accountability, migration is recognized both as a responsible action and as one that occurs within the messy 'hurly-burly' of everyday life. This perspective avoids the 'intellectual fallacy', part of the 'theoretical purification of practical orders' (Thrift 1996: 5) whereby human agents are treated as acting rationally according to a series of abstracted and decontextualized discrete prompts. A biographical approach takes a more longitudinal perspective (Green 1995; Warnes 1992), entangling any migration decision – to a greater or lesser extent – within the multiple concerns that any one individual (person or household) has at any time (for example, see Gutting 1996). Migration exists within what Giddens (1984) terms the 'practical consciousness', the level of thought which lies between unconscious decision making, which is unacknowledged by the participant, and discursive decision making, which includes that which is 'actively thought about'.

Together with a biographical perspective on the gendering of migration, we can also link the internal and external perspectives noted above through talking of this migration's structuration (after Giddens 1984). Here, the relationship between structure and agency is seen as a duality: 'agency produces structure produces agency produces structure in a never-ending recursive process' (Thrift 1985: 612). For gendered labour migration:

> Turning to the labour migration of a 'typical' . . . family, the various actors – principally the wife, husband and employers – draw upon the structures of patriarchy, represented by the exploitative position of the wife in both the home and waged workplace, in order to constitute the migration. However, this action, which tends to result in the wife being a secondary migrant, also serves to reproduce the 'original' patriarchal structures . . . through which the migration was constituted. In the domestic sphere, the wife's role and status as a co-provider is undermined and her labour market marginalization as a support for her husband is enhanced. In the sphere of waged labour, the secondary migration serves to legitimate the sex-typing of occupations by reinforcing both employers' and fellow male workers' perceptions of women as 'uncommitted'. Such lack of commitment also helps to undermine women's struggles against patriarchal oppression in both the home and the waged workplace. In sum, the structures of patriarchy, principally but not exclusively in the domestic and waged labour spheres, are both the medium and outcome of the gendering of 'labour migration'.
>
> (Halfacree 1995: 172–3)

In addition to the roles of the home and the waged workplace summarized above, the state, male violence, sexuality and culture in general may also serve to gender human migration. For example, what are the effects of state policies in maintaining 'traditional' gender divisions of labour (Morris 1990)? Attention could be given to the ways in which laws affecting security of employment have gendered effects in, say, not relating to part-time employment (women being especially concentrated into part-time work), thereby undermining women's career strategies. Smith (1997) notes how in Britain the social security system discourages women's employment; for example, through an implied expectation that they will earn little, with household benefits soon cut drastically if they earn very much (Morris and Irwin 1992). Elsewhere, Hanson and Pratt (1995: 62) argued, in respect of employment in Worcester, Massachusetts, how 'availability of benefits perfectly mirrors wage pattern'. From the perspective of culture – which can often provide an integrating lens through which to study the multidimensionality of migration (see Boyle, Halfacree and Robinson 1998: Ch. 9) – migration's ability to act as an escape route (Fielding 1992b) from oppressive patriarchal societies can be explored. For example, Walter (1991) has examined Irish women's emigration from such a perspective, with several authors (for example, Beale 1986; Nash 1993; Walter 1995) recognizing the highly cultural subordination of women within Irish society.

It is important to note here that we do not consider the type of qualitative techniques which are immediately associated with the biographical approach to migration to be the *only* way in which the subject is to be addressed. Instead, it should be possible to include both individual and structural level variables in modelling techniques. For example, Davies (1991; Davies and Pickles

1985) has incorporated a biographical angle in his modelling work on 'housing and migration careers'. Moreover, structuration may be a fundamentally dynamic process but it is mediated through 'social systems', the 'patterning of social relations across time-space, understood as reproduced practices' (Giddens 1984: 377). Thus, besides measuring attributes of the individual, satisfactorily theorized elements of the institutional structure can also be incorporated into modelling techniques.

Rethinking migration: is 'labour migration' a masculinist concept?

> The issue of modern working parents is, essentially, how to relegate the place of work in our lives. We need to reclaim private life and not just nuclear family time as something distinct and different from paid work . . . [not] rejecting the workplace but rather as a rejection of the place that work currently occupies in all our lives.
>
> (Benn 1998: 4)

In this final subsection we consider briefly the proposition suggested in the literature but not really researched in depth that 'labour migration' is not simply a category of human migration but is in practical terms a highly masculinist concept, its contemporary character relying upon unequal gender relations. This perspective can perhaps be best approached through examining the idea that the dual-career household provides a model for degendering labour migration (Halfacree 1995).

At one level, broadening and deepening the base of the dual-career household away from young professional couples might be expected to promote more 'supportive' husbands and 'androgynous' gender identities, thereby undermining the basis of the structuration of patriarchy through labour migration. However, at another level, this model can appear rather naïve, with so much of the dominant gender orders being sustained (Bondi 1993). These include:

- Daily domestic responsibilities: whilst recent years have seen a growing acceptance of the idea of wives having 'careers', they are still largely responsible for the domestic sphere (Bradley 1989), with women adapting their paid employment to accommodate traditional gender divisions/sex roles within the home (for example, Pinch and Storey 1992). From such failures in household equality we have the myth of the 'superwoman', juggling domestic responsibilities with those of a high-powered professional career (Benn 1998). Hence, the resurgence of waged domestic labour (Gregson and Lowe 1994).
- Breaks from employment: even when they have professional jobs, wives are still more likely than their husbands to take breaks from paid employment. Typically, of course, this is for childrearing purposes. Such breaks are likely to be harmful to an employment-orientated career

(Beechey and Perkins 1987), which may be undermined further if migration to a new labour market area takes place during this break.

- External entrenchment: we must be wary of concentrating solely on the way in which households are adjusting to the dual-career model, as this can overlook the entrenched patriarchal relations and attitudes identified from the external perspective discussed earlier. Major shifts in society as a whole are required to enable any household-led trends towards the dual-career household to flourish. Even such seemingly mundane issues as the adequacy of nursery provision still has far to go (for example, Benn 1998; Smith 1997).

In addition, from distinguishing dual-career from dual-income households, Bonney (1988: 91) notes how 'trends towards gender equality in the labour market may be contributing towards increasing labour market and social class inequalities *between* [households]' (our emphasis; see also Gregg and Wadsworth 1994).

From considerations such as these it is possible to suggest that exploring the gendered character of (labour) migration opens up a critique of taken-for-granted social systems such as labour migration. We are required to ask profound questions, such as those concerned with the meaning of work and the family (Hunt and Hunt 1982). Indeed, as the feminist scholar Hartmann argued nearly two decades ago, 'we will have to find ways to change both society-wide institutions and our most deeply ingrained habits' (Hartmann 1979: 232).

Adopting such a perspective in many ways requires us to rework the conclusions drawn from much of the research on the gendering of migration to date. In particular – and as a biographical approach implies – assessment of migration as being 'positive' or 'negative' for wives, for example, cannot just consider their labour market experiences. Instead, the fuller meaning of migration to each partner needs to be considered before it can be assessed. Monroe et al. (1985) observed spousal inconsistency in decision making which, when applied to migration, might suggest that each spouse seeks different things from a move. On the one hand, of course, this might lead one to conclude that this is just a pragmatic response to the reality of the patriarchal structures moulding women's (and men's) life chances. However, on the other hand, it does at least suggest a less deterministic 'different view' on migration as a socio-cultural practice. Thus, whilst in Jordan, Redley and James's (1994) study of family relations in South West England women lost out in battles over long-distance migration, the women acceded to this loss for the longer-term benefits it could bring them. In short, is it a hegemonically masculinist view to evaluate migration primarily through the lens of careers and the labour market when women may be more likely to take into account family commitments (Green 1994, 1995), notably the upbringing of children or the care of relatives (Green 1997), and other more 'grounded' (Marshall 1988) factors when explaining the rationale for moving or staying? Is it more than an internalized tolerance of domestication which makes tied migrants often appear happy with their status (for example, Bonney and Love 1991)?

At least questioning the normative status of career-related labour migration through the eyes of a gender-sensitive perspective leads to a revalorization of other types of migration. To some extent, it takes us away from a focus on 'work-rich' households to incorporate more of the experiences of the 'work-poor' (Gregg and Wadsworth 1994). There is not the space to outline these in detail, but it is again the case that these other migration strands have seen little work addressing their gendered nature. Included amongst these other migrations are those for:

- Quality of life. Of particular note here are moves to rural areas, where the recognized gender character of the 'rural idyll' (for example, Hughes 1997; Little 1997) suggests a need for further work on the gendering of rural restructuring (Agg and Phillips 1998). For example, given the often limited employment opportunities available in rural areas, we can appreciate how residential location decisions can feed back into restricting women's job prospects (see also Hanson and Pratt 1995).
- Retirement. There is a growing interest in the migration of the elderly (for example, Bean et al. 1994; Rogers et al. 1992; Warnes 1996), but little explicit focus has been given to gender issues, although evidence does suggest that such moves tend to be male-led (Law and Warnes 1982). In addition, with the majority of the population aged 60 years and over being female (Grundy 1996), we might wish to ask how gender relations interact with migration behaviour for this group.
- Ill health and caring. Migration associated with ill health may also have a gendered dimension, not least if men and women are seen to be able to cope differentially with the challenges posed. With the rise in the number of elderly people, such migrations are likely to be increasingly commonplace (for example, Grundy 1993).
- Other life-course transitions. The break-up of a marriage is understandably linked to migration, with men showing especially high rates (Hayes, Al-Hamad and Geddes 1995), as is remarriage, this time especially for women (Grundy 1985). More generally, life-course-related moves in early life are commonplace (Grundy and Fox 1985). In all of these transitions there is considerable scope for work on their gendered character to be explored further.

Gender, migration and the development of feminist scholarship

> ... the purpose of feminist geographers ... is to challenge the very nature and construction of that body of knowledge that is designated academic geography.
>
> (McDowell 1993a: 157)

An alternative look at the research reviewed above on gender and migration can be obtained through locating this work with respect to a 'map' of the evolution of feminist scholarship within geography. Such a perspective is useful, not only because it enables us to order work to date but also because it suggests something of the future directions that work on gender and migration in the developed world might take.

Clearly, there are a number of ways in which the contribution of feminism within geography might be ordered. Here, we draw primarily upon the structure suggested by McDowell – after Harding (1986) and Di Stefano (1990) – in two recent review articles (McDowell 1993a, 1993b). McDowell recognizes three broad families of feminist geography, although it is vital from the outset that we do not see these as being either mutually exclusive or as forming a clear chronology. The three categories of 'pedagogical convenience' (McDowell 1993b: 305) are feminist empiricism, standpoint feminism and postmodern feminism.

Feminist empiricism focuses on painting women (back) into the picture through an emphasis on showing the place of gender within social relations. This rationalist perspective stresses the 'unfairness' of excluding women from analysis (or, at best, assuming their experiences to be the same as those of men), advocating the implementation of similarly rational policies to counter any disadvantage experienced by women. Such work is by far the most common type of feminist geography scholarship; overall, the 'predominant emphasis has been on social relationships and activities as they vary across space' (McDowell 1993a: 162). Most of this work concentrates on gender experiences within paid employment and the home/waged work spatial split, the latter illuminating the so-called 'private sphere'.

One of the central weaknesses of feminist empiricism is its reliance on 'male' indicators to illustrate inequality. This critique can take us into the standpoint feminist approach, whereby greater attention is given to gender symbolism, the construction of gendered identities and the gendered nature of knowledge construction.[4] Moreover, it rejects the assumption of feminist empiricism that the aim should be to make both sexes' experiences the same, implicitly through making women's experiences like those of men. Instead, whilst of course attacking discrimination, it celebrates difference and valorizes women's perspectives – their standpoints.

Celebrating difference blends into the third category of feminist scholarship, that of postmodern feminism. Here, the singular category of 'woman' is subject to critical deconstruction. Ultimately, the salience of gender as an analytical category is questioned, with attention given to the construction of individual identities. Whilst such an approach can develop into the extreme relativism implied in much postmodern scholarship, this is not essential. Thus McDowell (1993b: 312), following Haraway (1991; see also Massey 1991), argues for 'embodiment as a node in a set of fields variously structured by social forces ranging from the global scale to the most intimate'. 'Gender', therefore, can assume various levels of significance related to time, space, activity, individuality, and so on. Some of this varied significance comes through from adopting a life-course perspective.

Using McDowell's mapping to interrogate work on gender and migration, it is clear that most falls into the feminist empiricism category – 'a tradition of intense empirical research through which the importance of women as actors in processes of economic, political and social change has been demonstrated' (Bondi 1993: 241). This work – as with feminist empiricism in general – has become increasingly sophisticated and underpinned by theory. Early studies were very much of the 'add women and stir' variety, whereby 'gender, if included at all, was merely another variable' (McDowell 1993a: 161), that of 'sex'. The limitations of this rather tokenistic work were soon acknowledged in studies of the gendering of labour migration, where attention moved from just considering the human capital 'spreadsheet rationality' of these moves for a household to outlining how gendered economic and social relations were produced and reproduced in the space economy and within the home, through developing concepts such as the 'tied migrant' and Finch's (1983) evocative notion of wives being 'married' to their husbands' jobs. The embeddedness of such unequal relations was challenged through analysis of the dual-career household, where it seemed that there had been *some* movement in women's experiences of employment, migration and domestic work which were bringing them closer to the experiences of their husbands. Barriers to such seemingly progressive trends were, moreover, to be explained through the legacy of patriarchy; the structuration of the latter coming through the sex-typing of occupations (each with their specific geographies), gender relations within the household (including issues of childcare responsibilities), discrimination in the workplace, and so on. Thus, we saw the 'emphasis on politically committed, critical, and place-based research' (Nast 1994: 56) which typifies contemporary feminist scholarship.

Increasingly, however, various authors have recognized that the degendering of migration raises other questions besides the need to counter and turn back the consequences of patriarchy as experienced by women. The influence of standpoint feminism has become increasingly apparent within studies of gender and migration. This is reflected not least in the questioning of the ability of the dual-career household to be seen as any kind of blueprint for a more egalitarian, non-discriminatory division of labour. Thus questions are asked, such as are the stresses and strains associated with the 'spiralist' career trajectory to be advocated for women as well as men? Moreover, is this even *possible*, if 'the land-use patterns of western cities reflect the patriarchal assumptions of an earlier era' (McDowell 1993a: 166), with all that this implies for the practice of everyday life? The construction of gender differences is also a central theme of much gender and migration work: for example, through the idea of the structuration of patriarchy or the contrasting meanings of 'migration' – based on experiences of moving – to husbands-as-careerists and wives-as-homemakers, for example.

From the work on the construction of these differences we begin to develop an alternative normative projection to that of women's behaviour coming to mirror that of men. With a concern to situate migration in a biographical context, we see that the seemingly clear-cut rational idea that migration is a simple, functional, instrumental act associated with 'income maximization' (however complexly defined) is overly reductionist. Other stories of migration come through, associated with people's priorities beyond the wage or salary, such as 'quality of life' issues and the desire to bring up one's family as qualitatively benignly as possible. Crucially, these latter issues are more often associated with a feminist perspective on migration (and non-migration). Particularly interesting here is recognition of gender within the concept of 'diaspora' (for example, Clifford 1994), where the importance attached to settlement and to links between communities exposes the masculinism inherent in studies of just the 'public' activity of the migration itself.

From the broader perspective introduced by standpoint feminism, we also see a more normative 'political' project. Issues such as spousal inconsistency in migration decision making and many women's seemingly willing acceptance of 'tied migrant' status need not necessarily be seen solely as being ideologically inscribed by patriarchy. Instead, perhaps these women's standpoints suggest an alternative future, where labour migration is not seen as some eternal, benevolent, almost day-to-day concept? Clearly, these are suggestions that we may wish to run with.

Most weakly developed within studies of gender and migration has been the postmodern feminist approach. This can be explained partly by the relative lack of work overall and, indeed, it is probably wise that academic interest has not sought to deconstruct gender in the context of migration before its significance at varying degrees of generalization for 'men' and 'women' has been demonstrated and detailed. Nevertheless, there are developments within migration research which are in line with some of the tenets of postmodern feminism. For example, there is increasing recognition of the sheer diversity of women's experiences regarding migration, explained not just in terms of conventional categories, such as class and age, but also acknowledging culture, personality and individuality. More generally, talk of the structuration of gendered migration speaks, on the one hand, of the construction of gendered identities associated with standpoint feminism but, on the other hand, of the variable significance of the socially constructed category 'gender' within migration. Furthermore, the biographical approach to migration research suggests much future potential work in this area, with emphasis being given to the highly contextual (in the non-trivial sense) development of actions such as migration.

Book outline

As has been emphasized above, migration and gender intersect in a wide variety of ways. This is reflected in this book in the diversity of issues covered, and by the various approaches used to study the precise topic of interest. As regards heterogeneous methodologies and theoretical roots, an interest in recognizing the value of studying migration from contrasting perspectives reflects in part a postmodern sensitivity to difference, diversity and context within population studies (Graham 1995). Choosing the 'theory and methodology to fit the task at hand' (Boyle, Halfacree and Robinson 1998: 82) needs to be done reflexively and critically but helps us to gain a more rounded picture of the gendering of migration than a more specialized collection which concentrates on one perspective, be that multilevel modelling, behavioural studies, qualitative research, and so on. For a subject still relatively under-researched by academics, we hope also that an emphasis on diversity will suggest as wide a range as possible for avenues of future research.

None the less, in spite of this call for a degree of eclecticism, the chapters in this book do reflect a number of biases. First, as already noted, migration tends to be analyzed most commonly through the lens of economic factors, and this is also the case here. Thus, after a couple of chapters that set the scene for the book as a whole, we have a large section that is concerned with exploring the gendered character of the costs and benefits of migration. This is, of course, a key material issue and, as argued earlier, is fundamentally enmeshed within the patriarchal structures of contemporary society. Second, there is also a bias within the book in favour of empirical studies rather than theoretical outlines. This reflects both the predominant research interests of many of the contributors but also, from the point of view of the editors, a desire to illustrate the practical implications of the gendering of migration. Furthermore, with the 'data feast' (Champion 1992: 224) presented to migration researchers through data sets such as the British Census's Longitudinal Study and the (postmodern) importance increasingly attached to local statistics (Fotheringham 1997), there is still the need for often quite preliminary empirical work to take place. Third, although the book purports to cover the 'developed world', the majority of the chapters refer to British material, although the United States and Ireland also feature quite strongly. In many respects this reflects the origins of this book within a British conference. Whilst such a bias is not ideal for geographical representativeness, it does allow a further clarification of the range of gendered experiences of migration within the same broad geographical context.

From the perspective of McDowell's mapping of feminist geography, the majority of the contributions here, as suggested in the previous paragraph, fall into the feminist empiricism category, although some of the critical reflections on things such as the 'costs' to women of migration could be said to be engaging with feminist standpoint theory, albeit largely implicitly.

There is little in this book that would fit under the postmodern feminism category and, as noted earlier, we think that this is probably quite positive in that it seems worth while to build up a feminist scholarship on migration before the postmodernists attempt to pick it apart! None the less, again as already noted, we are keen to encourage the postmodern valorization of mixed methodologies and difference in the collection.

The chapters in the book can be read together or separately as we have deliberately tried to keep them as self-contained as possible, avoiding cross-referencing. Overall, we feel that the collection gives a sound overview of the current state of play within research on gender and migration within developed countries, although there is clearly still much academic work to be done.

Chapter outlines

An important further sense of context for the book is set in chapter 2, where the value of applying a geographical perspective to the more established body of work on gender and social mobility is demonstrated. **Tony Fielding** and **Susan Halford** develop previous work using the British Census's Longitudinal Study (Fielding and Halford 1993) to examine gender-specific social and spatial mobilities within England and Wales over the 1981–91 period. This is approached first from a national perspective before the importance of space is drawn out. Nationally, a degree of social mobility for employed women is evident through a huge increase in professional employment and a degree of feminizing of both managerial and blue-collar jobs. However, a regional analysis is critical for exploring this conclusion further, since there is a complex geographical picture. Rates of upward mobility, in particular, are strongly regionally specific, with the South East standing out as a region of high rates. This geography is highly gendered, with women's movement into the managerial rather than the professional sector within the South East being of key note, although in general the regional geography of upward mobility is most marked for men. Finally, migration is a critical process for 'explaining' these patterns. A focus on migrants to the South East shows a strong association between women and social movement into the service class, including management, whilst out-migration for women is strongly linked to leaving the labour market.

The diversity of migration flows suggested by Fielding and Halford is explored further for thirty-seven British local authority districts through the modelling of 1991 census data by **David Atkins** and **Stewart Fotheringham** in chapter 3. Utilizing a 'competing destinations model' which is sensitive to the uneven spatial distribution of different potential locations for the migrant, single men and women are shown to act fairly similarly at the aggregate level. However, slightly more discriminatory analysis suggests single women are more deterred by distance, less attracted to high unemployment areas and more attracted to places with larger populations than single men. Looking at each parameter within the overall models separately also reveals a number of complex patterns of destination selection by gender, which suggest many future lines of potential research. The attractiveness of larger destinations to single female migrants is, however, reinforced.

Thus far, the book has concentrated on drawing out gendered patterns of migration and a next stage is to begin to explain these in more detail. The importance of employment factors has already been made clear and this becomes the centre of interest for the next few chapters. First, in chapter 4, after outlining the gendered character of the changing structure of employment, **Anne Green, Irene Hardill** and **Stephen Munn** provide an excellent summary of gender differentials with respect to the employment consequences of migration. On the one hand, it can be argued that migration for employment reasons has a highly gendered character, being almost always economically beneficial for the male partner in a household but often much less so for the female partner. This inequality is reflected in the concept of the 'trailing wife', outlined earlier. On the other hand, changes in attitudes towards women in employment, possibly reflected in and reflecting the increasing proportion of the workforce who are women, may be said to be contributing to a degendering of migration and employment. The complexity of the emerging picture is illustrated by Green and her colleagues through reference to their work on the in-migration of dual-career households to the English East Midlands. Whilst most employment 'adjustment' fell on the women, this need not have been negative for their careers, and the potential employment consequences for both partners must be considered. Outcomes are related to the employment structure of the migration destination and to the particular difficulties faced by migrants to more rural areas.

As already observed, the area's employment structure is critical to explaining the rather unique position of the South East of England, and this region is focused on again in chapter 5, where **Irene Bruegel** uses Tony Fielding's concept of the South East as an 'escalator' for upward social mobility. In particular, she is concerned to ask to what extent this escalator is gender-specific. A number of factors qualify the impression that the South East is inherently beneficial for women's careers. First, upward social mobility amongst migrants is much more associated with single women than with those in partnered situations. Second, and perhaps of most importance, we must note the selectivity of migration into the South East, with its bias in favour of the young, the well educated and the higher social classes. Third, we must always be careful in recognizing how definitions of different classes and so on can be gender-biased and how this can affect our conclusions. None the less, Bruegel ends on a more positive note with her suggestion of a growth in cross-class households, where the woman has the 'leading' role with respect to the migration decision.

In contrast to an emphasis on the costs of migration to partnered women, **Thomas Cooke** and **Adrian Bailey** have argued elsewhere (Cooke and Bailey 1996) that such migration can be beneficial to married women's careers. This issue is taken up and developed further in chapter 6, with its investigation of 'self-selection bias' as it applies to gendered migration, an issue also suggested by Bruegel's chapter. By self-selection bias, the authors argue that women who are more likely to migrate are more likely to be employed if they do migrate, whilst women who are more likely to stay are more likely to be employed if they do stay. Whilst there is a danger of circularity of argument here, self-selection could help explain why the authors' previous work suggested married women's migration led to an increase in their probability of employment in the destination, even though patriarchal relations may still influence the overall migration decision. Testing this issue, using probit models of Public Use Microdata Sample data for married women from the 1980 United States census, finds little evidence to support the hypothesis that women's employment was enhanced through migration. None the less, the idea of self-selection implies that the married woman's economic 'contribution' may be considered prior to the migration decision.

In chapter 7 **Paul Boyle, Keith Halfacree** and **Darren Smith** examine further the issue of the costs and benefits of migration to partnered women, using data from the 1991 British Census Sample of Anonymized Records. They argue that too little quantitative work has attempted to account for family characteristics, focusing instead on individual-level explanatory variables in the analyses. A consequence of previous studies is that few have made the effort to identify linked partners who move together and, as a result, the negative effect of family migration on women's employment characteristics has been underestimated. They also demonstrate that a number of explanatory variables, derived from the characteristics of linked partners, make a significant impact on model results. Studies that ignore these are prone to bias.

The mediation of migration behaviour due to entering into marriage is a central issue in chapter 8. **Norman Bonney, Alison McCleery** and **Emma Forster** deploy Becker's concept of 'commitment' to note the constraints on migration which affect individuals through their life course. The central importance of inertia for promoting (non-)migration behaviour is recognized through, *inter alia*, commitment to place. Bonney and his colleagues use British Census Sample of Anonymized Records data for 1991 to follow Cooke and Bailey in questioning the idea that the migration which occurs when such inertia is overcome tends to be disadvantageous in career terms for married women. Whilst recognizing that commitments such as marriage and having children depress inter-regional migration for both sexes, they suggest that relatively few women experience career disruption through their partner's migration. This is because most of these migrating women – as opposed to those younger and single – have already left established full-time career structures, notably through having children after marriage. The concept of commitment also draws attention to a household member's priorities other than employment status. For example, whilst from a short-term perspective female tied movers might be seen as losers through migration, this loss may be much less significant in the context of their longer-term goals within the household. Overall, therefore, the costs and benefits of migration must be seen within the context of the balancing acts that go on in daily life linked to one's varied commitments.

This emphasis on migration's place within the contexts of our daily lives – what we have termed elsewhere (Halfacree and Boyle 1993) migration's biographies – promotes research into the decision-making processes which give rise to migration. This issue is addressed head-on in chapter 9, where **Jenny Seavers** explores the joint decision-making process of migration for married couples based on sixteen detailed case studies from lowland England. The decision-making process is mapped using a 'decision plan net' which illustrates the range of issues involved in any one move. In terms of the overall dominance of seven of the decisions involved, three times as many moves are classified as 'female-dominated' as are 'male-dominated'. However, the gendering of dominance is shown to vary with the phase of the move, as illustrated by detailed examples taken from the fieldwork. Whilst the impetus for a move and the choice of area – notably job-related – may often be male-dominated, the overall decision to move typically brings the female partner firmly into the decision-making process. Again, therefore, the importance of multiple goals for almost any move should be recognized. Seavers then goes on to develop a conceptual framework for considering joint decision making in housing migration decisions.

Emigration, especially to a culture very different from that which one was brought up within, is likely to stimulate vigorous activity within Seavers's conceptual framework. Indeed, this does seem to be the case from evidence presented in chapter 10, where **Lin Li** and **Allan Findlay** outline the role of women in the emigration of professional couples from Hong Kong. Information from qualitative interviews is used to see whether – to draw on an often-stated Chinese saying – the women were 'following the chicken' (their husbands!) in their moves, as the patriarchal structure of traditional Chinese society might lead us to expect, or whether more individualistic reasons for emigration were given by the women. At first sight, the importance of the woman's career seemed clear, with the wives in many households applying for the work visas to come to Britain and Canada. However, this impression was frequently misleading, as this 'lead' was often made for strategic reasons concerning the likely success of obtaining a visa (the wives had more 'bankable' jobs). Instead, women's careers were in many instances played down, often through cultural constructions of gender characteristics, although once again the details of any migration could often be influenced by the priorities of both partners. In summary, for this group of migrants at least, traditional cultural values still underpin the migration decision-making process, even if hybrid modern-traditional identities are emerging.

The importance of culture in mediating the migration decision-making process was emphasized in Li and Findlay's chapter, and this broadens out the coverage of the book from primarily economic considerations. This expansion is taken further in the remaining chapters, starting with chapter 11. Here,

Ray Hall, Philip Ogden and **Catherine Hill** make use of Longitudinal Study data for 1981–91 and other sources to examine gender relations in the characteristics of working-age migrants living alone in England and Wales. As over a quarter of households in England and Wales in 1991 are one-person, and with migration being strongly associated with the life-course transition to such a status, this is clearly an important group to research. For some characteristics there are only minor gender differences, but there was a tendency for the single women migrants to be older (often associated with widowhood), to be biased in favour of skilled non-manual workers amongst non-professionals, and for them to stress housing reasons (reiterating the importance of non-job-related migration). As was suggested in earlier chapters, (Inner) London stood out as something of an unusual place, where migration (often long-distance) to living alone was linked clearly to a rise in social class for young women as well as for young men. This reinforces Bruegel's conclusion in her chapter on the selectivity of the urban escalator for women.

The movement into inner cities by younger professionals, recognized as something of a distinctive flow by Hall and her colleagues, has contributed to the gentrification of many city centres within the developed world. The gendered character of this process is considered by **Liz Bondi** in chapter 12, based on qualitative research in two areas of Edinburgh in Scotland, both associated with single people of both sexes getting mortgages. Bondi uses the metaphor of journey to express the process of gentrification, since this draws attention to the experiences, continuing change and personal narratives of those involved. The chapter explodes a few of the myths of gentrification. For example, not all of those interviewed were young professionals who had developed a taste for inner-city living as students. Instead, they were those seeking home ownership on a low budget. Also, the environment in the two areas was not especially supportive of 'non-traditional' households, although the predominance of 'modern' middle class views did lend the environment a 'pro-woman' character. The gentrification move itself was often regarded as a 'staging post' on the way to a suburban parenthood, with strong life-course associations between suburbs and having children coming through. However, we must be wary of generalizations since the diversity of the two areas' 'gentrifiers' supports Bondi's initial metaphor, with inner-city Edinburgh as a place-as-network (Massey 1991) intersected by many often-contrasting journeys.

The migration of younger adults is again considered in chapter 13, but here it is the out-migration rather than the in-migration process that forms the focus. **Catriona Ní Laoire** considers gendered processes of subordination with respect to qualitative work on youth migration from rural north Cork in Ireland. A useful starting point is to observe the gender relations within Irish society. A visibly hegemonic masculinity – reflected for example in a sports culture and a localized economic marginalization – promotes the higher rates of out-migration of the better-educated females, especially from rural areas such as north Cork. Personal experiences express the shock and strains apparent for both those who leave and those who stay, yet migration is seen as being especially empowering for rural women. Moreover, this gendering of migration is becoming more and more embedded, as the out-migration of young educated women reinforces the masculinization of Irish rural identity. In contrast, the twin discourses of migration as exile and migration as opportunity, which are both largely male, place the 'home' of rural Ireland as a feminine place. Unsurprisingly, the women's perspective on migration tends to be more complex than either of these twin discourses given, not least, their experiences of living in rural Ireland.

Martin Phillips in chapter 14 also takes up the intersection of constructs of rurality with the book's central themes of gender and migration. Starting with the viewpoint that service-class restructuring explanations of counterurbanization have been rather gender-blind, Phillips uses material gathered in five villages in the English Midlands to discuss the 'gender order of Middle England'. First, he shows how the service class itself in the villages is highly gendered, with a highly feminized 'service proletariat' class distinguished from those with more professional and managerial occupations. Furthermore, a gender order (after Connell 1987) is apparent in reproductive, economic and leisure activities, coming across in his interviews, if not without a degree of contestation. Turning to the place of migration, Phillips shows the female service proletariat to be associated with secondary migration; thus migration can be said to help structure the observed gender order, whilst a patriarchal gender order is an ingredient in moulding the migrations themselves (see also Halfacree 1995). Finally, Phillips reiterates how dominant cultural meanings of rurality and their material representations within the village (the landscapes of leisure, economies, and so on) serve to reinforce patriarchal relations within rural England.

The last few chapters have emphasized the association between key life-course transitions, often associated with changes in commitment, and certain types of gendered migration, and this issue is the focus of chapter 15. Here, in what is part of a much larger project, **Lynn Hayes** and **Alaa Al-Hamad** concern themselves with the migration behaviour of divorced men and women, using data from the 1991 British Census Sample of Anonymized Records. Recoding the data set, they identify six categories of divorced migrant, defined with respect to whether they move individually and/or whether they move to be with someone. There are some key gender differences in the migration behaviour of these differentially constrained groups, particularly associated with women's childcare commitments. For example, divorced men are biased in favour of the 'move alone' groups, whilst divorced women more often move with someone, usually a child; women are more associated with moves into social housing; and returning to live with parents was linked to lone men and to women with children. The evidence also suggests that divorced male migrants form relationships more frequently, and are thus less likely to be living alone or just with children. Consequently, there is something of an excess of older divorced women. Clearly, adopting a gender-sensitive perspective helps to undermine the saliency of describing any singular migration tendencies for divorced individuals.

The issue of age, a key variable in migration research, is considered directly in the next two chapters, which focus on the widely observed secondary peak – or 'late age slope' – in migration frequency associated with the elderly. First, in chapter 16, **Emily Grundy** and **Karen Glaser** again utilize the British Census Longitudinal Study to show elderly migration to be linked strongly to a change in household type, notably a move on one's own (widowhood) or to enter a non-private household. This relationship was especially strong for the most elderly and for women. Indeed, women made up 80 per cent of migrants aged 75 years or over. Clearly, this reflects women's longer life expectancies but it suggests how women will experience to a greater extent things such as the often negative economic consequences of widowhood. Data from the 1980s also provided evidence for an increase in 'granny farming', whereby frail elderly women became increasingly likely to move into a non-private institutional household rather than into a different type of private household. Finally, just looking at migration rates for 1991 showed very high rates of migration for both men and women changing their household type where long-standing illness was present, again pointing to a growing trend of institutionalization.

Some of the processes underpinning the differential rates of elderly migration are considered in chapter 17, where **Anthony Warnes** reports on findings from census data and a survey of the elderly in Britain. The increasing differentiation of the elderly generates dynamic residential requirements. Warnes identifies a 'gender inversion', where migration linked to bereavement increases for women as they get older, overtaking the rate for men which tends towards zero. Health decline is also a critical reason for women's migrations when they get very aged. Other factors prompting migration are also gendered, with men showing an association with locational factors of distance and women with the stresses of keeping up the house and garden. Finally, the importance of other people in encouraging the migration of the elderly is clear, which perhaps could also be gendered in respect of the different societal expectations and abilities invested in men and women.

The diverse chapters in this collection illustrate a multiplicity of ways in which migration in the developed world is highly gendered. Something of the legacy of such patterning is expressed in chapter 18, which employs the integrating style of 'cultures of migration' (Boyle, Halfacree and Robinson 1998) to provide a fitting conclusion to the book. In this chapter, **Bronwen Walter** explores some aspects of the often quite contrasting experiences of Irish women in Britain and the United States. As in Ní Laoire's contribution, these women are regarded as a vital yet often largely hidden half of an Irish diaspora. Moreover, concentration on just the migration process itself is strongly masculinist and neglects the importance of women in producing the diasporic identity. Hence, Walter shows the central role played by women in maintaining the ties with the homeland (Ireland) which help to define a diaspora, but also their central role in creating new 'roots'. Yet, from written evidence and other representations, these women and their roles are often absent or gratuitously stereotyped. Invisibility is particularly acute in Britain, with its much stronger legacy of anti-Irish racism. However, as Walter makes clear in this example – but which is of more general significance – the stories the women have to tell – whether through statistics, surveys, literary sources, narratives, and so on – add so much vital cultural richness to any understanding of the biographies of migration within the developed world.

Notes

1 Of course, gender and sex are not equivalent terms but the measured sex differences have clear implications for any conclusions that may be drawn regarding gender differences.
2 One of these implications, stemming from the need to locate somewhere where both partners' employment demands are met, is a rise in the number of dual-location households (Green 1997).
3 One possible criticism to make of this attempt to 'take geography on board' concerns the lack of attention that it has given to scale. In particular, the work of Fielding and others tends to focus on the (administrative) region, and the appropriateness of this scale perhaps needs some further consideration.
4 Associated with this work (and also with postmodern feminism) are considerations of masculinism inherent within 'fieldwork' (Nast 1994) and the issue of the possibility of 'feminist methods' (Moss 1993). Takes from these perspectives on the work on gender and migration are not considered here.

References

Abercrombie, N. and Urry, J. (1983) *Capital, Labour and the Middle Class*, London: Allen and Unwin.
Acker, J. (1990) 'Hierarchies, jobs, bodies: a theory of gendered organization', *Gender and Society* 5: 139–58.
Agg, J. and Phillips, M. (1998) 'Neglected gender dimensions of rural social restructuring', in P. Boyle and K. Halfacree (eds) *Migration into Rural Areas: Theories and Issues*, London: Wiley, pp. 252–79.
Beale, J. (1986) *Women in Ireland: Voices of Change*, Basingstoke: Macmillan Education.
Bean, F., Myers, G., Angel, J. and Galle, O. (1994) 'Geographic concentration, migration and population redistribution among the elderly', in L. Martin and S. Preston (eds) *Demography of Aging*, Washington DC: Committee on Population, United States National Research Council, National Academy Press, pp. 279–318.
Beechey, V. and Perkins, T. (1987) *A Matter of Hours: Women, Part-time Work and the Labour Market*, Cambridge: Polity Press.
Benn, M. (1998) 'Father Time and mother courage', *Guardian* (second section) 28 May: 2–4.
Bielby, W. and Bielby, D. (1992) 'I will follow him: family ties, gender-role beliefs, and reluctance to relocate for a better job', *American Journal of Sociology* 97: 1241–67.
Bird, G. and Bird, G. (1985) 'Determinants of mobility in two-earner families: does the wife's income count?', *Journal of Marriage and the Family* 47: 753–58.
Bondi, L. (1991) 'Gender divisions and gentrification: a critique', *Transactions of the Institute of British Geographers* 16: 190–98.

Bondi, L. (1992) 'Gender and dichotomy', *Progress in Human Geography* 16: 98–104.

Bondi, L. (1993) 'Gender and geography: crossing boundaries', *Progress in Human Geography* 17: 241–46.

Bonney, N. (1988) 'Dual-earning couples: trends of change in Britain', *Work, Employment and Society* 2: 89–102.

Bonney, N. and Love, J. (1991) 'Gender and migration: geographical mobility and the wife's sacrifice', *Sociological Review* 39: 335–48.

Boyle, P. and Halfacree, K. (1995) 'Service class migration in England and Wales 1980–1: identifying gender-specific mobility patterns', *Regional Studies* 29: 43–57.

Boyle, P., Halfacree, K. and Robinson, V. (1998) *Exploring Contemporary Migration*, Harlow: Addison, Wesley, Longman.

Bradley, H. (1989) *Men's Work, Women's Work*, Cambridge: Polity Press.

Bruegel, I. (1996) 'The trailing wife: a declining breed? Careers, geographical mobility and household conflict in Britain, 1970–89', in R. Crompton, D. Gallie and K. Purcell (eds) *Changing Forms of Employment*, London: Routledge, pp. 235–58.

Brun, J. and Fagnani, J. (1994) 'Lifestyles and locational choices trade-offs and compromises: a case-study of middle-class couples living in the Ile-de-France region', *Urban Studies* 31: 921–34.

Butler, T. and Hamnett, C. (1994) 'Gentrification, class and gender: some comments on Warde's "'gentrification of consumption"', *Environment and Planning D: Society and Space* 12: 477–93.

Camstra, R. (1994) 'The geodemography of gender: spatial behaviour of working women', *Tijdschrift voor Economische en Sociale Geografie* 85: 434–45.

Champion, A. (1992) 'Migration in Britain: research challenges and prospects', in A. Champion and A. Fielding (eds) *Migration Processes and Patterns. Volume 1. Research Progress and Prospects*, London: Belhaven, pp. 215–26.

Chant, S. (ed.) (1992) *Gender and Migration in Developing Countries*, London: Belhaven.

Clark, W. (1986) *Human Migration*, New York: Sage.

Clifford, J. (1994) 'Diasporas', *Cultural Anthropology* 9: 302–38.

Connell, R. (1987) *Gender and Power: Society, the Person and Sexual Politics*, Cambridge: Polity Press.

Cooke, T. and Bailey, A. (1996) 'Family migration and the employment of married women and men', *Economic Geography* 72: 38–48.

Courgeau, D. and Lelièvre, E. (1992) *Event History Analysis in Demography*, Oxford: Clarendon Press.

Davies, R. (1991) 'The analysis of housing and migration careers', in J. Stillwell and P. Congdon (eds) *Migration Models: Macro and Micro Approaches*, London: Belhaven, pp. 207–27.

Davies, R. and Pickles, A. (1985) 'The longitudinal analysis of housing careers', *Journal of Regional Science* 25: 85–101.

Dex, S. (1987) *Women's Occupational Mobility: a Lifetime Perspective*, Basingstoke: Macmillan.

Di Stefano, C. (1990) 'Dilemmas of difference: feminism, modernity and postmodernism', in L. Nicholson (ed.) *Feminism/Postmodernism*, London: Routledge, pp. 63–82.

Domsch, M. and Kruger-Basener, M. (1993) 'Personalplanung und entwicklung für dual career couples (DDCs)', in L. von Rosenstiel, E. Regnet and M. Domsch (eds) *Führung von Mitarbeitern. Handbuch für erfolgreiches Personalmanagement 2*, Stuttgart: Wege zur Chancengleichheit, pp. 469–80.

Dudleston, A., Hardill, I., Green, A. and Owen, D. (1995) 'Work rich households: case study evidence on decision making and career compromises amongst dual-career households in the East Midlands', *East Midlands Economic Review* 4: 15–32.

Duncan, S. (1991) 'The geography of gender divisions of labour in Britain', *Transactions of the Institute of British Geographers* 16: 420–39.

England, K. (1991) 'Gender relations and the spatial structure of the city', *Geoforum* 22: 135–47.

Ermisch, J. (1993) '*Familia oeconomica*: a survey of the economics of the family', *Scottish Journal of Political Economy* 40: 353–74.

Esping-Andersen, G. (1993) *Changing Classes: Stratification and Mobility in Post-industrial Societies*, London: Sage.

Fielding, A. (1989) 'Inter-regional migration and social change: a study of South East England based upon data from the Longitudinal Study', *Transactions of the Institute of British Geographers*, 14: 24–36.

Fielding, A. (1992a) 'Migration and social mobility: South East England as an "escalator" region', *Regional Studies* 26: 1–15.

Fielding, A. (1992b) 'Migration and culture', in A. Champion and A. Fielding (eds) *Migration Process and Patterns., Volume I. Research Progress and Prospects*, London: Belhaven, pp. 201–12.

Fielding, A. and Halford, S. (1993) 'Geographies of opportunity: a regional analysis of gender-specific social and spatial mobilities in England and Wales 1971–1981', *Environment and Planning A* 25: 1421–40.

Finch, J. (1983) *Married to the Job: Wives' Incorporation into Men's Work*, London: Allen and Unwin.

Fotheringham, A. (1997) 'Trends in quantitative methods I: stressing the local', *Progress in Human Geography* 21: 88–96.

Galster, G. and Killen, S. (1995) 'The geography of metropolitan opportunity: a reconnaissance and conceptual framework', *Housing Policy Debate* 6: 7–43.

Giddens, A. (1984) *The Constitution of Society*, Cambridge: Polity Press.

Graham, E. (1995) 'Population geography and post-modernism', Paper presented at the International Conference on Population Geography, University of Dundee, September.

Green, A. (1994) 'The geography of changing female economic activity rates: issues and implications for policy and methodology', *Regional Studies* 28: 633–39.

Green, A. (1995) 'The geography of dual career households: a research agenda and selected evidence from secondary data sources for Britain', *International Journal of Population Geography* 1: 29–50.

Green, A. (1997) 'A question of compromise? Case study evidence on the location and mobility strategies of dual career households', *Regional Studies* 31: 641–57.

Greenwood, M. (1985) 'Human migration: theory, models, and empirical studies', *Journal of Regional Science* 25: 521–44.

Gregg, P. and Wadsworth, J. (1994) 'More work in fewer households', *NIESR Discussion Paper* 72, London: NIESR.

Gregson, N. and Lowe, M. (1994) *Servicing the Middle Classes*, London: Routledge.

Grundy, E. (1985) 'Divorce, widowhood, remarriage and geographic mobility among women', *Journal of Biosocial Science* 17: 415–35.

Grundy, E. (1993) 'Moves into supported private households among elderly people in England and Wales', *Environment and Planning A* 25: 1467–79.

Grundy, E. (1996) 'Population review: the population aged 60 and over', *Population Trends* 84: 14–20.

Grundy, E. and Fox A. (1985) 'Migration during early married life', *European Journal of Population* 1: 237–63.

Gutting, D. (1996) 'Narrative identity and residential history', *Area* 28: 482–90.

Halfacree, K. (1993) 'Migration and gender part 1: a neglected area of research?', Paper presented at a meeting of the Migration Research Network, University of Bristol, July.

Halfacree, K. (1995) 'Household migration and the structuration of patriarchy: evidence from the U.S.A.', *Progress in Human Geography* 19: 159–82.

Halfacree, K. and Boyle, P. (1993) 'The challenge facing migration research: the case for a biographical approach', *Progress in Human Geography* 17: 333–48.

Halfacree, K., Flowerdew, R. and Johnson, J. (1992) 'The characteristics of British migrants in the 1990s: evidence from a new survey', *Geographical Journal* 158: 157–69.

Hanson, S. and Pratt, G. (1995) *Gender, Work, and Space*, London: Routledge.

Haraway, D. (1991) 'Situated knowledges: the science question in feminism and the privilege of partial perspective', in D. Haraway (ed.) *Simians, Cyborgs and Women: the Reinvention of Nature*, London: Free Association Books, pp. 183–201.

Harding, S. (1986) *The Science Question in Feminism*, Ithaca: Cornell University Press.

Hartmann, H. (1979) 'Capitalism, patriarchy and job segregation by sex', in Z. Eisenstein (ed.) *Capitalist Patriarchy and the Case for Socialist Feminism*, New York: Monthly Review Press, pp. 206–47.

Hayes, L., Al-Hamad, A. and Geddes, A. (1995) 'Marriage, divorce and residential change: evidence from the Household Sample of Anonymised Records', *Migration, Kinship and Household Change Working Paper* 3, Department of Geography, Lancaster University.

Hertz, R. (1987) *More Equal than Others: Women and Men in Dual-career Marriages*, Berkeley: University of California Press.

Hiller, D. and Philliber, W. (1982) 'Predicting marital and career success among dual-worker couples', *Journal of Marriage and the Family* 44: 53–62.

Holmstrom, L. (1973) *The Two Career Family*, Cambridge, MA: Schenkman.

Houseknecht, S. and Macke, A. (1981) 'Combining marriage and career: the marital adjustment of professional women', *Journal of Marriage and the Family* 43: 651–61.

Hughes, A. (1997) 'Rurality and "cultures of womanhood"', in P. Cloke and J. Little (eds) *Contested Countryside Cultures: Otherness, Marginalisation and Rurality*, London: Routledge, pp. 123–37.

Hunt, J. and Hunt, L. (1977) 'Dilemmas and contradictions of status: the case of the dual-career family', *Social Problems* 24: 407–16.

Hunt, J. and Hunt, L. (1982) 'Dual-career families: vanguard of the future or residue of the past?', in J. Aldous (ed.) *Two Paychecks. Life in Dual-Earner Families*, Beverly Hills, CA: Sage, pp. 41–59.

Jordan, B., Redley, M. and James, S. (1994) *Putting the Family First: Identity, Decisions, Citizenship*, London: UCL Press.

Kanter, R. (1977) *Men and Women of the Corporation*, New York: Basic Books.

Kiernan, K. (1992) 'The roles of men and women in tomorrow's Europe', *Employment Gazette* 100: 491–99.

King, R., Connell, J. and White, P. (eds) (1995) *Writing Across Worlds. Literature and Migration*, London: Routledge.

Law, C. and Warnes, A. (1982) 'The destination decision in retirement migration', in A. Warnes (ed.) *Geographical Perspectives on the Elderly*, Chichester: Wiley, pp. 53–81.

Lawson, V. (1998) 'Hierarchical households and gendered migration in Latin America: feminist extensions to migration theory', *Progress in Human Geography* 22: 39–53.

Lelièvre, E. and Bonvalet, E. (1994) 'A compared cohort history of residential mobility, social change and home-ownership in Paris and the rest of France', *Urban Studies* 31: 1647–65.

Little, J. (1997) 'Employment marginality and women's self-identity', in P. Cloke and J. Little (eds) *Contested Countryside Cultures*, London: Routledge, pp. 138–57.

Long, L. (1974) 'Women's labor force participation and the residential mobility of families', *Social Forces* 52: 342–48.

McDowell, L. (1991) 'Life without father and Ford: the new gender order of post-Fordism', *Transactions of the Institute of British Geographers* 16: 400–19.

McDowell, L. (1993a) 'Space, place and gender relations: Part I. Feminist empiricism and the geography of social relations', *Progress in Human Geography* 17: 157–79.

McDowell, L. (1993b) 'Space, place and gender relations: Part II. Identity, difference, feminist geometries and geographies', *Progress in Human Geography* 17: 305–18.

Marshall, J. (1988) 'Re-visioning career concepts: a feminist invitation', in M. Arthur, D. Hall and B. Lawrence (eds) *The Handbook of Career Theory*, Cambridge: Cambridge University Press, pp. 275–91.

Massey, D. (1990) 'Social structure, household strategies, and the cumulative causation of migration', *Population Index* 56: 3–26.

Massey, D. (1991) 'A global sense of place', *Marxism Today* June: 24–29.

Mincer, J. (1978) 'Family migration decisions', *Journal of Political Economy* 86: 749–73.

Molho, I. (1986) 'Theories of migration: a review', *Scottish Journal of Political Economy* 33: 396–419.

Monroe, P., Bokemeier, J., Kotchen, J. and McKean, H. (1985) 'Spousal response consistency in decision-making research', *Journal of Marriage and the Family* 47: 733–38.

Morris, L. (1990) *The Workings of the Household*, Cambridge: Polity Press.

Morris, L. and Irwin, S. (1992) 'Employment histories and the concept of the underclass', *Sociology* 26: 401–20.

Morrison, D. and Lichter, D. (1988) 'Family migration and female employment: the problem of underemployment among migrant married women', *Journal of Marriage and the Family* 50: 161–72.

Moss, P. (1993) 'Focus: feminism as method', *Canadian Geographer* 37: 48–49 (and following papers).

Nash, C. (1993) 'Remapping and renaming: new cartographies of identity, gender and landscape in Ireland', *Feminist Review* 44: 39–57.

Nast, H. (1994) 'Opening remarks on "Women in the field"', *Professional Geographer* 46: 54–66 (and following papers).

Nelson, K. (1986) 'Labor demand, labor supply and the suburbanization of low wage office work', in A. Scott and M. Storper (eds) *Production, Work and Territory*, London: Allen and Unwin, pp. 149–71.

Owen, D. and Green, A. (1992) 'Migration patterns and trends', in A. Champion and A. Fielding (eds) *Migration Processes and Patterns. Volume 2: Research Progress and Prospects*, London: Belhaven, pp. 17–38.

Pahl, J. and Pahl, R. (1971) *Managers and their Wives*, London: Allen Lane.

Pinch, S. and Storey, A. (1992) 'Who does what, where? A household survey of the division of domestic labour in Southampton', *Area* 24: 5–12.

Rank, M. (1982) 'Determinants of conjugal influence in wives' employment decision-making', *Journal of Marriage and the Family* 44: 591–604.

Rapoport, R. and Rapoport, R. (1976) *Dual-career Families Re-examined*, London: Martin Robertson.

Ravenstein, E. (1885) 'The laws of migration', *Journal of the Royal Statistical Society* 48: 167–235.

Reskin, B. and Hartmann, H. (1986) *Women's Work, Men's Work: Sex Segregation on the Job*, Washington DC: National Academy Press.

Reskin, B. and Roos, P. (1990) *Job Queues and Gender Queues: Explaining Women's Inroads into Male Occupations*, Philadelphia: Temple University Press.

Rives, J. and West, J. (1992) 'Worker relocation costs: the role of wife's labor market behavior', *Regional Science Perspectives* 22: 3–12.

Rogers, A., Frey, W., Rees, P., Speare, A. and Warnes, A. (eds) (1992) *Elderly Migration and Population Redistribution: a Comparative Study*, London: Belhaven.

Rose, D. (1989) 'A feminist perspective of employment restructuring and gentrification: the case of Montréal', in J. Wolch and M. Dear (eds) *The Power of Geography*, Boston: Unwin Hyman, pp. 118–38.

Rose, G. (1993) *Feminism and Geography. The Limits of Geographical Knowledge*, Cambridge: Polity Press.

Sandell, S. (1977) 'Women and the economics of family migration', *Review of Economics and Statistics* 59: 406–14.

Savage, M., Barlow, J., Dickens, P. and Fielding, A. (1992) *Property, Bureaucracy and Culture*, London: Routledge.

Shihadeh, E. (1991) 'The prevalence of husband-centered migration: employment consequences for married mothers', *Journal of Marriage and the Family* 53: 432–44.

Shotter, J. (1984) *Social Accountability and Selfhood*, Oxford: Blackwell.

Sjaastad, L. (1962) 'The costs and returns of migration', *Journal of Political Economy* 70: 80–93.

Smith, Y. (1997) 'The household, women's employment and social exclusion', *Urban Studies* 34: 1159–77.

Snaith, J. (1990) 'Migration and dual-career households', in J. Johnson and J. Salt (eds) *Labour Migration*, London: David Fulton, pp. 155–71.

Thrift, N. (1985) 'Bear and mouse or bear and tree? Anthony Giddens's reconstitution of social theory', *Sociology* 19: 609–23.

Thrift, N. (1996) *Spatial Formations*, London: Sage.

Van Den Berghe, P. (1970) 'The two roles of women', *American Sociologist* 5: 375–76.

Walby, S. (1986) *Patriarchy at Work*, Cambridge: Polity Press.

Walby, S. (1989) 'Theorising patriarchy', *Sociology* 23: 213–34.

Walby, S. (1990) *Theorizing Patriarchy*, Oxford: Blackwell.

Walter, B. (1991) 'Gender and recent Irish migration to Britain', in R. King (ed.) *Contemporary Irish Migration*, Dublin: Geographical Society of Ireland, pp. 11–20.

Walter, B. (1995) 'Irishness, gender and place', *Environment and Planning D: Society and Space* 13: 35–50.

Ward, C. and Dale, A. (1992) 'Geographical variation in female labour force participation: an application of multi-level modelling', *Regional Studies* 26: 243–55.

Warde, A. (1991) 'Gentrification as consumption: issues of class and gender', *Environment and Planning D: Society and Space* 9: 223–32.

Warnes, A. (1992) 'Migration and the life course', in A. Champion and A. Fielding (eds) *Migration Processes and Patterns. Volume 1. Research Progress and Prospects*, London: Belhaven, pp. 175–87.

Warnes, A. (1996) 'Migrations among older people', *Reviews in Clinical Gerontology* 6: 101–14.

2 A longitudinal and regional analysis of gender-specific social and spatial mobilities in England and Wales 1981–91

Tony Fielding and Susan Halford

Introduction

Feminist research over the past two decades has clearly demonstrated the significance of gender to the study and interpretation of social class stratification and social class mobility (Dex 1987; Marshall et al. 1988; Abbott and Sapsford 1987; Payne and Abbott 1990). At any one time, women and men display distinct patterns in terms of their distribution across the social class hierarchy, whilst over the course of individual lifetimes women and men tend to experience quite different patterns of movement between social classes. 'Snapshots' of the gendered composition of the class hierarchy reveal that women are disproportionately concentrated in semi-professional, clerical and semi-/unskilled manual work, whilst men have far greater presence in the professions, skilled manual occupations, and occupy senior managerial sections of the service class almost exclusively. Part of the explanation for this is that men are far more likely than women to experience upward social mobility during the course of their working lives, especially into the service class. Upward social mobility is less common for women. Even where this does take place, movements are frequently rather limited (Abbott 1990; Chapman 1990; Dex 1990), rather than the longer-range movements between social classes more typically experienced by men (Goldthorpe, Llewellyn and Payne 1980). More commonly, women experience downward mobility over their life course, with a great majority leaving the labour market at the end of their working lives from jobs of lower status and income than those they held as young adults.

These female mobility patterns have been closely linked to the effects of marriage and child-care responsibilities on women's careers. It has been widely reported that on marriage women's careers become secondary to those of their husbands (Kanter 1977; Finch 1983; Dex 1987) and that, in particular, the primary responsibility which women take for child care has devastating consequences for their chances of upward social mobility. Extensive research for the British Women and Employment Survey in 1984 revealed that, for women, 80 per cent of any upward mobility occurs either prior to childbirth or after major child-care responsibilities are over (Brannen

1989) and that after child-care breaks from the labour market few women ever regain the job status they held prior to their first career break (Beechey and Perkins 1987). However, even childless women who have taken no career breaks experience far less upward social mobility than men. Indeed as many as 45 per cent of this group of women had also experienced downward social mobility over their working lives (Dex 1990). This has been explained through reference to the types of jobs women tend to be employed in, many of which offer few opportunities for advancement, and to the 'gendered sub-structure' (Acker 1990) of workplace organizations, whereby organizational structures, images and practices enhance male careers whilst limiting female careers (Acker 1990; Halford, Savage and Witz 1997).

The insights contributed by this sociological research on women, class stratification and social class mobility have revealed the partial and 'malestream' nature of established work in the field, which was commonly based only on men but presented as the norm and the basis for universal explanations and theories. This is clearly not sustainable once gender is shown to be a central structuring process in class stratification and social mobility. Thus, in place of earlier malestream accounts a new orthodoxy has emerged emphasizing the differences between male patterns of social class and mobility on the one hand, and female patterns on the other. However, assumptions about the uniformity of women's experiences (of any nature) have been subject to sustained critique within feminist debate and a strong trajectory within current feminist research and theory is away from universal claims about 'women' and towards recognition of differentiation within the category 'woman' (and, indeed, 'man'). However, what continues to be absent from most feminist sociology – and particularly from debates about women, class and mobility – is a consideration of differences between places. Whilst race, ethnicity, sexuality and so on are established as axes of difference, these are presented as if they were the same everywhere.

Of course, we know from research on the geography of gender that gendered patterns and processes do indeed vary spatially. Qualitative studies of gender in particular places reveal substantial differences in terms of the types of work (paid and unpaid) which women and men perform, gendered cultural norms and expectations, and gendered identities and practices (McDowell and Massey 1984; Mark-Lawson, Savage and Warde 1986; Hanson and Pratt 1995). More quantitative studies have also begun to 'map' these differences within regions, across Britain, the EU and even the world (for example, Duncan 1991; Ward and Dale 1992; Congdon 1990). However, with a few notable exceptions, these geographical studies have been 'snapshots' describing the gendering of place and/or the geography of gender at any one particular time. Hence, these studies have failed to capture temporal changes, either over the course of individual lifetimes or changes in regional/national patterns over time. Yet, as we have described above, we know that gendered patterns do change over the life course. Furthermore, we also know that just as gender patterns and processes are not constant over space, nor are they constant over time (McDowell and Massey 1984). The picture is further complicated by the fact that many people do not remain in one place throughout their lives but move between regions or countries. Thus, whilst the gendering of place, region or nation shapes women's and men's opportunities, as individuals move from one place to another their opportunities and the pattern of their careers may change.

Of course, the link between spatial mobility and social mobility has long been recognized, at least for men moving up into or within professional and managerial occupations (Whyte 1957; Savage et al. 1992). For this group of 'spiralists', spatial mobility – be it in pursuit of education, a new job, or at the behest of employers – is a central feature of career development. Spatial mobility is thus explained by the opportunities available in a new place although the specifically gendered nature of these opportunities, or the places in which they are located, has rarely been explicitly considered. Indeed, women have rarely been considered at all in such studies of social and spatial mobility (Boyle and Halfacree 1995). Perhaps this is because of initial assumptions that women migrants would fit the malestream models constructed for men or because women have been assumed to be less mobile than men, on both a daily and a lifetime basis because of the constraints of marriage and child care (Hanson and Pratt 1995). Where female migration has been considered there is a widespread assumption that women are mainly 'secondary' migrants, following their husband's careers rather than their own which are seen to be less important or to have been eclipsed by a domestic career (Bonney and Love 1991). More metaphorically, some feminist writing in career theory suggests that women are in any case more grounded, or fixed, than men; tied to relationships and more concerned with 'being' than hierarchical career movement and spiralist behaviour (Marshall 1988; Gallos 1988).

In sum, whilst existing sociological and geographical research has made substantial contributions to our understanding of the articulations between gender, class and region, a number of crucial interlinkages remain relatively unexplored. Principally, studies of gender and social mobility fail to take into account the uneven gendering of space, whilst studies of gender and space do not consider patterns of social mobility and rarely consider changes to the gendering of space and place over time. In an earlier paper (Fielding and Halford 1993) we used material from the Longitudinal Study (see below for further details about this data set) to explore gendered patterns of social and spatial mobility in the period 1971–81. This showed quite clearly that the gendering of social mobility varies from one region to another. Taking the country as a whole, the regions in which women experienced upward social mobility were not necessarily the same as those regions in which men were upwardly mobile. We also showed that migration flows between regions show some distinctively gendered patterns, in particular that movement into the South East resulted in especially high rates of upward social mobility for women. In this chapter we aim to develop this analysis in a number of ways using material linking the 1981 and 1991 censuses. This enables a more up-to-date analysis of the national rates of social mobility for women and men. It also allows us to consider both the regional patterns of gendered mobility and the social mobilities of male and female inter-regional migrants during the 1980s.

Our earlier study of the 1970s provided some limited evidence that (certain) women were experiencing greater social mobility – particularly into professional occupations – than was the case a decade earlier. Explanations for this presumably include the rapid expansion during the 1960s of opportunities for higher education, which offered women far greater opportunities to gain qualifications than in earlier times, combined with the impact of the second-wave women's movement and the 'equality' legislation implemented by the British government in 1975. These influences have continued into the 1980s but we might also expect that the general trend towards more upward social mobility for women will have been exaggerated during the last decade by the economic boom in the service sector; by the increased adoption of equal opportunities initiatives by public and private sector organizations which, in many cases, moved beyond basic legal requirements; and by the galvanizing effect of fears about the 'demographic time bomb' which, for a while at least, prompted employers to examine new ways of recruiting, retaining and even promoting women. As far as men were concerned, whilst the expansion of the service sector may also have offered an array of new opportunities, the rapid contraction of skilled manual work might be expected to have had a more negative effect on the career opportunities of those in working class jobs.

In this chapter we concentrate on describing the gendered patterns of social and spatial mobility during this turbulent last decade. However, as we proceed we will also make comparisons with the 1970s, drawing out the most important patterns of change at the national level, between regions and in the fates of inter-regional migrants. In looking at migration flows we concentrate on movements into and out of the South East. Together these account for well over half of all moves into, within or out of the labour market, which were made during the 1980s. We would also expect gendered patterns of social mobility to be especially pronounced in this region, since many of the factors described above as features of the 1980s were significantly exaggerated in the South East.

The structure of our chapter is simple. In the following section we provide some brief details about the data set. Next, we summarize the overall national patterns of gendered social mobility in the 1980s, making brief comparison with the previous decade. In the following section we look at the regional patterns of social mobility, investigating which regions appear to offer especially good or especially bad chances of upward social mobility for women and men, and whether there have been any significant changes in this over the past twenty years. From this we move on to consider the social mobilities of inter-regional migrants, again making comparisons with the 1970s, before drawing together some conclusions.

The Longitudinal Study

Our analysis depends on the availability of information on the work histories of men and women in different regions, and of those who have moved between regions. This is a tall order, and in our opinion it is only the Office of Population Censuses and Surveys (now the Office of National Statistics) Longitudinal Study (LS) that has a sample size (of about 500,000) large enough to allow such calculations. The LS is a 1 per cent sample of the whole population achieved by selecting all those individuals recorded in the population census as having birthdays on four specified dates in the year. The census forms for the individuals in this sample (LS members) are then matched with the forms for those same individuals at the following census, allowing one to see the way in which the circumstances of those individuals have changed over the ten-year period. This matching process has now been carried out for two inter-censal periods – 1971–81 and 1981–91 – and the sample has been kept up-to-date between the censuses by extracting the deaths of LS members and by including new births occurring on those four dates. The product of this process is a very large sample of individuals for whom we have all the demographic, household, housing and employment facts that are collected in the census. So large is the sample that we can even analyse the social class changes that occur among a group of people migrating between one region and another over the inter-censal period, and the analysis of region-specific and gender-specific social mobility rates presents few problems.

In this chapter a number of practical decisions affecting the interpretation of the results have been taken. First, a distinction has been made between those regarded as being in the labour market and those deemed to be outside it. The former includes all full-time and part-time workers (employers, employees and the self-employed) and those seeking employment (the unemployed). For entries into the labour market we use the categories education (meaning in full-time education – school, university, college) and other. The category other is very small for men – it consists largely of those entering the labour market from the armed forces – but for women it is very large, consisting primarily of women who have left the labour market to raise children at home. Other is also, therefore, a destination category for women who are exiting the labour market. However, the main exit category for both men and women is, of course, retirement. Second, the basis of our classification of occupations is the socio-economic group (SEG). The census SEGs have been grouped into six categories:

- Professionals: this includes SEGs 3, 4 and 5.1 (and the virtually empty 1.1). The important point here is that SEG 5.1 is included. This means that the category includes many women who work as teachers and nurses – these women are often (unjustifiably, in our opinion) excluded from this category.
- Managers: SEGs 1.2 and 2.2.
- The petite bourgeoisie (the self-employed and owners of mostly small and medium-sized businesses): SEGs 2.1, 12, 13 and 14.

Table 2.1 Gender-specific intra-generational social mobility of non-migrants in England and Wales 1981–1991 (1.096 per cent sample)

Social class in 1981	*Social class in 1991*						
	PRO	*MAN*	*PeB*	*WhC*	*BlC*	*UNE*	*Total (labour market)*
(1) Males							
PRO (professionals)	7139	1997	437	461	775	356	11165
MAN (managers)	1465	5791	1254	816	906	502	10734
PeB (petite bourgeoisie)	297	554	6239	180	934	455	8659
WhC (white-collar)	1234	2398	825	4163	1353	584	10557
BlC (blue-collar)	1779	2542	4764	1561	28740	3964	43350
UNE (unemployed)	384	382	1130	376	2879	2670	7821
Education	3093	2096	1819	4755	11026	5825	28614
Other	123	108	131	158	359	207	1086
Total	15514	15868	16599	12470	46972	14563	121986
(2) Females							
PRO	5886	423	174	606	297	138	7524
MAN	383	886	190	681	144	84	2368
PeB	104	112	771	361	160	31	1539
WhC	1735	2346	1000	15697	3163	929	24870
BlC	420	305	314	2267	6134	618	10058
UNE	274	139	114	920	656	448	2551
Education	3147	1774	420	11723	3444	3287	23795
Other	1829	885	1194	7705	4512	966	17091
Total	13778	6870	4177	39960	18510	6501	89796

Social class in 1981 (1971)	*Social class in 1991 (1981)*		
(1) Males	*Middle class*	*Working class*	*Total (labour market)*
Middle class	25173 (20642)	5385 (6037)	30558 (26679)
Working class	15438 (12034)	46290 (57136)	61728 (69170)
Education & Other (entries)	7370 (5830)	22330 (23998)	29700 (29828)
Total	47981 (38506)	74005 (87171)	121986 (125677)
(2) Females			
Middle class	8929 (5459)	2502 (1944)	11431 (7403)
Working class	6647 (3662)	30832 (28450)	37479 (32112)
Education & Other (entries)	9249 (7064)	31637 (33766)	40886 (40830)
Total	24825 (16185)	64971 (64160)	89796 (80345)

- White-collar workers (lower-level non-manual employees): SEGs 5.2, 6 and 7.
- Blue-collar workers (manual employees): SEGs 8, 9, 10, 11, 15 and 17.
- The unemployed.

The first three categories are regarded as constituting the three middle classes.[1] The last three categories comprise the three working classes.

National patterns of gendered social mobility in the 1980s

In this section we examine national patterns of social mobility for women and men. Table 2.1 provides raw data on all those who were in the labour market in 1991, showing their occupational position in 1981. Table 2.2 transforms the figures in table 2.1 into rates of flow between the occupational categories by dividing each of the transitions 1981–91 by the row total for

Table 2.2 Gender-specific intra-generational social mobility rates of non-migrants in England and Wales 1981–91 (1.096 per cent sample)

	Social class in 1991						
	PRO	*MAN*	*PeB*	*WhC*	*BlC*	*UNE*	*Total*
Social class in 1981							
(1) Males							
PRO (professionals)	63.94	17.89	3.91	4.13	6.94	3.19	100.0
MAN (managers)	13.65	53.95	11.68	7.60	8.44	4.68	100.0
PeB (petite bourgeoisie)	3.43	6.40	72.05	2.08	10.79	5.25	100.0
WhC (white-collar)	11.69	22.71	7.81	39.43	12.82	5.53	100.0
BlC (blue-collar)	4.10	5.86	10.99	3.60	66.30	9.14	100.0
UNE (unemployed)	4.91	4.88	14.45	4.81	36.81	34.14	100.0
Education	10.81	7.33	6.36	15.62	38.53	20.36	100.0
Other	11.33	9.94	12.06	14.55	33.06	19.06	100.0
(2) Females							
PRO	78.23	5.62	2.31	8.05	3.95	1.83	100.0
MAN	16.17	37.42	8.02	28.76	6.08	3.55	100.0
PeB	6.76	7.28	50.10	23.46	10.40	2.01	100.0
WhC	6.98	9.43	4.02	63.12	12.72	3.74	100.0
BlC	4.18	3.03	3.12	22.54	60.99	6.14	100.0
UNE	10.74	5.45	4.47	36.06	25.72	17.56	100.0
Education	13.23	7.46	1.77	49.27	14.47	13.81	100.0
Other	10.70	5.18	6.99	45.08	26.40	5.65	100.0

Social class in 1981 (1971)	*Social class in 1991 (1981)*		
(1) Males	*Middle class*	*Working class*	*Total (labour market)*
Middle class	82.38 (77.37)	17.62 (22.63)	100.0 (100.0)
Working class	25.01 (17.40)	74.99 (82.60)	100.0 (100.0)
Education & Other (entries)	24.81 (19.55)	75.19 (80.45)	100.0 (100.0)
(2) Females			
Middle class	78.11 (73.74)	21.89 (26.26)	100.0 (100.0)
Working class	17.74 (11.40)	82.26 (88.60)	100.0 (100.0)
Education & Other (entries)	22.62 (17.30)	77.38 (82.70)	100.0 (100.0)

each occupational category. This shows us, in percentage form, the 1991 destinations of all the members of each occupational category in 1981. Together, these tables draw our attention to a number of important patterns.

Looking down the columns in table 2.1, which shows us the distribution of women and men between occupational categories in 1991, we can see that the numbers of women and men in the professional category are roughly equal (13,778 women and 15,514 men). Given the way in which we have constructed this category from SEGs we should be wary of the meaning of this, since pay and conditions in the professional occupations of teaching and nursing – both feminized areas of work – are generally far lower than in other professional categories (for example, doctors, lawyers, accountants) which continue to be male-dominated. None the less, these figures represent a remarkable change when compared with 1971 and 1981.[2] During this period the overall number of professional jobs has expanded rapidly, more than doubling from 1971 to 1981 and increasing again by over 50 per cent from 1981 to 1991. Not surprisingly then, the total numbers of both women and men in professional jobs have increased significantly. Most notable for us, though, is the enormous increase in the proportion of those jobs taken by women. Whereas women only accounted for 33 per cent of those in professional jobs in 1971, this had risen to 40 per cent in 1981 and again to 47 per cent by 1991. At first sight, this appears to confirm women's use of a 'qualifications lever' (Crompton and Sanderson 1990) as their most effective career move. The suggestion here is that since professional qualifications rely for their credibility on being gender-neutral (i.e. must be of equal value to both women and men who hold them), gaining such formal credentials may be women's best chance for competing with men in the labour market. Conversely, it is suggested that management posts will be particularly difficult for women to enter since qualifying credentials are far more nebulous and informal, and consequently more vulnerable to various forms of sexism.

Certainly, men continue to dominate in the managerial category. In 1991 men held over 70 per cent of all managerial jobs. However, even the small number of women in managerial posts in 1991 represents a 73 per cent increase on the numbers in 1981. Confirming this, the LS data reveals that the *rate* of increase for women moving into managerial posts is higher than that for men. None the less, as Crompton and Sanderson (1990) argue, movement from professional jobs to managerial jobs is lower for women than men (but not by very much). Instead, as table 2.2 shows us, the new women managers are coming from low-level white-collar work, where the rate increased from 5.95 to 9.43, and especially from education where the rate of flow increased from 3.15 to 7.46!

Like managerial occupations, the petite bourgeoisie is still predominantly male but, unlike the managerial category, male domination of the petite bourgeoisie strengthened during the 1980s. Predictably, table 2.2 shows that the highest rate of flow is from management to self-employment, confirming stereotypes of the independent entrepreneur and the significance of a small-business enterprise culture in the 1980s. However, there has also been a big increase in the rates of flow for those from working class origins to the petite bourgeoisie (rates of flow from the blue-collar working class rose from 5.66 to 10.99, and those from unemployment from 8.60 to 14.45). This seems likely to be a reflection of another side of 1980s Britain, namely a decline in the numbers of builders, mechanics and others employed on permanent contracts and a corresponding increase in the sub-contracting of nominally self-employed workers – often in fact the very same workers – in their place. In other words, there seems to be a sharp shift towards a proletarianization of the category male petite bourgeoisie during the 1980s.

In sum, two of the three middle class categories remain heavily male-dominated, although the management category shows signs of change. In the professional category, women have advanced to almost equal numbers (bearing in mind the caveat outlined above).

Turning to the working class categories, some patterns have remained constant. Lower-level white-collar jobs are still dominated by women, whilst the blue-collar working class remains predominantly male. However, there also appears to be an important pattern of change under way in the gendered composition of working class categories. Table 2.1 shows that whereas the male manual working class is extremely stable (i.e. individuals tend to be in this category in both 1981 and 1991), the female manual working class is in greater flux. Adding to this, table 2.2 shows that the rates of flow into the blue-collar working class are very distinctive. Those for men have decreased extremely sharply (for example, the flow from education has dropped from 50.14 to 38.53), but those for women have been maintained at fairly steady levels. This means that the manual working class has become much less male-dominated in its composition during the 1980s. This is not surprising given the contraction of heavy industry and the skilled manufacturing sector in Britain during the 1980s and the rapid expansion of routine assembly work, where semi- and unskilled jobs are held largely by women.

The rates of flow into unemployment have increased for both men and women, with the increase in the flow of young men from education into unemployment, and the increased size of the female unemployment to unemployment transition being particularly marked. None the less, table 2.2 reveals that long-term unemployment was far more prevalent amongst men (34.4 per cent of men who were unemployed in 1981 were in the same position in 1991, more than twice the rate for women[3]) and that there was a far higher level of male entry into unemployment from education.

Let us finally look more broadly at the patterns of change between the working classes and middle classes over the period 1981–91. Adding together the working class occupational categories in table 2.1, we can see that the female working class is now only slightly smaller than the male (65,000 against 74,000). However, adding together the middle class occupational categories, we can see that the female middle class is still much smaller than the male middle class (25,000 against 48,000). Over the period 1981–91, women's membership of the middle class also showed far less stability than did male membership (9,000 female middle class stayers against 25,000 male middle class stayers).

Regional patterns in gendered social mobilities

Next, we turn to look at gendered patterns of social mobility at the regional level, that is, to examine the mobility patterns of those women and men who were living in the same region in both 1981 and 1991. This entails disaggregating the data in tables 2.1 and 2.2 into their regional components. To do this, tables for each of the nine regions in England and Wales (there is, unfortunately, no LS for Scotland or Northern Ireland) equivalent in format to tables 2.1 and 2.2 were produced. (These are not reproduced here due to lack of space.) For each region, the percentage figures equivalent to those in table 2.2 were divided by the national percentages in table 2.2 to produce 'location quotients'. These location quotients express the regional figures as a ratio of the corresponding national (i.e. England and Wales) figure. In this way we can see in what respects and to what degree the regional situations differ from the national, and whether or not these differences are the same for men and for women. A summary is presented in table 2.3.

Two patterns are clear. The first is the highly privileged status of the South East region. It has by far and away the highest location quotients for both upward social mobility from the working class categories to the middle class ones (1.21 for men and 1.14 for women) and from education to the middle classes (1.28 and 1.22). Downward social mobility location quotients for the South East are correspondingly low. Other regions in southern and eastern England have marginally higher than national average location quotients (notably the South West region), but the regions of northern England and Wales show far less upward mobility than the national average and far bigger downward mobility rates. The second important pattern is the strong propensity for the male and female location quotients across the regions to differ from one another. Linked to this, we can see that the degree of variability is far less for women than for men. For example, in the case of upward mobility from the working class to the middle class, the values for men are higher than those for women for all of the regions of southern and eastern England, but are lower for men than for women in the regions of northern England and Wales. This means that the advantage for a man to live in the south of England, compared with men living other regions, is greater than that for a woman, where the gap is less pronounced. This appears to indicate that, in terms of long-range movements between classes, space makes more of a difference for men than it does for women. We shall show below how this comes about – the important point for the moment is the fact that there is a divergence between male and female rates.

Considering this regional divergence in greater detail reveals a complex pattern. Table 2.4 records a selection of key transitions broken down by

Table 2.3 Summary of social mobility rates by gender by region: location quotients (England and Wales = 1.00)

	Middle class to middle class		*Middle class to working class*		*Working class to middle class*		*Working class to working class*		*Education to middle class*		*Education to working class*	
	Male	*Female*	*Male*	*Female*	*Male*	*Female*	*Male*	*Female*	*Male*	*Female*	*Male*	*Female*
South East	1.01	0.99	0.98	1.05	1.21	1.14	0.93	0.97	1.28	1.22	0.91	0.94
East Anglia	0.99	1.04	1.05	0.85	1.00	0.96	1.00	1.01	0.93	0.87	1.02	1.04
South West	1.00	0.99	0.98	1.05	1.11	1.09	0.96	0.98	1.03	0.91	0.99	1.03
East Midlands	1.00	0.97	0.98	1.10	0.95	0.89	1.02	1.02	0.81	0.91	1.06	1.03
West Midlands	1.00	1.02	1.02	0.92	0.87	0.84	1.04	1.03	0.89	0.86	1.03	1.04
Wales	0.99	1.00	1.04	0.98	0.85	0.92	1.05	1.02	0.82	0.85	1.06	1.04
North West	1.00	1.01	1.01	0.98	0.93	0.97	1.02	1.01	0.88	0.97	1.04	1.01
Yorkshire & Humberside	1.00	1.03	1.02	0.89	0.85	0.86	1.05	1.03	0.77	0.82	1.07	1.05
No region	0.98	1.02	1.11	0.93	0.71	0.93	1.10	1.01	0.67	0.80	1.11	1.06

Source: OPCS 1991 Longitudinal Study (Crown Copyright Reserved).

Table 2.4 Social mobility rates by gender and region: (a) into middle class positions, location quotients (England and Wales = 1.00)

	WhC to MAN		*Education to PRO*		*Education to MAN*		*BlC to PeB*	*UNE to PeB*	*PRO to MAN*	*Other to PeB*	*Other to MAN*	*Other to PRO*
	Male	*Female*	*Male*	*Female*	*Male*	*Female*	*Male*	*Female*	*Male*	*Female*	*Male*	*Female*
South East	1.07	1.17	1.31	1.16	1.40	1.40	1.15	1.25	1.10	0.93	1.25	1.13
East Anglia	1.06	0.93	0.83	0.73	0.79	1.06	1.03	1.40	0.92	1.21	0.97	0.94
South West	0.91	0.98	0.83	0.89	0.89	0.83	1.31	1.23	1.06	1.25	1.13	0.92
East Midlands	1.08	0.99	0.79	0.90	0.72	0.82	0.96	1.16	0.83	1.03	0.77	0.87
West Midlands	1.10	0.84	0.90	0.85	0.84	0.89	0.89	0.84	0.98	0.80	0.89	0.71
Wales	0.89	0.76	0.84	0.94	0.70	0.56	0.91	0.99	0.73	1.28	0.79	0.95
North West	0.88	0.89	0.95	1.08	0.85	0.80	0.92	0.75	1.01	1.00	0.96	1.03
Yorkshire & Humberside	0.87	0.92	0.74	0.83	0.78	0.75	0.91	0.99	0.86	1.06	0.80	1.00
No region	0.76	0.88	0.76	0.96	0.64	0.58	0.61	0.56	0.80	0.75	0.55	1.11

Source: OPCS 1991 Longitudinal Study (Crown Copyright Reserved).

region, focusing on upward social mobility. The first six columns refer to transitions where there are sufficient numbers of both women and men to make a direct comparison viable. The first case concerns the important transition between low-level white-collar work and managerial positions. Although for both men and women there is a bias towards high rates in the South East region, it is noticeable that the rate *for women* in this region is considerably higher than that for men, and that the geography of high and low rates for the remainder of the regions differs markedly between the sexes. Thus, whereas there is a distinct bias towards high upward mobility for women in the South East region alone (at least, from routine white-collar to managerial jobs), for men this South Eastern bias is only slight and it affects all but one of the southern and midland regions of the country. As far as this particular transition is concerned, space makes more of a difference to women than it does to men. The second case concerns the transition from full-time education to professional work. On this occasion the bias towards high upward mobility in the South East is particularly strong in the case of men. This may well reflect the spatial concentration of private sector professional jobs, which are more male-dominated, in the South East and the greater spatial dispersion of public sector professional jobs (including teaching and nursing, included in our 'professional' category). The third case concerns the transition from education to management. In this case the bias towards upward mobility in the South East is extremely strong for both men and women (40 per cent above the England and Wales average for both sexes), but the figures for other regions are not always alike. In particular, this path to upward mobility is not important for men in East Anglia (where it is for women), and is very unimportant for women in Wales (where it is of modest importance for men). Furthermore, in the case of this particular transition, women's rates are even more variable (ranging from 1.40 to 0.56) than men's (ranging from 1.40 to 0.64).

The remaining columns in table 2.4 refer to those upward transitions that have a strong gender bias, in so much as there are vastly higher numbers of women – or men – making those transitions. As we have already seen, an overwhelmingly male transition is the movement into the petite bourgeoisie with the 1980s showing significant increases in the inward movement of men from blue-collar work and unemployment. In both cases, the South East region, along with other southern and eastern regions, has witnessed a very rapid flow of men into the petite bourgeoisie from these origins. In contrast, the flows in the north and west are very small, especially so in the Northern region. For very different reasons, the transition from professional to managerial jobs is also interesting. This move is customarily seen as a promotion (although both categories are, according to our classification, middle class). Again, the South East region has a value for men of over 1.00, indicating a higher than average rate of promotion.

Turning to transitions of particular importance to women, table 2.4 shows transitions from the category 'other' (comprising mainly women undertaking unpaid domestic responsibilities on a full-time basis). Here the flows into the petite bourgeoisie are completely out of line with the other two patterns of upward movements out of the 'other' category. Whereas upward movement into managerial and professional jobs is higher amongst women in the South East, movement into the third middle class category is below the average, whilst in East Anglia, the South West and Wales rates of flow are well above the average. A possible explanation for this is that these women are middle class wives (some of whom were out-migrants from the South East) who are becoming the proprietors of small businesses in the retail and catering trades when demands on their time from their children have diminished.

Table 2.5 continues this analysis but focuses this time on a selection of downward transitions into working class categories. An interesting broad impression can be gained by comparing the values across the rows for the South East region at the top of the table and the Northern region at the bottom. For both women and men, the South East values are mainly below 1.00 showing low levels of downward movement into the working classes, whilst comparable values for the North are mainly above average. Looking at the regional patterns in more detail, the first two columns of table 2.5 show downward mobility from middle class to working class jobs (i.e. the converse of the first two columns in table 2.4). The rates for both men and women are slightly below average in the South East, the West Midlands and Yorkshire and Humberside but mostly above average elsewhere. A simpler and clearer picture occurs with the transition from education to blue-collar work. Here the values for the South East are well below the average and are above average almost everywhere else. Particularly noticeable are the high values for female entry into manual work in those regions with modern branch plants such as the East Midlands, Wales and the North. The movement from education to unemployment is equally clear-cut, only this time the low values in the South East are joined by the other southern and eastern regions, and it is only the North and West that have high values. As with the previous two cases, the values for men and women are rather similar, suggesting that gender-specificity might be a more marked feature of upward social mobility than that of downward social mobility (a result which complies with our analyses for 1971–81 – Fielding and Halford 1993).

The remaining columns in table 2.5 cover cases where a comparison between male and female rates is not appropriate, usually because of the dominance of either male or female labour in the particular transition. For men the pattern of downward mobility from being petit bourgeois to becoming part of the manual working class is biased towards regions of rural industrialization (East Anglia, East Midlands and Wales). The next two columns show flows into unemployment. The patterns are remarkably similar to those for the shift from education into unemployment. The transition for women from education to white-collar jobs shows very low variability with all the values lying between 0.90 and 1.05 (notwithstanding a bias for higher values in the South and East). Women who re-enter the

Table 2.5 Social mobility rates by gender and region: (b) into working class positions: location quotients (England and Wales = 1.00)

	MAN to WhC		*Education to WhC*		*Education to UNE*		*PeB to BIC*	*BIC to UNE*	*UNE to UNE Male*	*Educ'n to WfhC*	*Other to BIC*	*WhC to UNE*
	Male	*Female*	*Male*	*Female*	*Male*	*Female*	*Male*	*Male*	*Male*	*Female*	*Female*	*Female*
South East	0.96	0.98	0.83	0.64	0.83	0.84	0.96	0.95	0.89	1.05	0.79	1.07
East Anglia	1.10	0.99	1.18	1.19	0.77	0.85	1.31	0.68	0.65	1.05	1.11	0.69
South West	1.20	1.17	1.06	1.04	0.84	0.91	0.95	0.83	0.87	1.05	0.97	0.86
East Midlands	0.75	1.13	1.21	1.53	0.95	0.95	1.17	0.87	0.94	0.90	1.16	0.79
West Midlands	0.87	0.97	1.10	1.15	0.99	1.20	1.03	1.04	1.02	0.96	1.24	0.92
Wales	1.05	1.04	1.10	1.41	1.14	1.09	1.11	1.11	1.02	0.92	1.16	0.95
North West	1.16	1.07	0.98	1.01	1.28	1.12	1.01	1.11	1.21	0.98	0.98	1.26
Yorkshire & Humberside	0.92	0.86	1.11	1.20	1.15	1.05	0.85	1.12	0.98	1.01	1.07	0.79
No region	1.40	0.90	1.05	1.22	1.43	1.41	0.90	1.17	1.20	0.91	1.23	1.09

Source: OPCS 1991 Longitudinal Study (Crown Copyright Reserved).

labour market by taking manual jobs tend to do so at rates which are very similar to flows from education into manual work, with high rates in regions of rural and branch plant industrialization (East Anglia, East Midlands and Wales, plus West Midlands and the North). Flows from the female white-collar working class into unemployment tend to be rather higher than the average in some of the northern regions, but the figure for the South East region is also above the average (reflecting perhaps a broad tendency for the stability of class membership to be rather lower in the South East than in other regions of the country – see table 2.3, row 1).

Taking tables 2.3–2.5 together (including information on transitions not selected for these tables) we can construct regional profiles which reveal the regional distinctiveness of gender-specific social mobility regimes. A comparison between the South East region and the North can usefully be made to convey something of this distinctiveness. The South East has high flows into its service class categories (professionals and managers) but not into the third component of the middle class – the petite bourgeoisie. This is possibly related to the difficulty of entry into the ownership of small and medium-sized businesses in this region due to the high costs of land and buildings, and the high cost of labour. But there are two exceptions – the rates of flow of men into the petite bourgeoisie from unemployment and from the blue-collar working class are very high. This might be related to the strong presence in the South East of ethnic minority businesses, but it is also feasible that this trend represents a national development which has shown its first effects in the South East region (spreading perhaps to the rest of the country during the 1990s). A second feature which qualifies the picture of high upward social mobility in the South East is the difference between male and female rates of entry into the professional and managerial middle classes. In general, the rates of entry into managerial jobs are higher for women, and into professional jobs rather higher for men. We have already suggested the discrepancy in entry to professional jobs may be linked to a regionalized and gendered public/private distinction in professional work. The discrepancy in entry to managerial jobs may reflect the presence of some more progressive employers in the South East and/or the greater representation of the types of work in which women are more likely to move into managerial roles, such as advertising, journalism and financial services (Boyle and Halfacree 1995). Amongst working class categories, entry into the white-collar jobs is slightly higher than the average (reflecting no doubt the service-sector orientation of the South East economy), but entry into the blue-collar section of the working class is uniformly low for both men and women. The situation with unemployment is rather different, however. One would expect, perhaps, that rates would be uniformly low (or perhaps not, given the concentration in the southern half of Britain of the damage done by the bursting of the 'bubble economy' of the late 1980s!). In fact, for both men and women, while the unemployment to unemployment and education (and other) to unemployment transitions are lower than the national average, the rates of flow of most other transitions are above the average. This suggests a higher vulnerability to redundancy in the South East than in other regions.

The profile for the North is very different. Here the rates of flow into middle class jobs are very low, but there are interesting differences between men and women. The rate of flow for men is so low that in the case of the education to middle class transition, a higher percentage of women (18.0 per cent) enter middle class positions than men (16.3 per cent) (overturning the national situation). The reasons for this lie largely in the higher rate of flow of women into the professions (possibly related to the greater significance of the public sector in the regional economy), and in the lower rates of entry of men into the petite bourgeoisie. Conversely, and as expected, the rates of flow into the blue-collar working class are very high. But here again gender differences arise. The rate of flow of men from education into manual work is only slightly above average (1.05) but the rate for women is very much above average (1.22). It seems that the feminization of the manual working class is at work here. However, the most distinctive feature of the Northern region profile is the propensity for men and women to enter unemployment. The overall location quotient for entry into unemployment is 1.35, which can be interpreted as meaning that this flow is 35 per cent above the national average. Within this, there are marked gender differences. Although the rates of flow from education into unemployment are high for both men and women (1.43 and 1.41), the unemployment to unemployment transition is 1.20 for men and only 0.78 for women. Thus, the North has many more men who have been unable to avoid unemployment at both census dates than women. There are also differences in the rates of flow from professional and managerial positions into unemployment; men are much more liable to this transition in the North (relative to the national average) than are women.

What are the possible reasons for the gendered regional differences illustrated here? First, contrasting regional labour markets (services versus manufacturing; high pay and job security versus low pay and redundancy; private versus public sector; large establishment and single industry versus small establishment and diverse ownership) result in gender-differentiated work histories and experiences. Second, contrasting regional social histories and cultures lead to gender-specific work and domestic roles and gender-specific possibilities of career advancement. Specifically, a strong cosmopolitanism (individualism and an emphasis on rights, self-fulfilment, and so on) contrasts with strong localism (collectivism and emphasis on obligations to family, and so on).

Differences in the social mobilities of male and female inter-regional migrants

Not everyone spends the whole of their working lives in one region. Some 9.3 per cent of those in the labour market in 1991 lived in another region of England and Wales in 1981. The importance of this group, however, lies not so much in their numbers but in the fact that these people were more socially mobile in both directions, but especially upwards, than the rest of the population. In the period 1981–91, for example, while 41.1 per cent of those who were in the labour market at both dates but were non-migrants were socially mobile, the equivalent figure for migrants was 52.8 per cent. As might be expected, the figures for those entering and leaving the labour market are even more striking. Some 30.5 per cent of those moving from education to a professional job, and 21.9 per cent of those moving from education to a managerial job, were inter-regionally mobile between 1981 and 1991 (this compares with an 8.9 per cent migration rate for all those living in England and Wales at both dates). The same situation (though in smaller measure) applies to those leaving the labour market – 12.2 per cent of professionals and 11.8 per cent of managers migrated as they entered retirement. The extreme contrast to these figures is provided by those in blue-collar jobs in both 1981 and 1991. Only 2.8 per cent of these men and women migrated inter-regionally during this period. The remainder of this section explores the ways in which the social mobility associated with migration is gendered.

As with non-migrants, we are fairly selective in this section, focusing attention on just two migration streams – that from the rest of England and Wales to the South East region, and that from the South East region to the rest of England and Wales. In fact, so dominant is London in the inter-regional system that these two migration streams account for over half of all inter-regional migrants in England and Wales. The question then becomes whether or not there are differences between men and women in the social class changes that accompany these migrations to and from the South East region.

To do this we need data not only on those within and entering the labour market but we also need data covering those who left the labour market between 1981 and 1991, since many migrations accompany retirement and the movement of women into full-time domestic work. The format chosen uses location quotients again, but this time the location quotient registers the ratio of the selected transition from one class to another among those migrating to (or from) the South East (expressed as a percentage of all migrants to or from the South East), to the equivalent percentage for all inter-regional migrants in England and Wales. For example, men making the transition from education outside the South East to professional job within the South East represent 12.73 per cent of in-migrants to the region. The equivalent figure for all inter-regional migrants is 7.81 per cent. The ratio of the first to the second gives us a (dyadic) location quotient of 1.63. Two tables result from this exercise. The first, table 2.6, shows the social mobilities of those women and men moving into the South East during the 1980s and the second (2.7) shows the patterns for women and men who left the region during this period.[4]

Table 2.6 contains a great deal of interesting detail, but we begin by pointing out some of the broader features. The final column of each part of the

Table 2.6 Social composition of the migration stream to the South East region 1981–91

	PRO	*MAN*	*PeB*	*WhC*	*BlC*	*UNE*	*Total (labour market)*	*Retire*	*Other*	*Total*
Males										
PRO	0.89	1.09	0.53	0.70	0.90	1.16	0.93	0.32	1.15	0.83
MAN	1.13	1.03	0.36	0.88	0.86	0.28	0.91	0.32	0.40	0.78
PeB	0.78	0.82	0.53	0.72	1.08	0.94	0.74	0.41	0.57	0.66
WhC	1.13	1.01	0.66	1.13	0.94	0.81	1.01	0.27	0.57	0.85
BlC	1.26	1.13	0.81	1.12	0.79	0.74	0.90	0.24	0.55	0.73
UNE	1.80	1.65	1.39	1.51	1.40	1.07	1.39	0.51	0.83	1.15
Total (labour market)	1.03	1.07	0.70	1.07	0.91	0.85	0.96	0.31	0.69	0.82
Education	1.63	2.01	1.43	1.80	1.31	1.13	1.59	0.75	1.47	1.23
Other	0.87	1.16	1.10	0.94	0.80	1.21	0.97	0.75	1.17	1.02
Total	*1.29*	*1.29*	*0.82*	*1.40*	*1.03*	*0.97*	*1.17*	*0.58*	*1.24*	*1.00*
Females										
PRO	1.11	1.52	0.65	1.22	1.01	0.90	1.13	0.31	0.84	0.97
MAN	1.27	0.78	0.39	0.95	0.25	0.70	0.84	0.24	0.84	0.70
PeB	0.65	1.62	0.24	1.31	1.40	1.58	1.00	0.55	0.63	0.81
WhC	1.27	1.35	0.82	1.03	0.77	1.00	1.05	0.27	0.78	0.83
BlC	1.20	1.20	0.34	1.39	0.62	1.22	0.96	0.25	0.59	0.67
UNE	1.51	1.58	1.20	1.53	1.31	1.60	1.48	0.43	0.79	1.13
Total (labour market)	*1.16*	*1.29*	*0.66*	*1.12*	*0.80*	*1.13*	*1.08*	*0.29*	*0.76*	*0.85*
Education	1.75	2.09	1.29	1.57	1.16	1.24	1.62	0.81	1.28	1.29
Other	1.08	1.05	0.58	0.92	0.80	0.46	0.88	0.62	0.80	0.77
Total	*1.43*	*1.62*	*0.73*	*1.26*	*0.91*	*1.09*	*1.27*	*0.63*	*0.88*	*1.00*

Source: OPCS 1991 Longitudinal Study (Crown Copyright Reserved).

table shows us that the migrants to the South East are biased towards (young) people leaving education and towards those in unemployment. Not surprisingly, the attraction of the South East region, with its higher wage rates and lower unemployment tends to be strongest for those at the beginning of their working lives and those who most need to find employment. The differences between men and women are slight but it can be seen that the middle class migration of women to the South East is rather higher than that of men (see below). The bottom rows of the two parts of table 2.6 show us that the in-migrants display a strong tendency to end up in the labour market, especially in professional jobs and even more so in managerial jobs.[5] Here the values for women are significantly higher than those for men (1.43/1.29 for professionals, and 1.62/1.29 for managers). This is further evidence of the particular advantage of the South East region for the formation of women's middle class careers (especially managerial careers). By contrast, male in-migrants are unusually likely to end up in routine white-collar work, an interesting reversal of the national patterns discussed above. More generally, it is important to note that for both women and men who moved into the

South East two of the three working class categories have location quotients over 1.00. This indicates that not all of the in-migrants succeed in obtaining middle class jobs – many finish up in low-level white-collar jobs, while some do manual work and others enter unemployment.

Turning to some of the detail in table 2.6, we can see the specific patterns making up the broad features described above. The high rates of entry into the service class categories arise partly from the extremely high rates of flow, for both men and women, from education. On top of this, however, but *only for women*, there are high rates of transfer into professional and managerial jobs in the South East from other parts of the labour market outside the South East (1.16 and 1.29 respectively). Particularly noteworthy is the upward mobility of women white-collar workers migrating into managerial jobs in the South East (1.35) – a less usual transition at the national level. Also very interesting, though based on small numbers, is the high rate of movement of women professionals into management as they migrate to the region (1.52). As we have already seen, this is a generally a particularly unusual move for women professionals who appear often to be segregated into what Crompton and Sanderson (1990) have called 'practitioner niches', whilst male professionals get more opportunities to move into managing other professionals. Another very interesting gender difference arises in the case of entry into working class categories. The reason why the values for men migrating into white-collar and blue-collar jobs are higher than those for women is largely accounted for by the tendency for young men to migrate from full-time education outside the South East into these jobs in the South East. This, added to what was said above about professional and managerial women migrants suggests that there is something of a contrast between the sexes when it comes to this migration stream. The women tend to be from more middle class backgrounds and they enter working class jobs less than the men (relative, of course, to the norm set by all inter-regional migrants).

Examining the figures separately for men and for women, and ranking the location quotients from the highest values to the lowest, it can be seen that the five highest values for men all have education outside the South East as their origin category. Three of these are into the three middle classes, but the fourth is odd – education to other. Since other for men means the armed forces it looks as though there is a migration of young men from the provinces into the armed services in the South East. The next two values are flows from unemployment, but the destinations are blue-collar work and the petite bourgeoisie. It is conceivable that this reflects the migration of Asians from the Midlands and North into London. Of the remaining ten values above 1.10, three relate to promotion from working class jobs to middle class ones, so there is ample evidence of the upward mobility benefits enjoyed by male migrants to the South East region.

For women, education as an origin for inter-regional migrants to the South East region dominates rather less than for men, and the white-collar occupations as a destination rather more. But high up in the ranking is the very significant transition from white-collar working class to manager. Like men, the white-collar to professional transition is also represented, as are movements from education in the provinces to working class jobs in the South East. A distinctive feature of the female flow matrix is the presence of high values for the professional to professional transition (1.11) and from professional to white-collar work (1.22). It is interesting to speculate on these gender differences.

It has been said that middle class male careers are built upon the presence of a supportive wife and family, whereas middle class female careers are best built alone, unencumbered by family commitments. This could possibly give rise to gender differentiation in migration flows, since a migration to the South East is more easily undertaken by those 'travelling light' (due to the high costs of housing, difficulties with schooling, stress of commuting from suburban homes) who are entering a labour market particularly well suited to their career development (i.e. unattached professional women), than by those possessing the opposite characteristics (i.e. married men with routine middle class jobs and with suburban family commitments).

Table 2.7 shows the equivalent transitions for migrants from the South East region to the rest of England and Wales. Again, it is sensible to look at the broader features before inspecting the detail. The final column shows the social class origins of these out-migrants. The main features are that the out-migrants come from labour market origins, and especially from managerial and white-collar occupations, and that gender differences are slight (but notice the much higher figure for the male petite bourgeoisie than for the female). The situation with respect to the class destinations of these out-migrants is very different. The bottom rows of the figures for men and women show that retirement is the most distinctive destination for both men and women, and to almost the same degree. The second most important destination may come as a surprise – the petite bourgeoisie location quotients are high for both men and women (though in the case of women the numbers involved are very small). Migrants from the South East region have a strong tendency to set up in business on their own when they leave the region. The third most attractive destination category is highly gender-specific. The category other is of importance as a destination for women (where the numbers are very large), but not for men (where the numbers are very small). Finally, it can be seen that the values of service class destinations differ between men and women. The men tend to go to professional and managerial jobs to a far greater extent than women. We can summarize these results by saying that while retirement is important for both sexes, men tend to get better jobs than women do when they migrate away from the South East, and women often leave the labour market altogether when they make this migration.

If the individual location quotients are ranked from highest to lowest we can add some detail to this picture. For men, the top four rankings all have retirement as their destination; two of them from middle class categories,

Table 2.7 The social composition of the migration stream from the South East region 1981–91

	PRO	*MAN*	*PeB*	*WhC*	*BlC*	*UNE*	*Total (labour market)*	*Retire*	*Other*	*Total*
Males										
PRO	0.99	0.98	1.39	1.44	1.15	1.03	1.03	1.69	0.22	1.14
MAN	1.18	0.96	1.40	1.26	1.56	1.47	1.13	1.82	1.25	1.28
PeB	1.51	0.79	1.18	0.78	1.02	1.30	1.12	1.46	0.89	1.20
WhC	1.02	1.05	1.56	0.89	1.56	1.44	1.12	1.93	1.34	1.29
BlC	0.86	0.89	1.17	0.68	0.97	1.11	0.98	1.77	0.92	1.18
UNE	0.50	0.57	0.90	0.63	0.61	0.82	0.70	1.16	0.48	0.80
Total (labour market)	*0.99*	*0.96*	*1.24*	*0.92*	*1.01*	*1.12*	*1.02*	*1.70*	*0.79*	*1.17*
Education/Retirement	0.53	0.42	0.68	0.62	0.77	0.88	0.62	1.05	0.56	0.79
Other	1.22	1.15	0.91	1.23	1.05	0.64	1.03	1.11	0.77	0.95
Total	*0.80*	*0.84*	*1.15*	*0.80*	*0.94*	*1.00*	*0.89*	*1.31*	*0.66*	*1.00*
Females										
PRO	0.96	0.84	1.40	1.01	0.84	0.98	0.97	1.62	1.26	1.11
MAN	1.15	1.59	1.33	1.20	1.07	1.52	1.34	1.54	1.25	1.32
PeB	0.70	0.47	1.56	0.74	1.01	0.76	0.97	1.10	0.84	0.98
WhC	0.92	0.90	1.44	0.96	1.14	1.21	1.02	1.91	1.32	1.28
BlC	0.65	0.54	1.23	0.75	0.89	0.40	0.78	1.70	1.11	1.12
UNE	0.43	1.01	1.01	0.65	0.60	0.87	0.68	1.52	0.78	0.83
Total (labour market)	*0.92*	*0.99*	*1.39*	*0.93*	*0.96*	*1.01*	*0.98*	*1.74*	*1.20*	*1.18*
Education / Retirement	0.50	0.38	0.65	0.56	0.79	0.77	0.56	1.08	0.70	0.76
Other	0.92	0.79	1.09	0.87	0.93	0.99	0.91	1.29	1.11	1.10
Total	*0.72*	*0.70*	*1.20*	*0.77*	*0.90*	*0.88*	*0.80*	*1.30*	*1.06*	*1.00*

Source: OPCS 1991 Longitudinal Study (Crown Copyright Reserved).

managers and professionals, and two from working class categories, white-collar and blue-collar workers. Then, at 1.56, comes the first of the five cases of significantly high figures for entry into the petite bourgeoisie (if the professional to petite bourgeoisie category were a little larger this would make a sixth). This first one is particularly interesting because it would perhaps normally be regarded as a case of social promotion – the transition from low-level white-collar work in the South East to membership of the petite bourgeoisie outside the South East. The transition from blue-collar employment to the petite bourgeoisie also figures in the list of transitions with location quotients of over 1.10. Thus the male figures show both retirement and the petite bourgeoisie as important destinations but with signs of upward social mobility in addition to those listed above.

For women out-migrants the picture is complicated by the strong presence of other as a destination category. Once again, retirement tops the list of ranked figures, but although the white-collar to petite bourgeoisie is there, most of the remaining high values are for shifts from labour market positions in the South East to full-time housewife or carer outside the South East.

Whether by choice or by constraint, many women who leave the South East region, leave the labour market at the same time.

Summary and conclusions

This chapter examined some of the differences which adopting a geographical perspective makes to the gendering of social mobility. We began with a brief selective review of relevant literature from both sociology and geography and suggested that some important aspects of the articulation between gender, class and region remain relatively unexplored. In particular, studies of gender and social mobility rarely consider whether their findings are representative of all places, whilst studies into the geography of gender rarely consider whether or how gendered experiences, practices and relations change over time. The rest of the chapter contributed to the investigation of these themes using new empirical material from the 1981–91 Longitudinal Study. The main results were:

- During the 1980s women made substantial inroads into the expanding managerial and professional categories, as well as moving into blue-collar work at an increased rate, whilst men showed particular increases of movement into the petite bourgeoisie.
- Space makes a difference to the patterns of social mobility which women and men experience. There are distinct differences in the geographies of opportunity between men and women. Specifically, while the South East region displays exceptionally high rates of social promotion for both men and women, it is distinguished by its particularly high rates of upward mobility of women into managerial positions. There is also a tendency for the spatial patterns of upward mobility to be more gender-specific than the spatial patterns for downward mobility.
- Social class changes that accompany inter-regional migration both reflect and reinforce these regional differences in social mobility rates, and these are also gender-differentiated. Both men and women who migrate to the South East region tend to experience social promotion, but the rate of upward mobility for women is higher (relative to other regions) than that for men. Both male and female migration flows away from the South East are biased towards those leaving the workforce through retirement, but the female flows are also biased towards exit from the labour market before retirement and towards downward social mobility. A remarkable similarity in the geography of gender-specific social mobility rates for both migrants and non-migrants between the 1971–81 and 1981–91 periods is found, suggesting that these structures and differences are very deep-rooted and were not fundamentally altered by the Thatcher revolution of the 1979–90 period.

Notes

1 This is in line with the class analysis developed in Savage et al. (1992) – i.e. classes based upon (a) credentials and qualifications, (b) organizational skills and experience, and (c) the ownership of capital assets.

2 See Fielding and Halford (1993), where the LS tables for 1971–81 are presented in identical format to the 1981–91 tables presented here.

3 The figures for unemployment must be interpreted with some care. In particular, it is possible that there is a degree of under-reporting of female unemployment, given benefit rules which exclude many married women, even though the census material used here is quite separate and not supposed to be affected the rules determining access to benefit.

4 In these tables we have normally only included those transitions which exceed 0.5 per cent of all male or female inter-regional migrants. Exceptionally, we have included smaller flows but these are shown in brackets.

5 The rates of entry into the third middle class category, the petite bourgeoisie, are low for both female and male migrants to the South East.

References

Abbott, P. (1990) 'A re-examination of "three theses re-examined"', in G. Payne and P. Abbott (eds) *The Social Mobility of Women*, Basingstoke: Falmer Press, pp. 37–46.

Abbott, P. and Sapsford, R. (1987) *Women and Social Class*, London, Tavistock.

Acker, J. (1990) 'Hierarchies, jobs, bodies: a theory of gendered organization', *Gender and Society* 5: 139–58.

Beechey, V. and Perkins, T. (1987) *A Matter of Hours: Women, Part Time Work and the Labour Market*, Cambridge: Polity Press.

Bonney, N. and Love, J. (1991) 'Gender and migration: geographical mobility and the wife's sacrifice', *Sociological Review* 39: 335–48.

Boyle, P. and Halfacree, K. (1995) 'Service class migration in England and Wales 1980–1: identifying gender-specific mobility patterns', *Regional Studies* 29: 43–57.

Brannen, J. (1989) 'Childbirth and occupational mobility', *Work, Employment and Society* 3: 179–201.

Chapman, T. (1990) 'The career mobility of women and men', in G. Payne and P. Abbott (eds) *The Social Mobility of Women*, Basingstoke: Falmer Press, pp. 73–82.

Congdon, P. (1990) 'Graduation of fertility schedules', *Regional Studies* 24: 311–26.

Crompton, R. and Sanderson, K. (1990) *Gendered Jobs and Social Change*, London: Unwin Hyman.

Dex, S. (1987) *Women's Occupational Mobility: a Lifetime Perspective*, Basingstoke: Macmillan.

Dex, S. (1990) 'Occupational mobility over women's lifetime', in G. Payne and P. Abbott (eds) *The Social Mobility of Women*, Basingstoke: Falmer Press, pp. 121–38.

Duncan, S. (1991) 'The geography of gender divisions of labour in Britain', *Transactions of the Institute of British Geographers* 16: 420–39.

Fielding, A. and Halford, S. (1993) 'Geographies of opportunity: a regional analysis of gender-specific social and spatial mobilities in England and Wales 1971–1981', *Environment and Planning A* 25: 1421–40.

Finch, J. (1983) *Married to the Job: Wives' Incorporation into Men's Work*, London: Allen & Unwin.

Gallos, J. (1988) 'Exploring women's development: implications for career theory, practice and research', in M. Arthur, D. Hall and B. Lawrence (eds) *The Handbook of Career Theory*, Cambridge: Cambridge University Press, pp. 110–32.

Goldthorpe, J., Llewellyn, C. and Payne, C. (1980) *Social Mobility and Class Structure in Modern Britain*, Oxford: Oxford University Press.

Halford, S., Savage, M. and Witz, A. (1997) *Gender, Careers and Organizations: Current Developments in Banking, Nursing and Local Government*, Basingstoke: Macmillan.

Hanson, S. and Pratt, G. (1995) *Gender, Work and Space*, London: Routledge.

Kanter, R. M. (1977) *Men and Women of the Corporation*, New York: Basic Books.

McDowell, L. and Massey, D. (1984) 'A woman's place?', in D. Massey and J. Allen (eds) *Geography Matters*, Buckingham: Open University Press, pp. 128–47.

Mark-Lawson, J., Savage, M. and Warde, A. (1986) 'Gender and local politics: struggles over welfare policies, 1918–39', in L. Murgatroyd, M. Savage, D. Shapiro, J. Urry, S. Walby, A. Warde and J. Mark-Lawson *Localities, Class and Gender*, London: Pion, pp. 195–215.

Marshall, G., Newby, H., Rose, D. and Vogler, C. (1988) *Social Class in Modern Britain*, London: Hutchinson.

Marshall, J. (1988) 'Re-visioning career concepts: a feminist invitation', in M. Arthur, D. Hall and B. Lawrence (eds) *The Handbook of Career Theory*, Cambridge: Cambridge University Press, pp. 275–91.

Payne, G. and Abbott, P. (eds) (1990) *The Social Mobility of Women*, Basingstoke: Falmer Press.

Savage, M., Barlow, J., Dickens, P. and Fielding, T. (1992) *Property, Bureaucracy and Culture*, London: Routledge.

Ward, C. and Dale, A. (1992) 'Geographical variation in female labour force participation: an application of multi-level modelling', *Regional Studies* 26: 243–55.

Whyte, W. (1957) *The Organization Man*, New York: Touchstone.

3 Gender variations in migration destination choice

David Atkins and Stewart Fotheringham

Introduction

Internal migration is the major factor shaping the population distribution of Great Britain and an understanding of this process is essential to many aspects of central and local government planning and to the business strategies of private enterprises. An effective means of developing such understanding is through the mathematical modelling of migration choice. The migration process itself has often been deconstructed into two phases of decision making; whether to move and where to move (Stillwell and Congdon 1991; Nam, Serow and Sly 1990). These phases need not be independent: for instance, the decision to make a short-distance 'housing-related' move may be prompted by a particular destination becoming available. However, longer-distance moves such as those examined in this chapter are more often 'employment-driven' and in these cases the two decision phases can be broadly considered to be independent.

Many techniques have been employed to model the aggregate movement of migrants (Stillwell and Congdon 1991) but the most popular is the compensatory framework, often termed gravity modelling, based on the principles of distance-decay and a variety of place attraction and/or information surrogates. The model applied here is in essence a gravity model but with the important addition of a variable which measures the spatially varying degree of competition between potential destinations resulting from their uneven spatial distribution. Destinations in close proximity to other destinations and which face greater spatial competition appear to be chosen less frequently than destinations which are relatively isolated, *ceteris paribus* (Fotheringham 1983, 1986). The development of this model, termed a competing destinations model, owes much to the field of cognitive science and in particular investigations of spatial perception (Fotheringham 1991; Fotheringham and Curtis 1992, 1997).

When examining migration behaviour through the application of models, traditionally one set of model parameters is obtained and these are used to draw inferences about the determinants of the migration process. It is not unreasonable, however, to assume that such determinants might vary over space. Indeed, recently it has been increasingly recognized that global statistics can hide a great deal of spatial variation. Essentially, global statistics give an 'average' value that hides potentially interesting differences over space. Consequently, there has been a general movement away from global and towards local statistics (Fotheringham 1997). In order to examine the extent of such spatial variation in the determinants of destination choice the competing destinations model was calibrated independently for migration from each origin. This not only gives an indication of those areas where certain variables are more important to destination selection but also allows identification of those areas where gender variations are most marked.

In this chapter, using the British example, the competing destinations model of destination choice is calibrated separately for single male and single female migrants, enabling gender variation in the determinants of destination choice to be examined. In order to isolate single migrants, and hence to increase the independence of migrants, the 1991 Census Special Migration Statistics were used. These provide ward- and district-level flow data disaggregated by a number of socio-economic variables, including marital status and gender. This analysis examines movements between 37 selected districts. Additional data about these districts were obtained from the census and other sources.

The competing destinations model

The general gravity model

The most widely applied model of migration behaviour is the gravity model of spatial interaction (Haynes and Fotheringham 1984). In its general form this model can be expressed:

$$M_{ij} = k\ O_{i1}^{\alpha 1}\ O_{i2}^{\alpha 2} \ldots O_{im}^{\alpha m}\quad D_{j1}^{\phi 1}\ D_{j2}^{\phi 2} \ldots D_{jn}^{\phi n}\ d_{ij}^{\beta} \qquad (1)$$

where: M_{ij} is the migration flow from area i to area j;

k is a scaling parameter which ensures that the sum of all predicted flows equals the sum of actual flows (i.e. ensuring that $\Sigma_j \hat{M}_{ij} = \Sigma_j M_{ij}$);

O_i represents an origin propulsiveness variable of which there are m;

D_j represents a destination attractiveness variable of which there are n;

d_{ij} represents the spatial separation between areas i and j;

α is a parameter representing the sensitivity of M_{ij} to variations in a particular origin attribute;

ϕ is a parameter representing the sensitivity of M_{ij} to variations in a particular destination attribute;

β is a distance-decay parameter which describes the sensitivity of M_{ij} to variations in spatial separation.

When considering destination choice alone, the array of origin propulsiveness variables is omitted from the above equation and the model is then written:

$$M_{ij} = \frac{D_{j1}^{\phi 1}\, D_{j2}^{\phi 2} \ldots D_{jn}^{\phi n}\, d_{ij}^{\beta}}{\Sigma_j \quad D_{j1}^{\phi 1}\, D_{j2}^{\phi 2} \ldots D_{jn}^{\phi n}\, d_{ij}^{\beta}} \tag{2}$$

This ensures that the constraint $\Sigma_j \hat{M}_{ij} = \Sigma_j M_{ij}$ is met for each origin.

A wide variety of destination variables can be and have been included in this model (Fotheringham and Pitts 1995; Boyle 1993, 1994). As well as origin-destination separation and destination accessibility (discussed below), the current analysis includes variables representing population, unemployment rate, house prices, tenure and social class structure at each of the destinations.

Competition between potential destinations

The standard model of migration described above (equation 2) has been questioned by Fotheringham (1983, 1986, 1987, 1991) because its theoretical derivation depends on the assumption that individuals consider every alternative before selecting the one which maximizes their utility. In most spatial choice situations, and particularly in migration, the number of alternatives is so large as to make such an assumption untenable. If the behavioural assumption embedded in the derivation of the standard migration model is invalid, the application of the model to understand aspects of the migration process must be highly questionable.

As an alternative, Fotheringham (1986, 1991) has suggested that individuals do not evaluate every alternative but instead process spatial information hierarchically. It is envisaged that individuals store information in spatial clusters and initially make decisions based on these clusters rather than on the individual destinations within them. Hence, not all destinations are evaluated and suboptimal choices may well result.

One type of modelling framework which has been developed for hierarchical choice such as this is the nested logit model (Ben-Akiva and Lerman 1987), which can be viewed as a multi-stage extension of the standard multinomial framework described above. However, several problems exist with the nested logit model when applied to migration destination choice. One is that it is assumed that the modeller knows the spatial hierarchy viewed by an individual. In cases such as shopping choice where the hierarchy consists of a set of shopping centres and shops within each centre, the spatial hierarchy is well defined and the assumption is a reasonable one. However, in migration destination choice there is generally no well-defined hierarchy that can be assumed by the modeller. Consequently, the operation of the nested logit model in such a circumstance is fraught with subjectivity in the definition of the spatial hierarchy.

A second problem is that the nested logit model assumes a spatially invariant hierarchical structure that again is unreasonable for large-scale migration studies. While it seems reasonable to assume that migrants view space hierarchically and have views on 'clusters' such as 'The North' or 'The North East', it is entirely unreasonable to assume that all migrants' views of space are alike. What constitutes 'The North' is likely to vary according to the location of an individual, and a person's spatial hierarchy is likely to reflect his/her knowledge surface which will vary across space. Individuals are more likely to subdivide mentally the area around where they live, and with which they have a greater familiarity, than the areas more distant, and more unfamiliar, *ceteris paribus.*

A third problem with the application of the nested logit model to migration studies is that the destinations within a particular cluster are all assumed to be equally substitutable, whereas it is more reasonable to assume that closer places are more likely to be substitutes for each other than are more distant ones. Similarly, destinations on the edges of clusters might well be seen as substitutes for destinations in close proximity but which lie in a neighbouring cluster – a facet of the choice process not recognized within the nested logit framework.

Due to these problems, the nested logit model seems inappropriate to most spatial choice situations. Fotheringham (1991) has developed the competing destinations model as an alternative framework in which to model spatial choice resulting from hierarchical information processing. The general form of the competing destinations model is:

$$M_{ij} = \frac{D_{j1}^{\phi 1} D_{j2}^{\phi 2} \ldots D_{jn}^{\phi n} d_{ij}^{\beta} P_i(j \in \Lambda)}{\Sigma_j D_{j1}^{\phi 1} D_{j2}^{\phi 2} \ldots D_{jn}^{\phi n} d_{ij}^{\beta} P_i(j \in \Lambda)} \quad (3)$$

where: $P_i(j \in \Lambda)$represents the probability that the destination j is in the cluster Λ evaluated by the individual at i.

The formula allows that suboptimal choices might be made because not every alternative is evaluated: there might be 'better' alternatives not chosen simply because they are not evaluated.

Fotheringham has shown that it is not necessary to define the exact nature of the clusters viewed by individuals. It is only necessary to include an attribute describing the general location of a destination with respect to other 'competing' destinations that result from a hierarchical choice process. The argument is as follows. Suppose that individuals do cognize destinations in clusters. It is reasonably well accepted that individuals tend to underestimate the number of objects in large groups (Stevens 1975), so that they are less likely to be cognizant of, and therefore evaluate, destinations that occur in large clusters. To represent this relationship within a migration model it is only necessary to include a term that measures the probability of a destination being within a large cluster as perceived by an individual. Although there are several measurements that could be used, a potential accessibility measure is often used and makes intuitive sense. The relationship assumed is:

$$P_i(j \in \Lambda) = f(A_j) \tag{4}$$

where: A_j represents the accessibility of destination j to other destinations.

Equivalently, A_j can be viewed as a 'centrality' measure – the closer j is to other destinations, the larger A_j will be and vice-versa. A commonly accepted measure of accessibility is:

$$A_j = \Sigma_k P_k / j_k \tag{5}$$

where: k indexes a destination other than j.

By substitution, a specific version of the competing destinations model is then:

$$M_{ij} = \frac{D_{j1}^{\phi 1} D_{j2}^{\phi 2} \ldots D_{jn}^{\phi n} d_{ij}^{\beta} A_i^{\lambda}}{\Sigma_j D_{j1}^{\phi 1} D_{j2}^{\phi 2} \ldots D_{jn}^{\phi n} d_{ij}^{\beta} A_i^{\lambda}} \tag{6}$$

where: λ is a parameter to be estimated.

If λ is zero, then the competing destinations model is equivalent to the standard logit model given above; increasingly negative values indicate increasing intensities of hierarchical decision making. Fotheringham and O'Kelly (1989) have shown that this framework no longer has the undesirable 'independence from irrelevant alternatives' property embedded in the standard logit framework and therefore represents a substantive change in the formulation.

Numerous empirical examples have supported the suggestion that the competing destinations model provides a more accurate framework for the modelling of spatial flows than does the traditional logit framework (Fotheringham 1983, 1986, 1987; Fotheringham and O'Kelly 1989; Curtis 1991; Pellegrini 1996). The evidence mostly falls into three types:

- The parameter on the additional accessibility variable is significant and negative;
- The model replicates known flow patterns more accurately, even adjusting for the extra degree of freedom in the model;
- Because the standard logit formulation is inappropriately derived, it is a gross mis-specification of reality and parameter estimates obtained from it contain a potentially severe mis-specification bias. This is particularly noticeable in origin-specific estimates of the distance-decay parameter for reasons described by Fotheringham (1991).

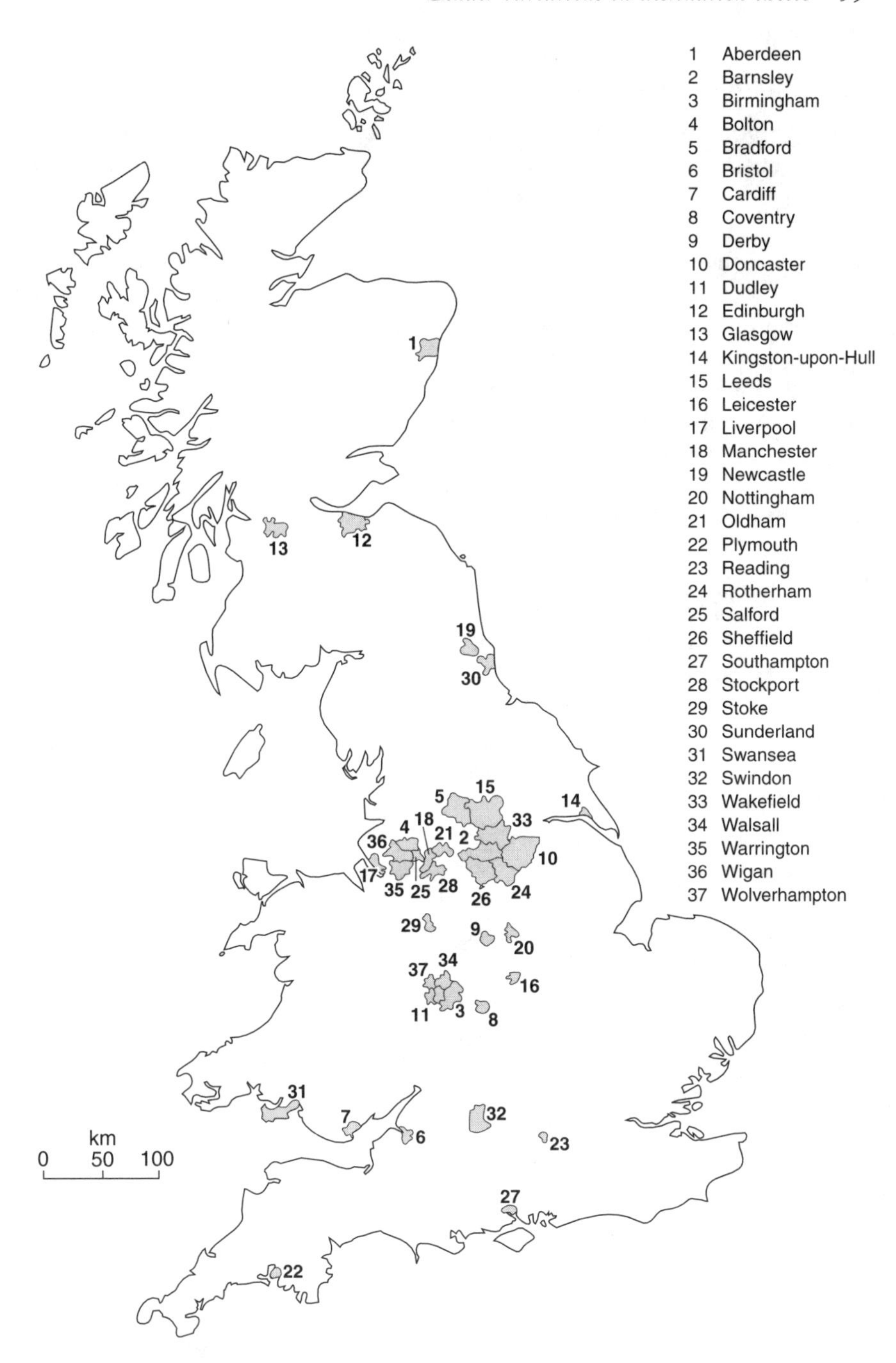

Figure 3.1 The location of 37 districts selected for analysis

The data

The migration system

The migration system under study consists of 37 selected local authority districts. These are the largest districts, in terms of population, which the authors considered to have a distinct 'identity', where the administrative boundary coincides with a more widely recognized 'place' boundary. Where this is not the case, for instance when a district contains a number of similar-sized settlements but no major focal point, the district level was felt to be an inappropriate scale for migration destination choice analysis. Although various scales could be combined within the same migration analysis, this is not attempted in the present study. For similar reasons, all London boroughs were excluded from the analysis.[1] A few districts were also included to improve the spatial coverage over the rest of the country (notably Aberdeen, Falmouth, Southampton and Swansea). Figure 3.1 shows the locations and names of the 37 selected districts.

The migration flow data

The flow data used to calibrate the models were extracted from the 1991 Special Migration Statistics (SMS) derived from the Census of Population (OPCS 1992; Rees and Duke-Williams 1994). These are held on the MIDAS datasets server at Manchester University and can be accessed from there using the SMSTAB extraction software (MIDAS 1997; Duke-Williams 1995). The SMS consist of two sets of migration matrices: Set 1 at ward-level disaggregated by gender and 5 year age groups; and Set 2 at district-level, disaggregated by broad age groups, gender, ethnic group, marital status, limiting long-term illness, economic position and tenure. More specifically, the flow data used here were derived from table 4 of the SMS Set 2 data. This table comprises six mutually exclusive migration matrices which have here been aggregated to three matrices describing the movements of married, single male and single female migrants. By using the other tables of the SMS Set 2 data, a variety of migrant subgroups can be isolated and their destination choice behaviour examined independently.

Destination characteristics

Data for the destination attractiveness variables – population, unemployment rate, tenure structure and social class structure – were obtained from the Small Area Statistics (SAS) of the 1991 British Census (OPCS 1991). Average house price data, weighted by property types, were obtained from a building society for house sales during 1988 (Dorling 1989). Straight-line origin-destination separations were calculated from local authority district centroids derived from population weighted ward centroids (Atkins et al. 1993). The district population data are the 1991 definition of the 'normally resident' populations, as reported in table 1 of the SAS. The unemployment rate was calculated as the percentage of the economically active population which was reported as either unemployed or on a government scheme in table 8 of the SAS. The tenure variable represents the percentage of all households in a district that are owner-occupied. This was extracted from table 20 of the SAS. The social class structure variable is calculated as the percentage of a district's economically active heads of households who are professionals and managers or, more specifically, the proportion of those heads of households in social classes I–V who are in social classes I and II. This variable was extracted from the SAS and excludes those in the armed forces or on government schemes.

Gender variation: spatially aggregate results

Using the spatially aggregate data (the type usually used in migration analysis) a comparison of the parameter estimates indicates few apparent differences between single males and single females in their migration destination choice behaviour. Table 3.1 compares spatially aggregate parameter estimates for single male and single female migrants. These parameter estimates indicate the nature of the relationships that the model's explanatory variables are found to have with the independent variable once variations in the other variables have been taken into account.

A negative distance parameter estimate indicates that more distant destinations are less likely to be chosen by migrants *all other things being equal*. Furthermore, the more negative this parameter estimate the stronger this relationship, i.e. the more strongly migrants are deterred from moving to more distant destinations rather than to closer ones. Both the percentage of owner-occupiers and the accessibility of areas also have negative relationships with migration flows. Higher levels of owner occupancy imply lower levels of rented accommodation which is often the first step for in-migrants (especially single in-migrants), as they may not want to commit themselves to the owner-occupied housing market in a new location which they may not like, or where they may not plan to stay long. The negative relationship between migration and destination accessibility supports the hypothesis of hierarchical destination choice. Higher accessibility represents increased likelihood of an area being perceived as being in a large cluster of potential destinations. The size of large clusters is generally underestimated so the cluster as a whole

Table 3.1 Gender comparison of parameter estimates from aggregate analysis

	Distance	*Population*	*Unemployment*	*Class*	*House prices*	*Tenure*	*Accessibility*	R^2
Single males	–1.48	0.64	0.64	0.94	1.07	–0.84	–1.10	0.885
Single females	–1.53	0.71	0.43	0.83	0.95	–0.97	–1.04	0.890
Z stat.	4.28	–2.77	2.64	1.24	1.93	1.35	–1.18	—
95% sig.	Yes	Yes	Yes	No	No	No	No	—

will receive less attention than its actual size merits. This means the members of large clusters are less likely to be individually assessed or therefore selected than areas in smaller clusters. For instance, Aberdeen, which is quite isolated and therefore has a low accessibility, is more likely to receive individual scrutiny than Salford, a high-accessibility area that is surrounded by other metropolitan areas.

Although there is almost always a negative relationship between accessibility and immigration, the value of the accessibility parameter estimate varies. If the model is calibrated for separate population subgroups and, as we shall see later, for different origins it is useful to draw conclusions from comparisons of the resulting accessibility parameter estimates. A more negative accessibility parameter indicates a more intensely hierarchical process of destination choice. In other words, the accessibility parameter estimate is indicative of the extent to which migrants' mental spatial representations of potential destinations are hierarchically constructed.

The parameter estimates in table 3.1 show that the population, social class, house prices and, perhaps surprisingly, unemployment of potential destinations have positive relationships with migration into these areas. So, *ceteris paribus*, areas where these variables are higher will experience higher levels of in-migration. Population is a well-documented attractor of migrants, largely because of the amenities and employment opportunities which larger urban areas have to offer, but also because migrants are more likely to be aware of areas of high population and have information about them (Haynes and Fotheringham 1984). Similarly, a high percentage of the higher social classes is a well-documented positive determinant of migration. Higher house prices might intuitively be expected to have a deterrent effect on purely economic grounds, but house prices are also an indicator of the general affluence of an area, and most people like to live in the most affluent area they can afford. The positive relationship between an area's unemployment rate and its likelihood of selection as a migration destination is also counterintuitive. This might be a result of the rapid changes in industrial structure, which are occurring in areas of declining traditional industries. Areas such as the North East, Merseyside and Yorkshire where mining, shipbuilding and steel-working were once the main employers are now characterized by high numbers of middle-aged unemployed. New employers (often in high-tech or service sectors) enter these areas because of low costs but often prefer to attract younger blood from further afield rather than to retrain and employ the locally unemployed.

Whilst, individually, all of these parameter estimates are statistically significant, difference of means tests indicate that gender variation is only statistically significant (at the 95 per cent level) for the Distance, Population and Unemployment parameters. From this it can be concluded that female migrants are:

- more deterred by distance;

- more attracted to areas with larger populations;
- less attracted to areas with higher unemployment rates.

However, whilst these differences in migration behaviour between males and females appear relatively slight at the aggregate level, an examination of the parameter estimates from origin-specific model calibrations reveals a more complex picture.

Gender variations: spatially disaggregate results

Parameter estimates

Performing separate calibrations for each of the 37 origins in the migration system provides 37 estimates for each parameter. Figures 3.2–3.8 show these parameter estimates plotted for single male versus single female migrants. For some origins there appear to be considerable gender variations in the determinants of migration flows. In some cases this may result in part from sample size problems, as flows of single males and single females between some study areas are very small. Figures 3.2–3.8 distinguish such cases by plotting parameter estimates which are 95 per cent significant as black circles and cases where parameter estimates were not 95 per cent significant as white squares. This does not refer to the statistical significance of the gender comparison but reflects the parameter estimate standard errors calculated by the calibration procedure. Although in many cases where gender variation is apparent at least one of the parameter estimates is individually insignificant, there are also many cases of individually significant parameter estimates showing quite marked gender variation. The figures illustrate a wide range of parameter estimate values for migrants from different origins and also

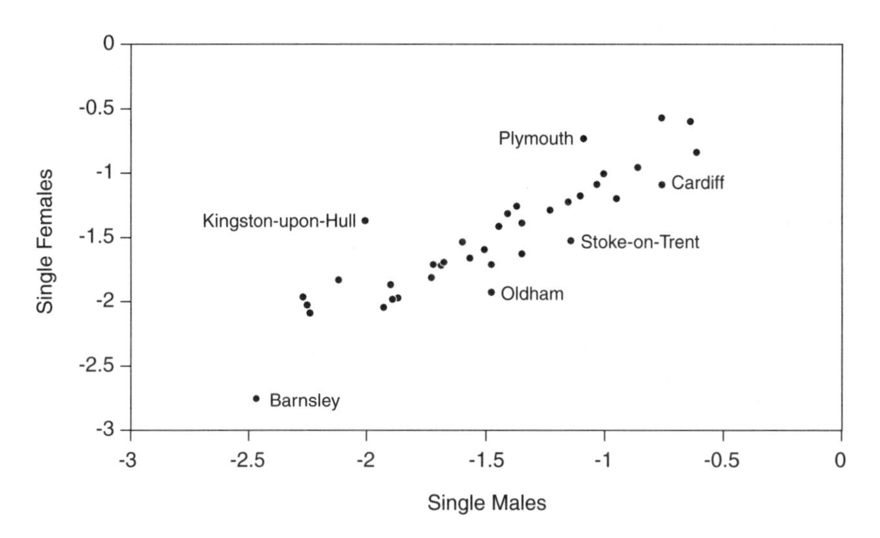

Figure 3.2 Gender variation in the distance parameter estimates

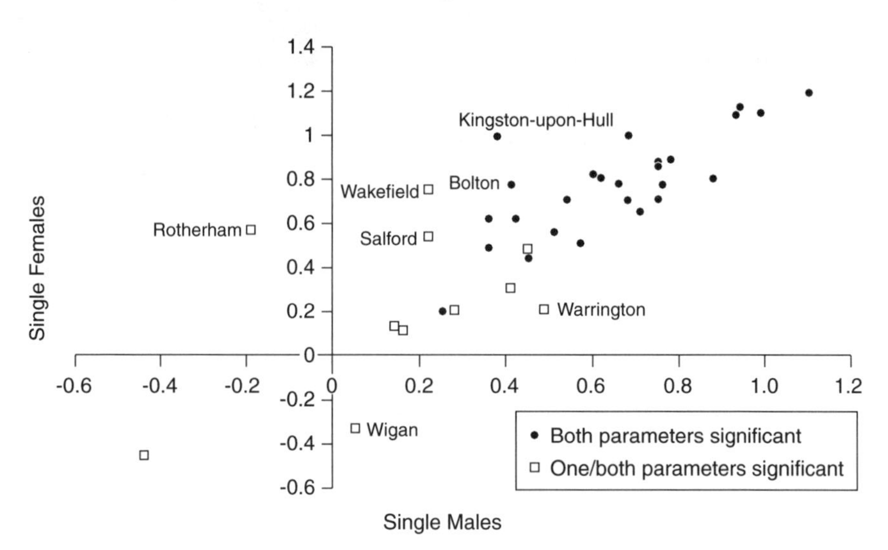

Figure 3.3 Gender variation in the population parameter estimates

show the degree to which gender variation is spatially variable, something which is impossible from the results of global model calibrations.

The motivation behind most migration is employment – people usually move home to start a new job or to try and find a new job. Thus, the industrial structure of the various areas in a migration system has a very significant effect upon migration within that system. Because of the difficulty of obtaining data on industrial structure and usefully incorporating it in a modelling framework, the competing destinations model applied here takes no account of the differing employment opportunities available in different areas. It is believed that this is the major cause of the high degree of spatial and gender variation in migration destination choice behaviour evident from figures 3.2–3.8.

Figure 3.2 shows the distance parameter estimates for single male and single female migrants from each of the 37 origins. This parameter represents the extent to which increasing distance from a potential destination deters migrants from selecting that destination. This is the only explanatory variable for which all parameter estimates were individually significant (at the 95 per cent level), suggesting that despite the wide range of values of the parameter estimate spatial separation played an important role in the decision-making process of migrants from all origins. There is limited gender variation, however. Hull and Plymouth are above the diagonal, meaning that male migrants from those places are more deterred by distance than female migrants. Conversely, in Oldham, Stoke and Cardiff it is the female migrants who are more deterred by distance. This could reflect the gender bias in the nature of the employment opportunities available in and around these areas.

The range of parameter estimate values will in part result from vacancies in some industries being more spatially concentrated than others. If an area has a number of migrants with jobs or seeking jobs in an industry which is only located in distant areas then distance is likely to be less of a deterrent than in an area where all out-migrants are, say, plumbers, for whom work is very evenly distributed over space. This could explain why areas such as Aberdeen (oil industry), Hull (shipping) and Sheffield (steel) have high distance parameter estimates, whereas service- and high-tech industry-dominated places such as Bristol, Reading and Swindon have lower distance parameter estimates.

Figure 3.3 shows an interesting pattern of gender variation in estimates of the population parameter, with female migrants from almost all areas being more attracted to areas of higher population than male migrants from the same areas. This could result from the higher demands which young adult women and women with young children have for health and education services, and hence their desire to move to areas of high population which can supply these facilities.

The gender variation in class parameter estimates, which can be seen in figure 3.4, could result from variations in the employment structure and hence social class structure in origin areas. If, for instance, professional employment in an area was biased towards males then social classes I and II would be dominated by males, and this could increase the relative desire of males to move to areas of high social class. Aberdeen, home to the well-paid and male-dominated oil industry, however, appears to oppose this theory as it is female migrants from Aberdeen who are more attracted to higher class areas. This could be the result of return migration of male oil-workers who previously moved to Aberdeen from areas with fewer people in the higher social classes.

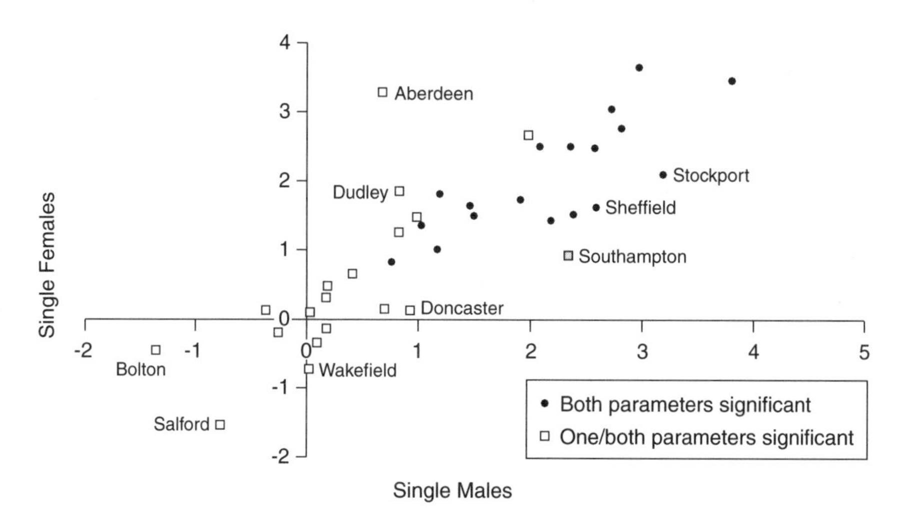

Figure 3.4 Gender variation in the class parameter estimates

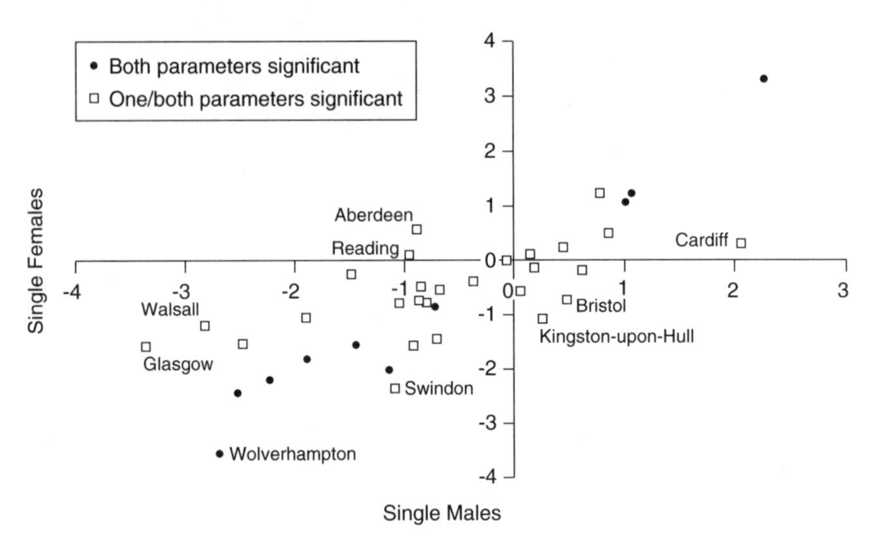

Figure 3.5 Gender variation in the house price parameter estimates

Gender bias in employment and earnings might also explain some of the gender variation in the house price parameter estimates, shown in figure 3.5. Male migrants from both Aberdeen and Hull, for instance, are more attracted to areas of higher house prices than their female counterparts. This could be because Aberdeen and Hull are dominated by male-biased oil and shipping industries, such that male migrants are on average better off and want to move to more affluent areas, with higher house prices.

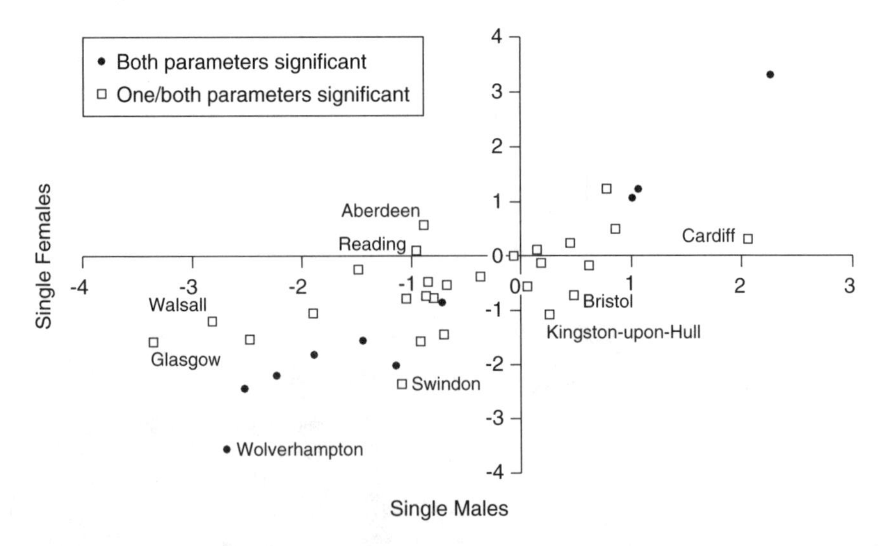

Figure 3.6 Gender variation in the tenure parameter estimates

An interesting pattern of gender variation can be seen in the tenure parameter estimates, shown in figure 3.6. From those origins where male migrants seek areas of high owner occupancy, female migrants are, broadly speaking, deterred by it, but from those origins where male migrants are deterred by high levels of owner occupancy, female migrants are less deterred. Indeed, male migration from two-thirds of origins was more responsive to destination areas' percentage of owner occupancy, despite the fact that in the global model single females had the larger tenure parameter estimate (–0.97 to males' –0.87). The attraction or otherwise of areas of high owner occupancy will be related to the stability of employment which a destination can offer, which is a variable which may well be different for males and females. It may be that areas such as Wolverhampton, Swindon and Hull have many women employed in part-time, contract or other 'low job security' employment. When these women migrate they may be going to similar employment elsewhere and may therefore be less likely to want or be able to commit themselves to owner occupancy. Such female migrants may be more inclined to avoid areas of high owner occupancy, whereas their male counterparts might well not be restricted to employment opportunities with the same degree of job security.

The trend in the spacially disaggregate unemployment parameters shown in figures 3.7 mirrors that of the global model in that most parameter estimates are significant positive. For most of the origins, there are no significant differences in the relationship between destination unemployment rate and migration, although for some, notably Cardiff, Southampton, Hull, Swindon, Plymouth, Aberdeen, Glasgow and Oldham, there are some large differences. In the case of the first four origins, the relationship between migration and destination unemployment rate is significantly more positive for single males

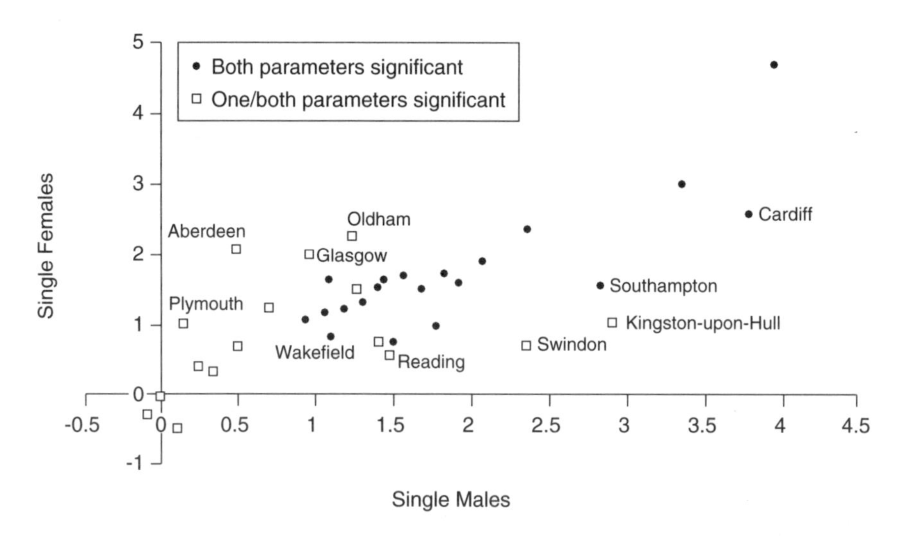

Figure 3.7 Gender variation in the unemployment price parameter estimates

than for single females. In the case of the latter four, the relationship is significantly more positive for single females. It is not immediately apparent why single male migrants from Cardiff, Southampton, Hull and Swindon and why single female migrants from Plymouth, Aberdeen, Glasgow and Oldham should be more attracted to areas of high unemployment.

The plot of accessibility parameter estimates in figure 3.8 shows some cases of quite marked gender variation. Recall that the accessibility parameter estimate can be thought of as representing the degree to which migrants from a particular origin are selecting their migration destinations hierarchically. An important determinant of how hierarchical a migrant's mental spatial representation of potential destinations will be is the total amount of information about those destinations that is available to migrants. Thus, there is a relationship between centrality and the accessibility parameter estimate as migrants from more central areas generally have more information available since they are situated closer to more potential destinations. Gender variation in information availability could result from employment patterns, with certain types of jobs, such as truck drivers and travelling salespeople, involving more travel and contact with people from other areas. Indeed, figure 3.8 shows that, for the majority of origins where parameter estimates were significant, male migrants had larger (negative) parameter estimates and were therefore making destination choices more hierarchically than female migrants. This suggests that there may indeed be some gender bias in 'information gathering' occupations creating an information imbalance and a consequent gender variation in accessibility parameter estimates.

Although much of the explanation of the patterns found in figures 3.2–3.8 is somewhat tentative, it will hopefully be recognized that analysis of this kind provides a valuable means of uncovering variation which global models cannot. More importantly it uncovers the spatial patterns in this variation,

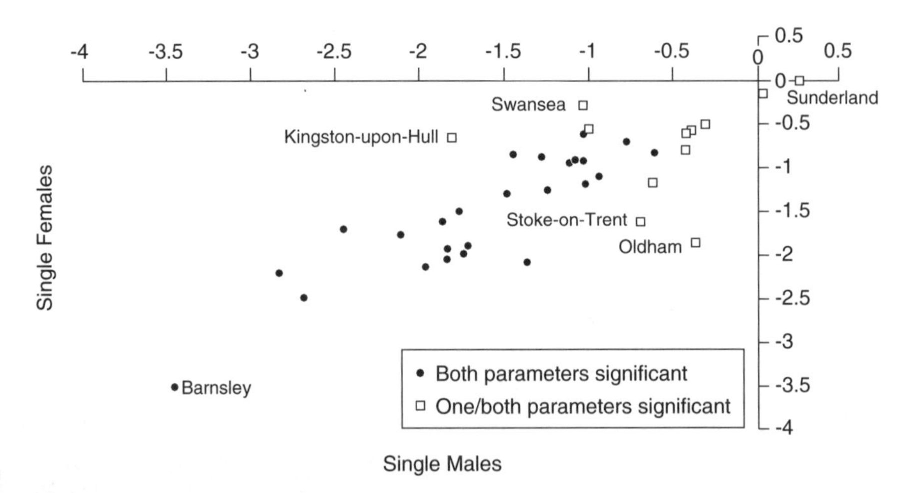

Figure 3.8 Gender variation in the accessibility parameter estimates

pinpointing specific areas where variations are more pronounced and where more in-depth analysis might prove valuable.

Goodness-of-fit variations

As with the aggregate case, the origin-specific calibrations show very little gender variation in goodness of fit, with the single males' and single females' R^2 statistics differing by less than 0.05 for over 80 per cent of origins. This can be seen in figure 3.9 which plots the R^2 values from the origin-specific model calibrations for single males against those for single females, indicating the range within which male and female R^2 values for an area vary by less than 0.05. The correlation coefficient between male and female R^2 values is 0.89.

It is interesting that there are similarities within the two sets of 'outliers' in figure 3.9. Of those with higher R^2 for single males, Swindon and Bristol are on the M4 corridor and, together with Southampton, all are southern in location. The three areas where single female R^2 values are more than 0.05 greater than single males' (Liverpool, Cardiff and Hull) are all coastal cities, but are widely spatially distributed. This suggests that goodness of fit may be related to an area's current and/or recent industrial structure as both the shipping industries of Liverpool, Hull and Swansea and the thriving high-tech industries of the M4 corridor are highly gender-biased employers.

Finally, in the light of recent research on the phenomenon of counterurbanization (Champion 1991; Atkins 1996; Champion and Atkins 1996), it might be tempting to assume that the negative relationship between migration and the accessibility statistic is an inevitable consequence of the accessibility statistic being a surrogate for urban–rural status. However, this assumption is not borne out by figure 3.10, which plots the accessibility statistic against net out-migration (1990–1) for the 37 districts in the migration system. It shows no clear relationship.

Conclusion

Separate calibration of the destination choice model for each origin in the migration system has provided an insight into a gender-variable process that is otherwise obscured by the 'averaging' effect of aggregate parameter estimates. Applications such as this provide further support for recent shift away from aggregate and towards local statistics (Fotheringham 1997).

Little gender variation in destination choice behaviour is evident from the global statistics but a more complex situation becomes apparent when migration decisions are modelled separately for each origin. Although in-depth interpretation and explanation of the gender variation evident from the origin-specific results is beyond the scope of this chapter, our analysis demonstrates the utility of calibrating spatial interaction models independently for subgroups of the population and for different origins. It is probable that the spatial variation in gender differences evident from our

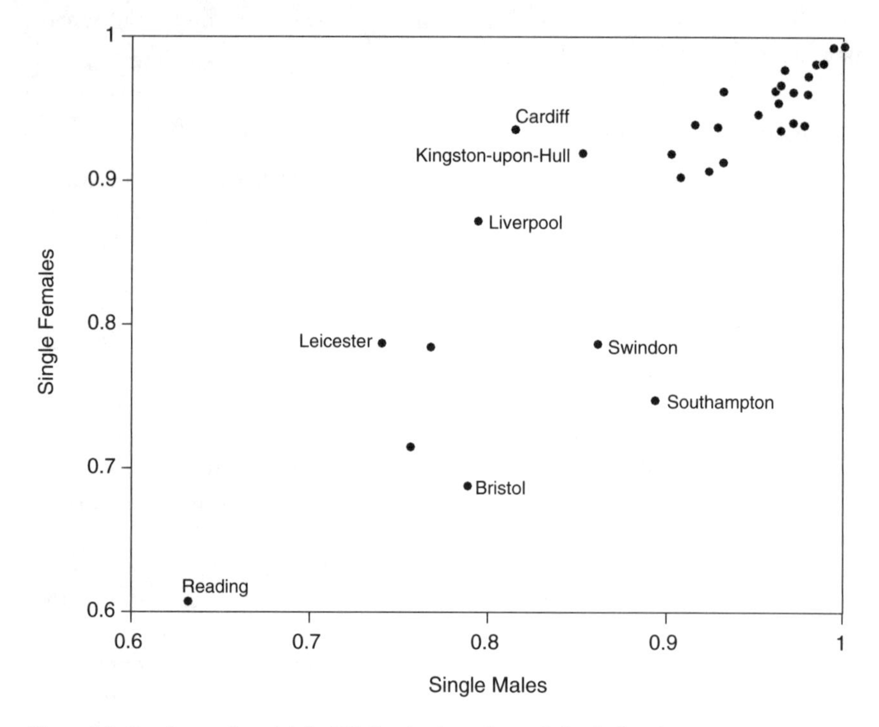

Figure 3.9 Goodness of model fit (R^2) for single males and single females

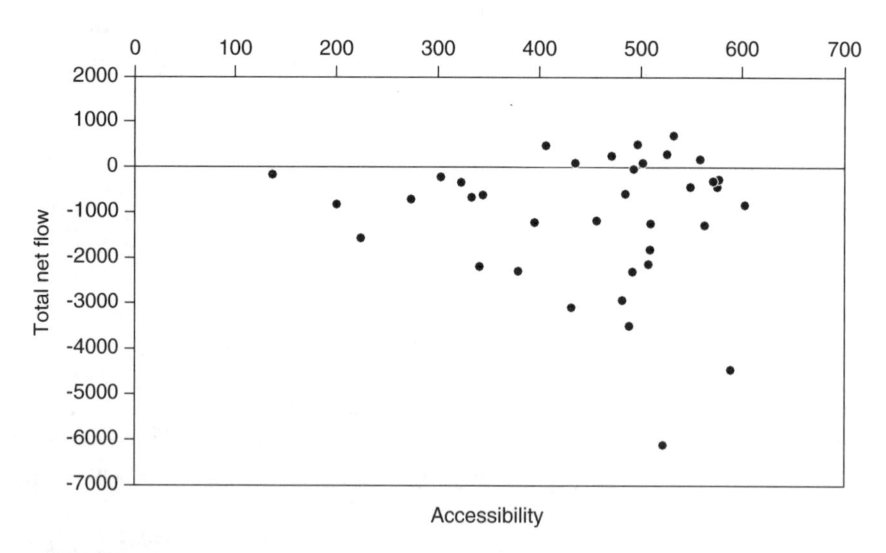

Figure 3.10 The accessibility statistic plotted against net out-migration

analysis will in many cases reflect the demography and employment structure of the various areas. Origin-specific analysis such as this can be used as a means of identifying 'unusual' areas – in this case in terms of gender variation – which can then be examined in more detail in subsequent research.

Finally, the almost universally negative and significant accessibility parameter estimates generated from the competing destinations model provide further evidence to suggest that migration destination choice is indeed a hierarchical process. Furthermore, this suggests that the inherently hierarchical competing destinations modelling framework is a significant and beneficial development of the traditional gravity model, especially when applied to the process of migration destination choice.

Note

1 Whilst Londoners consider boroughs independent places, migrants from further afield will have a less well defined picture of London so that aggregations of London boroughs might be the most appropriate scale of analysis were London to be included in the analysis.

References

Atkins, D.J. (1996) 'Visualising migration: an interactive exploration of 1991 Census data on migration', Unpublished MPhil thesis, Department of Geography, University of Newcastle upon Tyne.

Atkins, D.J., Charlton, M., Dorling, D. and Wymer, C. (1993) 'Connecting the 1981 and 1991 censuses', *North East Regional Research Laboratory Research Report* 93/9, Department of Geography, University of Newcastle upon Tyne.

Ben-Akiva, M. and Lerman, S.R. (1987) *Discrete Choice Analysis – Theory and Application to Travel Demand,* Cambridge, MA: MIT Press.

Boyle, P. (1993) 'Rural in-migration in England and Wales 1980–81', *Journal of Rural Studies* 11: 65–78.

Boyle, P. (1994) 'Metropolitan out-migration in England and Wales 1980–81', *Urban Studies* 31: 1707–22.

Champion, A.G. (1991) *Counterurbanisation*, London: Edward Arnold.

Champion, A.G. and Atkins, D.J. (1996) 'The counterurbanisation cascade: an analysis of the 1991 Census Special Migration Statistics for Great Britain', *Seminar Paper* 66, Department of Geography, University of Newcastle upon Tyne.

Curtis, A. (1991) 'An investigation into the hierarchical processing of spatial information by using the competing destination model', Unpublished MA thesis, Department of Geography, State University of New York, Buffalo.

Dorling, D. (1989) 'Visualisation of spatial social structure', Unpublished PhD thesis, Department of Geography, University of Newcastle upon Tyne.

Duke-Williams, O. (1995) *SMSTAB UNIX Manual Page*, Internet URL:gopher://cs6400.mcc.ac.uk:70/00/midas/datasets/census.dir/sms91.dir/man.help

Fotheringham, A.S. (1983) 'A new set of spatial interaction models: the theory of competing destinations', *Environment and Planning A* 15: 15–36.

Fotheringham, A.S. (1986) 'Modelling hierarchical destination choice', *Environment and Planning A* 18: 401–18.

Fotheringham, A.S. (1987) 'Hierarchical destination choice: discussion with evidence from migration in the Netherlands', *Working Paper* 69, Netherlands Interuniversity Demographic Institute, The Hague.

Fotheringham, A.S. (1991) 'Migration and spatial structure: the development of the competing destinations model', in J. Stillwell and P. Congdon (eds) *Migration Models: Macro and Micro Approaches*, London: Belhaven Press, pp. 57–72.

Fotheringham, A.S. (1997) 'Trends in quantitative methods I: stressing the local', *Progress in Human Geography* 21: 88–96.

Fotheringham, A.S. and Curtis, A. (1992) 'Encoding spatial Information: the evidence for hierarchical processing', Paper presented at the International Conference on 'GIS – from space to territory: theories and methods of spatio-temporal reasoning', Pisa, Italy.

Fotheringham, A.S. and Curtis, A. (1997) 'Spatial information processing: some evidence from the recall of US cities', manuscript available from the authors.

Fotheringham, A.S. and O'Kelly, M.E. (1989) *Spatial Interaction Models: Formulations and Applications*, Dordrecht: Kluwer Academic Publishers.

Fotheringham, A.S. and Pitts, T.C. (1995) 'Directional variation in distance decay', *Environment and Planning A* 27: 715–29.

Haynes, K.E. and Fotheringham, A.S. (1984) *Gravity and Spatial Interaction Models*, London: Sage.

MIDAS (1997) *SMS Gopher Page*, Internet URL: gopher://midas.ac.uk:70/11/midas/datasets/census.dir/sms91.dir

Nam, C., Serow, W. and Sly, D. (1990) *International Handbook on Internal Migration*, London: Greenwood Press.

OPCS [Office of Population Censuses and Surveys] (1991) *Local Statistics: Small Area Statistics: Prospectus*, 1991 Census User Guide 3, London: OPCS.

OPCS (1992) *Topic Statistics: Special Migration Statistics: Prospectus*, 1991 Census User Guide 35, London: OPCS.

Pellegrini, P.A. (1996) 'Migration modelling using the US public use micro data samples, 1985 to 1990', Unpublished PhD thesis, Department of Geography, State University of New York, Buffalo.

Rees, P. and Duke-Williams, O. (1994) 'The Special Migration Statistics: a vital resource for research into British migration', *Working Paper* 94/20, School of Geography, University of Leeds.

Stevens, S.S. (1975) *Psychophysics: Introduction to its Perceptual, Neural and Social Prospects*, New York: Wiley.

Stillwell, J. and Congdon, P. (eds) (1991) *Migration Models: Macro and Micro Approaches*, London: Belhaven Press.

4 The employment consequences of migration

Gender differentials

Anne Green, Irene Hardill and Stephen Munn

Introduction

This chapter is concerned with gender differentials in the employment consequences of migration. The context for the chapter is the increase in female participation in the formal labour market in the United Kingdom over the last thirty years. By the mid-1990s women comprised nearly half of all employees, and projections point to a continuing increase in the share of total employment accounted for by women. Married women, particularly married women with young children, account for the majority of the increase in female participation. Although many of these women work on a part-time basis, in industries and occupations traditionally characterized by a large share of women employees, increasing numbers of women – particularly highly qualified women – are pursuing careers in industries and occupations which were previously considered overwhelmingly the preserve of men. With more women in work and wanting to work, it is pertinent to examine whether the notion of a 'trailing wife' conforms to the current realities of migration behaviour.

The increase in the number of women in employment is not the only significant feature of labour market change over recent decades. There have also been important changes in the balance between full-time and part-time working, in the industrial and occupational structure of employment, and in employment relationships – including an increase in self-employment and flexible working. The nature of these labour market changes, and their differential impacts on women's, men's and households' experiences of employment, are examined in the first main section of this chapter.

The labour market changes referred to above have been accompanied by changes in household structures. Key trends over the last thirty years include increases in single-person households, in cohabitation rates, in the age of marriage, in divorce rates and in lone parent families (Green et al. 1997). There has been an overall fall in fertility, but perhaps equally (if not more) important for a study of the employment consequences of migration has been the tendency for delayed fertility. This means that greater numbers of women have more work experience, and have advanced further in their careers, before leaving the workforce for childbirth. Moreover, these same women also tend to take shorter and fewer breaks for childbirth and child-rearing.

Together, the restructuring of the labour market and of households has culminated in an increase in 'no earner' and 'multi-earner' households (Gregg and Wadsworth 1995; Williams and Windebank 1995), and a decline in the number of 'traditional' two-parent households with dependent children. These changing household structures have implications for the forms of tension and compromise inherent in household migration decisions, how such decisions are made, and their employment consequences.

The majority of residential moves involve 'wholly moving' households and are prompted by non-job-related reasons. Many moves are over short distances, and the labour market situation of household members remains unchanged. The main focus of this chapter, however, is on longer-distance moves – often prompted by a change in employment for one household member. In the second main section of this chapter theoretical perspectives on job-related migration are considered, and the relevance of conventional perspectives for current realities of migration behaviour is investigated with specific reference to the findings of recent research on the location and mobility decisions of dual-career households. Key factors in household migration decisions are explored, and the actual and perceived employment consequences of moves – for men and for women – are examined.

It is apparent from analysis of secondary data sources on the structure of employment and the evidence presented on migration decision making that there are actual and perceived geographical variations in labour market 'opportunity structures'. These geographical variations form the focus of the third main section of the chapter. Since there is a continuing counterurbanizing tendency in the United Kingdom, with in-migration being an important component of population increase in many small towns and rural areas, particular attention is paid to the quantity and quality of employment opportunities open to members of in-migrant households in rural areas, and how the employment consequences of migration might be different for men and women.

The key findings from the reviews of the changing structures of employment, migration decision making and geographical variations in 'opportunity structures' are synthesized in the final section of the chapter. The emphasis here is on lifestyle 'trade-offs' – between men and women at the household level, between 'work' and 'home' aspirations, between migration and commuting, between 'working at home' and 'working at work', and between continuity and change in family and working lives.

The changing structure of employment

The changing structure of employment is a function of the interactions between changes in labour supply and demand. In terms of labour supply, a key feature of change is the increase in female participation rates – most particularly amongst married women aged 25–44 years (Green 1994). While female participation rates have risen, a less pronounced fall in male participation rates is evident – although for men aged 55 years and over the decline in participation rates over the last ten years has been striking (Collis and Mallier 1996). As women have come to play an increasing role in the formal labour market, so the share of women in employment has risen. This rise has been felt across most industries and most occupations and in virtually all areas across the United Kingdom. From the mid-1960s an almost continuous increase in the proportion of women in paid work is evident (Joshi 1985; McRae 1997). In 1981 women filled two out of every five jobs in the United Kingdom. By 1996 the proportion had rise to 46.5 per cent, and projections suggest that by 2006 the share will have risen to 48 per cent (Institute for Employment Research 1997).

Those working on a part-time basis account for the majority of the increase in women in employment (Hewitt 1996). Across the United Kingdom economy as a whole the share of part-time employees has risen from 14 per cent in 1971 to a quarter by 1996, and this proportion is projected to increase further. Women account for the vast majority of part-time workers (Fagan and Rubery 1996), although the share of such jobs taken by men has increased markedly in recent years. For both men and women the majority of new jobs projected to be created over the medium term are part-time. Over the economy as a whole a smaller increase in self-employment is forecast. The major area of job loss in recent years has been full-time jobs for men, and over the medium term this trend is expected to continue with full-time job losses for men easily outstripping increases in full-time jobs for women.

The changing structure of employment by gender and full-time/part-time status is in part driven by the changing industrial and occupational profile of employment. In general, men dominate primary, manufacturing and construction industries, while women are concentrated in service industries. Men account for approximately nine out of every ten workers in craft and skilled manual occupations and eight out of every ten plant and machine operatives. By contrast, women are disproportionately concentrated in clerical and secretarial occupations (where they account for four out of every five workers) and in personal and protective service and sales occupations (where they form two-thirds of the total workforce). However, it is in the higher level non-manual occupations where the increase in the share of women in employment has been most pronounced. Women filled less than one-quarter of managerial and administrative positions in 1981, but by 1996 the share had risen to over a third. Over the same period, women's share of employment in professional occupations rose from less than a third to over two-fifths.

The long-term shift from primary and manufacturing industries to services has favoured women's employment opportunities at the expense of those for men. In occupational terms the picture has been less clear cut, although in most occupational groups projected gains in employment for women exceed those for men. Occupational groups witnessing the largest job gains are managerial and administrative, and professional and associate professional and technical occupations, along with protective and personal service occupations. In the higher-level non-manual occupations there has been a marked increase in the proportion of women employed. Although part-time working in such occupations has increased, highly qualified women engaged on a full-time basis account for much of the growth. In contrast, many of the personal service occupations – where the share of women employed exceeds the share of men – are organized on a part-time basis and are characterized by low pay levels. Sales occupations share similar characteristics. The loss of skilled manual and craft jobs has impacted disproportionately on men, while women have been the main losers in the face of the demise of other (unskilled) occupations (Institute for Employment Research 1997).

Despite these important changes in the industrial and occupational profile of employment, perhaps the word used most frequently to describe labour market restructuring is 'flexibility' (Beatson 1995; Meadows 1996). The term subsumes increases in part-time working (including very short working hours and zero hours contracts), in self-employment, in contracting out, in fixed-term contracts and in the use of incentives and bonuses. The obverse of enhanced flexibility is the demise of permanent employment relationships and of lifetime salaried conditions. The result is increasing job insecurity and changing organizational forms and career patterns. As many organizations have undergone a process of 'delayering', and some activities have been 'spun off' to contractors, so the scope for 'bureaucratic' career patterns within single large organizations has diminished (Savage 1988).

One facilitator of flexibility is the spread of information and communications technologies (ICT), which has meant the loss of some jobs, changes to the content of many of those remaining, and the creation of new jobs. ICT has also enabled the physical separation of many activities – sometimes on a global scale – which had previously been undertaken on the same site or at nearby locations. It has become technically feasible to undertake a greater range of work-related tasks in non-work locations, and it is this particular feature of the impact of ICT that is of particular relevance for migration studies.

While all of these labour market changes outlined above have placed a premium on 'flexibility', it is less clear that a willingness to be 'flexible' is necessarily accompanied by an increasing willingness amongst households and individuals to be 'mobile'. Individual and household attitudes to 'mobility' are discussed in greater detail in the next section.

Migration decision making

As noted in the introduction, the majority of residential moves are made for non-job-related reasons. Although only a minority of moves are made for job-related reasons, they are significant and distinctive in that they tend to be over longer distances than average. It is such longer-distance moves made primarily for job-related reasons that form the focus of attention here.

Since the household is the main unit around which many people organize their lives, and the majority of moves involve 'wholly moving households', it is appropriate to consider migration decisions in a household context. Most existing migration theories seem particularly apposite to the 'traditional' male breadwinner/female homemaker household (Mincer 1978), although they are also applicable to dual-earner households with a 'conventional' division of labour. In such two-adult households there are allocative efficiency gains from a division of labour in which one adult (traditionally the man) specializes in paid work in the formal economy (in order to benefit from the increasing returns created by accumulation of human capital), and the other (traditionally the woman) specializes primarily in home production (Ermisch 1993). In such households, the man's career/job would be expected to take precedence in migration decisions (Bonney and Love 1991; Bruegel 1996). However, in the face of increasing participation of women in paid work over extended periods of their working lives, growing insecurity in the labour market, and increasing rates of household dissolution, the gains from intra-household specialization (and the cooperation this entails) are increasingly uncertain. With such 'traditional' household structures in decline, migration theories need to take account of the current realities of increased participation of women in the labour force, their penetration into professional and managerial jobs and changing household arrangements (Green 1995).

The increase in employment rates amongst women, together with growing non-employment amongst men, has contributed to more complex, and also more fluid, household arrangements. Leaving aside the increasing number of no-earner households, the obverse of the 'traditional' male breadwinner/female homemaker household is the 'egalitarian' household in which both male and female partners have (more or less) equally absorbing careers, and in which household tasks – including caring for children – are shared (more or less) equally. In between the 'traditional' and 'egalitarian' types are a variety of other possible household structures. All of these types may be viewed from the perspective of migration decision making as being dynamic networks of household relations. The nature of intra-household relations is likely to differ between households, as exemplified by Bruegel's (1996) distinction between unified households, households as gendered but unified collectives, households as coalitions, and households as arenas of potential conflict. The way in which migration decisions are made, and their consequences, are likely to differ according to the nature of such relations.

Migration decision making in dual-career households

In households in which both partners are pursuing careers, in any migration decision the advantages of a move for one career have to be considered against the possible adverse consequences on the employment prospects of the other.

In the remainder of this section, some of the main findings from research on key factors in the location and mobility decisions of a sample of dual-career households in the East Midlands are explored (for further details see Dudleston et al. 1995; Green 1997), in order to provide some insights into the difficulties and challenges in coordinating the joint progress of careers.

From a theoretical perspective it might be expected that in order to maximize household income the more highly paid career in a dual-career household (whether it be the man's or the woman's) would tend to dominate in migration decisions. However, there may be countervailing factors at work. For example, it might also be expected that the more locationally constrained career in a dual-career household (which may or may not also be the highest-paid career) would tend to take first priority in migration decisions – in order to maximize overall employment options. Alternatively, it might be hypothesized that in order to minimize disruptions to individual careers a dual-career household may seek to avoid job-related migration and seek to optimize household well-being by staying in one particular location. Indeed, although professional workers display a greater migration propensity than most other socio-economic groups, and although certain groups of workers have 'mobility clauses' in their contracts of employment, there is evidence that fears of disruption to partners' careers is a key factor underlying employee objections to relocation. Moreover, the results of research undertaken by Black Horse Relocation (1993) suggest that concerns about the impact of relocation on a working spouse/partner are felt much more keenly by employees than employers recognize.

The research amongst dual-career households in the East Midlands indicated that most individuals considered that some degree of 'mobility' between jobs (and often between employers) – although, interestingly, not necessarily mobility involving long-distance migration necessitating a change of residence – was desirable or essential for career advancement. Indeed, in the face of growing employee resistance to, and the costs incurred in, relocation, there is evidence that some organizations have attempted to structure career development on a regional, rather than on a national, basis. Within such a context, a dual-career household may maximize employment options and reduce the need for migration by locating in a relatively central/accessible location.

Nevertheless, staying in one place is not always an option, and migration poses different sets of threats and challenges to different individuals and different households. In nineteen out of thirty dual-career households, with whom individual in-depth interviews were conducted in the East Midlands, the 'male career' could be said to 'take the lead' in migration decisions. In five instances the 'female career' was most influential, and in six cases both partners had careers of equal weighting. Hence, in the majority of households included in the case study, satisfying the demands of the 'male career' was the dominant factor in migration decisions, in accordance with 'traditional' migration theory. While some of the female partners felt they had sacrificed their own careers in order to 'follow' the 'male career', for what they considered to be the well-being of the household, in other instances the female partners felt that they had some influence over migration decisions. In particular, being able to exercise a 'veto' over locations perceived as particularly 'difficult' in employment terms. Some female partners commented on the advantages of selecting or following occupational paths which are more easily transferable geographically – such as teaching or nursing – in the expectation of becoming a 'tied/trailing spouse'.

While relocation surveys reveal that the 'typical' relocator is a 37-year-old man in a managerial position with a partner and two children, the same surveys also reveal an increase in female relocatees (Black Horse Relocation 1996). Although female relocatees are on average younger and more likely to be single than their male counterparts, it is recognized that there is a growing number of male trailing partners. In reality the concerns over finding alternative employment in the destination area, the impact of the move on career and promotion prospects and the potential loss of contact and support networks faced by male trailing partners are the same as those faced by women in the same position. However, societal and cultural norms may make it harder for male trailing partners to adapt and settle in the new area than for women in the same position, and this can bring added pressures to a move. Survey evidence suggests that the ability of a working 'trailing partner' to find suitable alternative employment in the destination area is an important influence on the overall success of the relocation. A survey conducted in early 1996 of people who had relocated early in 1995 revealed that of those individuals with working partners whose careers had prompted a household relocation, approximately two-thirds of those whose partners had found alternative employment relatively easily rated the move a 'success'. On the other hand, amongst those whose partners had experienced difficulty in obtaining suitable employment only 38 per cent felt the relocation had been 'successful' (Black Horse Relocation 1996).

Clearly, the potential employment consequences *for both partners* of migration are a key factor in the migration decisions of dual-career households. With the 'delayering' of organizations, fewer job-related moves than formerly – particularly for men – involve promotion, and it remains to be seen whether a greater proportion of household relocations prompted by 'horizontal' moves results in a decrease in the share of all moves deemed by the movers themselves to be 'successful'. However, in those households with school-age children, the age and stage of education of the children is a further key factor in decisions regarding whether, when and where to migrate. In the United Kingdom the introduction of the National Curriculum and school league tables has made parents more cautious about moving children between schools – particularly in the middle of examination courses. The larger the number of children in a household, the more limited are the suitable 'windows of opportunity' for undertaking moves.

For those households without children educational considerations are just one less factor to consider.

Housing considerations are another important factor facing households contemplating or undertaking migration. While relocation packages – including a 'guaranteed sale' of a household's current home – can and do take some of the anxiety out of moving home, there remain concerns over buying (or renting) property in the new location and about possible loss of equity in the housing market. After all, there are both 'gainers' and 'losers' on the housing market 'roller coaster' (Green 1997), and, even leaving monetary gains and losses aside, moving house is recognized as a stressful event in an individual's/household's lifecourse.

Geographical variations in 'opportunity structure'

The concerns expressed by individuals in households considering or contemplating migration about the 'difficulty' of some areas in terms of employment opportunities and possible losses of equity in the housing market highlight the influence of geography in shaping and constraining decision making and actions. There are 'objective' spatial variations in many components of the 'opportunity structure' (for example, the labour market situation, the housing market, the transport system, the educational system, the social infrastructure, and so on) between regions and local areas (Galster and Killen 1995). Clearly, these 'objective' variations may be a key factor in determining different 'choices' and 'outcomes' facing migrating households in different areas (Fielding and Halford 1993). There are also 'subjective' spatial variations in the values, aspirations and preferences of different individuals and households and in perceived opportunities in different regions and local areas, which are likely to influence the decisions regarding migration and actions taken. In understanding migration behaviour and its consequences for employment, it is important to take account of both 'objective' and 'subjective' variations in opportunity structures.

In 'objective' terms it would be expected that employment opportunities for all household members wanting work would be maximized in large urban areas. Such areas are characterized not only by the sheer 'quantity' of employment opportunities, but also the higher 'quality' of employment opportunities – in terms of the range and specialized nature of jobs available. Hence, as noted by one of the dual-career households referred to in Green (1997: 650): 'I can see anybody who is actually in a dual-career – if you were fairly equal about living in London as opposed to anywhere else it would be so sensible. You are not stuck to one firm.' However, while residential preferences vary between individuals and households, the main migration trend in Britain over the last thirty years has been a counterurbanizing one, with greatest net out-migration from some of the largest urban areas and net in-migration to smaller towns and rural areas (Champion 1996). Indeed, a key feature emerging from the study of dual-career households in the East Midlands was a strong residential preference amongst some of the households interviewed for an old 'character' house/cottage in a rural location (Green 1997). While alongside counterurbanization of population there has also been decentralization of employment to areas at successively lower levels of the urban hierarchy (Townsend 1993), it remains the case that in quantitative and qualitative terms employment opportunities are maximized in large urban areas.

In-migration to rural areas

The remainder of this section draws on findings of recent research concerned with employment opportunities and constraints faced by in-migrants to rural areas (for further details see Hardill et al. 1997; Green 1998). At face value, headline labour market statistics portray a positive picture of rural labour markets: in particular, employment growth in recent years has been higher than the national average and average unemployment rates are lower than nationally. However, the reality is often less favourable (Monk and Hodge 1995). Seasonal and casual unemployment structures – particularly in remoter rural areas – tend to be associated with marked seasonal variations in unemployment, job insecurity and low levels of training and staff development. Hidden unemployment and under-employment are often more rife in rural areas than elsewhere (Errington 1988; Beatty and Fothergill 1997). Although agriculture – historically the mainstay of most rural economies – is in long-term decline in employment terms, there often remains an 'agricultural legacy' in low wage levels (Osborne 1997), which tends to be further compounded by larger than average shares of part-time and casual employment. The low wages are not merely a function of the industrial and occupational structure of rural areas, since even when these have been accounted for, a disparity between rural areas and urban areas remains (Wilson et al. 1996).

However, perhaps the most significant feature of rural labour markets from the perspective of a consideration of the employment consequences of migration to rural areas is the limited nature of employment opportunities – in terms of both the quantity, and perhaps even more crucially, the quality, of those opportunities. Obviously, whatever the respective economic vibrancy (or otherwise) of rural and urban labour markets, in sheer quantitative terms the pool of job opportunities available in rural areas is smaller than in larger urban areas. In qualitative terms, the range of employment opportunities tends to be limited in terms of the relative lack of specialized jobs available, and the curtailment of promotion prospects. This position is often further exacerbated by the low turnover in 'good-quality' jobs in rural areas. So, not only are many rural areas characterized by a high proportion of low-paid and insecure jobs in the secondary labour market, but entry to the relatively low share of jobs available in the primary labour market tends to be more difficult than in large urban areas where levels of labour turnover tend to be higher.

Hence, many in-migrants to rural areas from urban areas face a limited range of local employment opportunities. In rural areas relatively easily accessible to larger urban labour markets, any problems associated with the limited nature of employment opportunities in rural areas may be overcome by commuting to jobs in larger urban centres. However, physical communications difficulties and lack of public transport mean that for all members of in-migrant households in rural areas wanting work, access to a car for journey-to-work purposes is likely to be crucial. Those unable to drive and/or without access to a car, are more likely to face severe employment constraints. It is notable that women – along with young people – are more likely to fall into this category of the 'transport-poor' (Wibberley 1978) than men.

Research amongst in-migrants to rural areas in eastern England (Green 1998) indicates that some households sought to overcome employment problems facing individual household members as far as possible by deliberately choosing to reside in 'accessible' rural areas – so as to maximize the quantity and quality of employment opportunities available. If all adult members wanting work in such in-migrant households have access to a car the employment disadvantages of a rural location may be overcome. However, in more inaccessible rural areas problems may be more severe and the research highlighted the fact that some household members – particularly women – had to 'trade down' and 'make do' with jobs which did not utilize their skills/qualifications. Of course, such situations are not unique to women in-migrants to inaccessible rural areas; rather the key point is that the geography of 'opportunity structures' is such that female in-migrants seeking work are particularly likely to face employment constraints in such locations. Moreover, there is some evidence from the study that the nature and severity of the employment constraints faced by at least some in-migrants to rural areas were often disguised by 'idyllic' perceptions of rural life (Little 1997).

Synthesis

From the evidence presented in this chapter it is apparent that migration may have differential employment consequences for men and women. Employment consequences may be positive or negative depending on the characteristics of the destination area (in terms of accessibility, nature and range of employment opportunities available, and so on), household resources, and the attitudes, aspirations and behaviour of the individuals concerned. As the number of two-earner/multi-earner households rises, the trade-offs involved in household migration decisions may become more complex, as the number of factors (i.e. the potential gains and losses from a move) to be considered in the migration 'balance sheet' increases.

It is clear that at the household level a move is more likely to be judged 'successful' if all those household members wanting work are able to find 'suitable' employment. However, the research referred to on the location and mobility decisions of dual-career households suggests that more such households may be placing a greater premium on locating in accessible areas – with good transport links – so as to maximize commuting potential and minimize the need for migration (and associated disruption to partner's careers, children's education, and so on). Hence, it would seem that longer-distance commuting is being traded off against residential migration. At the extreme, 'dual-location' households – where one partner undertakes long-distance commuting on a weekly basis, returning to the family home at weekends – may be formed.

The trend towards greater flexibility in working practices may also be a force in reducing the need for migration. Interviews conducted with employers in 1997 on employee location flexibility suggest that there is a trend amongst many large employers – particularly in the service sector – towards allowing people to work at home on some days on an occasional or regular basis. ICT is obviously a key enabling factor here. If an individual 'works at home' – for at least some of the time – he/she may be willing to commute further on those occasions when working at the workplace. The 'blurring of boundaries' between 'work' and 'home' (explored in more detail in Hardill, Green and Dudleston 1997), as more 'work' is or can be undertaken in the home, may also lead to a decline in levels of residential migration, and the associated (possibly detrimental) employment consequences for the 'following' partner's career. By the same token, increasing flexibility in ways, times and places of working, coupled with the spread of ICT, may enable more 'trailing spouses' to keep their jobs in the origin location when migrating with their partner to a new area.

Alternatively, moving to a new area may have positive employment consequences for both men and women. A move to a new area may provide an ideal opportunity to 'change direction' in the labour market, to embark on a new career or to take 'time out'. Despite the restructuring in the labour market in favour of women and in the form and nature of household relationships in favour of 'individualization' (Bumpass 1990), it remains the case that most of the employment adjustments fall upon women. Nevertheless, such adjustments need not be negative.

Acknowledgements

This chapter draws on the results of research projects funded by:

- the ESRC (reference R000236072) on 'Employment opportunities and constraints for in-migrants in rural areas', undertaken by the present authors plus Anna Dudleston (Nottingham Trent University) and David Owen (University of Warwick);
- the Leverhulme Trust (reference F/740) on 'The location and mobility decisions of dual-career households', undertaken by Dudleston, Green, Hardill and Owen;
- the Leverhulme Trust (reference F/215/AV) on 'Adapting to increased

insecurity: the emergence of dual-location households', undertaken by Green, Terence Hogarth and Ruth Shackleton (University of Warwick).

References

Beatson, M. (1995) 'Labour market', *Employment Department Research Paper* 48, Sheffield: Employment Department.

Beatty, C. and Fothergill, S. (1997) 'Unemployment and the labour market in RDAs', *Rural Research Report* 30, Salisbury: Rural Development Commission.

Black Horse Relocation (1993) *Company Loyalty Versus Family Values – Relative to Relocation*, Windsor: Black Horse Relocation.

Black Horse Relocation (1996) *The View from There: a Survey of Relocated Employees 1996*, Windsor: Black Horse Relocation.

Bonney, N. and Love, J. (1991) 'Gender and migration: geographical mobility and the wife's sacrifice', *Sociological Review* 39: 335–48.

Bruegel, I. (1996) 'The trailing wife: a declining breed?', in R. Crompton, D. Gallie and K. Purcell (eds) *Changing Forms of Employment*, London: Routledge, pp. 235–58.

Bumpass, L. (1990) 'What's happening to the family? Interactions between demographic and institutional change', *Demography* 27: 483–98.

Champion, A. G. (1996) 'Migration between metropolitan and non-metropolitan areas in Britain', *Economic and Social Research Council End of Award Report* (H507255132).

Collis, C. and Mallier, T. (1996) 'Third age male activity rates in Britain and its regions', *Regional Studies* 30: 803–9.

Dudleston, A. C., Hardill, I., Green, A. E. and Owen, D. W. (1995) 'Work rich households: case study evidence on decision making and career compromises amongst dual-career households in the East Midlands', *East Midlands Economic Review* 4: 15–32.

Ermisch, J. (1993) '*Familia oeconomica*: a survey of the economics of the family', *Scottish Journal of Political Economy* 40: 353–74.

Errington, A. (1988) 'Disguised unemployment in British agriculture', *Journal of Rural Studies* 4: 1–7.

Fagan, C. and Rubery, J. (1996) 'The salience of the part-time divide in the European Union', *European Sociological Review* 12: 227–50.

Fielding, A. and Halford, S. (1993) 'Geographies of opportunity: a regional analysis of gender-specific social and spatial mobilities in England and Wales 1971–81', *Environment and Planning A* 25: 421–40.

Galster, G. C. and Killen, S. P. (1995) 'The geography of metropolitan opportunity: a reconnaisance and conceptual framework', *Housing Policy Debate* 6: 7–43.

Green, A. E. (1994) 'The geography of changing female economic activity rates: issues and implications for policy and methodology', *Regional Studies* 28: 633–39.

Green, A. E. (1995) 'The geography of dual-career households: a research agenda and selected evidence from secondary data sources for Britain', *International Journal of Population Geography* 1: 29–50.

Green, A. E. (1997) 'A question of compromise? Case study evidence on the location and mobility strategies of dual-career households', *Regional Studies* 31: 641–57.

Green, A. E. (1999) 'Employment opportunities and constraints facing in-migrants to rural areas in England', *Geography* 84: 34–44.

Green, A. E., Elias, P., Hogarth, T., Holmans, A., McKnight, A. and Owen, D. (1997) *Housing, Family and Working Lives*, Coventry: Institute for Employment Research, University of Warwick.

Gregg, P. and Wadsworth, J. (1995) 'Gender, households and access to employment', in J. Humphries and J. Rubery (eds) *The Economics of Equal Opportunities*, Manchester: Equal Opportunities Commission, pp. 354–63.

Hardill, I., Green, A. E. and Dudleston, A. C. (1997) 'The "blurring of boundaries" between "work" and "home": perspectives from case studies in the East Midlands', *Area* 29: 335–43.

Hardill, I., Green, A. E., Owen, D. W., Dudleston, A. C. and Munn, S. J. (1997) 'Employment constraints and opportunities for in-migrants in rural areas', *Economic and Social Research Council End of Award Report* (R000236072).

Hewitt, P. (1996) 'The place of part-time employment', in P. Meadows (ed.) *Work Out–or Work In?*, York: Joseph Rowntree Foundation, pp. 39–58.

Institute for Employment Research (1997) *Review of the Economy and Employment: Labour Market Assessment 1996–97*, Coventry: Institute for Employment Research, University of Warwick.

Joshi, H. (1985) 'Motherhood and employment: change and continuity in post-war Britain', *Office of Population Censuses and Surveys Occasional Paper* 34, London: HMSO.

Little, J. (1997) 'Employment marginality and women's self-identity', in P. Cloke and J. Little (eds) *Contested Countryside Cultures*, London: Routledge, pp. 138–57.

McRae, S. (1997) 'Household and labour market change: implications for the growth of inequality in Britain', *British Journal of Sociology* 48: 384–405.

Meadows, P. (1996) *Work Out–or Work In?*, York: Joseph Rowntree Foundation.

Mincer, J. (1978) 'Family migration decisions', *Journal of Political Economy* 86: 749–73.

Monk, S. and Hodge, I. (1995) 'Labour markets and employment opportunities in rural Britain', *Sociologica Ruralis* 35: 153–72.

Osborne, K. (1997) 'Small area estimates from the 1996 New Earnings Survey', *Labour Market Trends* 105: 69–71.

Savage, M. (1988) 'The missing link? The relationship between spatial mobility and social mobility', *British Journal of Sociology* 39: 554–77.

Townsend, A. (1993) 'The urban-rural cycle in the Thatcher growth years', *Transactions of the Institute of British Geographers* 18: 207–21.

Wibberley, G. (1978) 'Mobility in the countryside', in R. Cresswell (ed.) *Rural Transport and Country Planning*, London: Leonard Hill, pp. 3–16.

Williams, C. and Windebank, J. (1995) 'Social polarization of households in contemporary Britain: a "whole economy" perspective', *Regional Studies* 29: 723–28.

Wilson, R., Assefa, A., Elias, P., Green, A. E., McKnight, A. and Stilwell, J. (1996) *Labour Market Forces and NHS Provider Costs*, Coventry: Institute for Employment Research, University of Warwick.

5 Who gets on the escalator?

Migration, social mobility and gender in Britain

Irene Bruegel

Introduction

After a long silence, gender is beginning to feature in the analysis of labour migration (Boyle and Halfacree 1995; Green 1997). Recent quantitative analyses, however, pose something of a paradox. While most theoretical approaches and much qualitative analysis suggest that women will gain less from migration than men, some quantitative analyses in Britain (Fielding and Halford 1993; Fielding 1995; Savage 1988) and the United States (Cooke and Bailey 1996) suggest that women can do as well as men out of labour market migration. In this chapter I attempt to untangle this paradox, concentrating on moves to London and the South East of England in the 1980s and early 1990s. Fielding and Halford (1993) found that flows to this region in 1971–81 were relatively to the advantage of women's careers compared with those of men and that promotion to managerial posts through migration to the South East between 1981 and 1991 was also more common for women than for men (Fielding 1995). Women appear as firmly footed on the London and South East 'escalator' (Fielding 1992) as men.

This chapter utilizes two complementary data sets that provide some account of employment change and its association with geographical mobility in order to explore the impact of migration and job change on the jobs held by men and women. Location and job change variables from the Labour Force Survey (LFS) have been analysed in addition to the Office for National Statistics Longitudinal Survey (LS), which Fielding relies on for both 1971–81 and 1981–91. The LS covers a ten-year period and therefore involves more moves. By the same token, the long time period makes the actual relationship between geographical mobility and employment change difficult to pinpoint; the change of region could have been nine years before or even nine years after the change of occupation. Hence much of the change in class position of migrants may relate to job changes within the region of origin, or within the region of destination, rather than being a direct result of a move between regions. The LFS provides year-on-year information about job changes – characterizing them in a number of ways – and year-on-year information about changes of residence. This allows a number of different types of mobility to be investigated. In addition to inter-regional moves, moves can be classified as inter-regional or intra-regional, as job-orientated or not, and as company-based transfers or involving a change of employer.

Using these data sets for South East England, four possible explanations for the paradox are explored in this chapter:

- The theorization is outdated, with women increasingly resisting the status of 'constrained migrant', being no longer as easily moved about as 'any other piece of furniture' (woman respondent quoted in Snaith 1990: 170).
- The measures of social mobility used in much migration analysis fail to take account of the problems involved in comparing the social mobility of men and women.
- The negative effects of tied migration are counterbalanced by the selectivity of migration. Women who migrate, whether under their own steam or not, are a select group of women who have a higher propensity to be upwardly socially mobile. Migration is not so much enhancing their job mobility as reflecting their potential, and the apparent social mobility would be lower if other women were to migrate.
- Women's access to the escalator depends on very particular circumstances. Only migration to a large and varied labour market with large numbers of high-status jobs enables women to improve their status with migration. This is Fielding's own explanation, but it still leaves open to question the 'escalator' mechanism and why migrants do better than long-term residents.

Migration and power relations within the household

Although the empirical finding that women may be more frequent migrants than men can be traced back to Ravenstein (Bartholomew 1991), the theoretical analysis of gender and inter-regional migration is relatively underdeveloped. The growth of the dual-earner household has certainly been recognized by geographers, including those concerned with migration (Boyle and Halfacree 1995; Green 1997). However, for the most part, women have been added as a data set, while gender relations, particularly those within the household, remain largely invisible. This is not to say that households are the sole institution structuring gender inequalities in returns from migration (Halfacree 1995); it is simply to argue that household relations have been under- rather than overestimated in much of the literature (Jarvis 1997). There is a danger, as Halfacree (1995) notes, of treating household structures and domestic arrangements as invariate, and failing to recognize how these can be constituted and reconstituted in the process of migration (Wilson and Tienda 1989).

Two contrary forms of invisibility are evident in the geographical literature:

- Household relationships are simply ignored; women are treated in effect as if they were just another kind of man. This can be characterized as the individualistic model of household behaviour, in which the household is whisked/wished away.
- The primacy of a male career is still taken for granted; the labour migration behaviour of households is then analysed purely on the attributes of the man. This may be because the household is reified as a single actor, with one common preference function, but can also occur where the household is taken to be made up of individuals with separate attributes. In the latter case male primacy is contingent rather than necessary. In such accounts, of which Mincer's (1978) is the best known, decisions to migrate are 'family decisions' to maximize total utility irrespective of its distribution. Moves are made only where gains to one partner compensate for losses to the other. Migration decisions reflect the husband's interests only in so far as his higher investment in human capital and higher earnings raise overall returns; they do not arise directly and unequivocally from his gender.

Neither of these approaches adequately reflects the gendered nature of power relations within the household (Bruegel 1996). Some sense of these comes through qualitative analysis of household decision making. For example, in Jordan, Redley and James's (1994) study of family relations in thirty-six higher-class households in South West England, in at least a quarter of these a 'battle of the sexes' was played out over long-distance migration. In the main, women lose out in this battle:

> to read the accounts of women facing unexpected changes is to become aware of their relative powerlessness in the face of these men's control over the household destiny.
>
> (Jordan, Redley and James 1994:162).

Jordan and his colleagues nevertheless argue that women accede to this loss of control for the longer-term benefits it brings them. In an adaptation of Becker's model of the family as a realm of rational decision making, the study recognizes that women's 'original preferences are modified because of the high value that they put on living together' (1994: 105). This is important in that tastes are not given, as in Becker or Mincer's account: power differentials serve to alter aspirations.

Both variations of the rational household model of migration tend to predict that married women will do worse in labour market terms from household migration. This is taken so much for granted – by women and men, and by geographers and sociologists – that it has hardly been analysed in Britain since the 1970s.[1] Two recent studies both provided evidence that such a position is taken as the natural state of affairs even in 1980s Britain. Faced with a hypothetical question about a move to a new location to improve their partner's job, a majority of women (63 per cent) thought that a woman should 'encourage the partner to take the job and look for any job she could do there'. Only 12 per cent thought she should 'ask the partner not to accept before she could be sure of finding as good a job there as she has now' (Rose and Fielder 1988: 10). Likewise, in their retrospective study of actual moves to Aberdeen in the 1980s, Bonney and Love (1991) found that a majority of the women forced to give up their jobs with the move to Aberdeen were quite happy with the outcome. Questioning the ubiquity of experiences such as these provides the opening to the first attempt to resolve our paradox.

Resolving the paradox

Constrained migration as an outdated concept

The first issue to address in trying to explain the apparent paradox of women both gaining and losing as a result of migration is whether or not the concept of the woman-as-constrained-migrant is now outdated. This can be approached from the perspective of whether there remain clear career 'costs' from women's migration at an aggregate level. Such an analysis must be sensitive to the marital (or equivalent) status of the women concerned.

There are, of course, differences between households relating both to the reasons for any move and to differences between the aspirations of women and their potential earnings. From the woman's perspective, there is a continuum between constrained and unconstrained forms of migration. The least constrained form is likely to be where she is an independent person with limited family ties. In 1990–1, for example, single women in the LFS sample were more than three times more likely than married or cohabiting women to have moved house for 'job-related' reasons. The differential between single and married men was half as large (3.0 per cent of single men and 2.0 per cent of married men made job-related moves). Indeed single women made more job-related moves than married men. Given this, we can expect that the effects of migration on single women's employment status will be more positive than that of married women, taken as a whole. This point is confirmed in table 5.1, which measures entry into service class employment of different groups of women. Irrespective of migration experience, upward social mobility is shown to be greater for women who were not married in either 1981 or 1991, than for those who were married or cohabiting in both years. A proportion of this second group will have married or cohabited in between, just as a proportion of the 'married' women will have been divorced/separated for some years in the interim, but we treat them as though they were either continuously married or single throughout. What is noticeable – and important to the subsequent discussion – is that social mobility differences between married and non-married women are increased by migration, and especially by migration to the South East.

Table 5.1 Upward social mobility of women by migration and marital status 1981–91

		Proportion of women not in service class jobs in 1981 who had service class jobs in 1991*			
		Women married in 1981 and 1991 (%)		*Women not married in 1981 and 1991 (%)*	
All women		4.1		6.9	
	n=		*81817*		*24601*
Women migrants:					
All inter-regional		6.2	5297	17.2	
	n=				*1900*
To South East only		7.9	930	23.6	
	n=				*636*

Source: LS, Crown Copyright.

Note
* SEGs 1.1, 1.2, 2.2, 3, 4 and 5.1.

The most constrained situation for a woman is where she is the dependant of a male partner subject to some kind of internal transfer with his current employer. The move and its destination are relatively fixed. Other moves associated with a husband's job, for example if he is unemployed, are less constrained, if only because the job search area is more open to joint decision making. Mincer (1978), however, found that employed women were as likely to be tied migrants as full-time housewives when their partners were subject to a company transfer or where they were unemployed, although, in general, households moved less readily when the wife was in employment.

To consider how the type of move affects the end result from the perspective of a partnered woman, data relating to some 90,000 couples were extracted from Labour Force Surveys for 1989, 1990 and 1991, covering some 2,500 moves made between 1988 and 1991. Four types of move, reflecting different levels of constraint on woman were identified:

- Transfers associated with the man's job. These were taken to be moves reported by the male partner as 'for reasons of a job' and where he remained with the same employer over the year in question.
- Other moves classified by the male partner as being job-related.
- Inter-regional moves as a whole.
- Moves classified by the female partner as being job-related.

Many of the last group of moves were ones which the male partner also classified as a move for a job reason. Women may therefore have taken the question to refer to their partner's job, illustrating women's continued identification with their partner's career. We therefore distinguished between moves identified by both partners as job-related and the small number where only the women said the move was job-related. Even then, many of the women concerned were not in employment, either before or after the move; when they were, these tended to be cross-class households in which, for whatever reason, the women were the breadwinners or had higher occupational status than the men. The ambiguous interpretation of this question by women living with male partners made it particularly difficult to identify women's involvement in company transfers. Where women remained with the same employer on migrating for a job-related reason they rarely reported a change in the type of work they did. It would seem that these 'transfers' were ones requested by women after a 'household' decision to move. Whilst these were not uncommon, married women were far less likely than either married men or single women to move house as part of an internal labour market career; hence, fewer of their moves would be directly associated with upward social mobility.

We can hypothesize that the more constrained the woman in the decision to move, or in the choice of destination, the higher the cost of a move to her job status. Some evidence for this is provided in table 5.2, which looks at unemployment rates and levels of professional and managerial employment by type of move. Compared to the 'no move' women who stayed in the same area throughout, women migrants of all types were rather more likely to have become unemployed in the year the move was made. Only 3.8 per cent of women non-migrants, employed a year before the survey date, were unemployed at the time of the survey, but over 12 per cent of all inter-regional movers and those who moved for reasons associated with their husband's job,

Table 5.2 Impact of move on female partner by type of move made by couples, 1988–91

		Women economically active in Year 1		
		% unemployed	*% professional/ managerial*	*% professional/ managerial*
*Type of move**	*n=*	*Year 2*	*Year 1*	*Year 2*
Husband's company transfer	418	15.3	15.3	12.1
Other job move by husband	1467	12.2	15.2	12.8
Other inter-regional moves	2406	12.6	19.0	17.3
Wife-only job move	502	3.6	16.8	17.4
No move in period	87313	3.8	13.3	12.9
All residents of South East	27236	3.6	16.1	15.9
Husband's job move to South East	440	12.0	18.3	15.2

Source: LFS 1989–91.

Note
* Where moves took place: Year 1 is before the move, Year 2 after the move.

were unemployed a year later, almost all having become unemployed in the intervening period. As predicted, women involved in the most constrained type of move, those where the man was being transferred, were most often unemployed following the move. Some occupational downgrading is evident for women involved in constrained moves, including those to the South East. Only where the moves were made in relation to a woman's job was any upward mobility immediately associated with the move. In the case of married women, much of the social mobility found over the ten-year span of the LS may, therefore, be less an effect of migration and more a reflection of the types of households that make inter-regional moves. The high rate of unemployment of migrant wives relative to non-migrants suggests that tied migration still has negative consequences for many of the women involved.

The outcomes of moves for women living in conventional couples then depends partly on the underlying reasons for the move. The analysis suggests that inter-regional moves are in some respects distinct from moves made for job-related reasons. Part of the reason why Fielding may have found moves from outside the South East region to the advantage of women may indeed be that these were not primarily moves associated with a partner's job. For the moment we can note that the effect of household migration on women's employment is susceptible to economic and organizational change as this affects both the rationale for migration and the demographic composition of migrants. In so far as a decline in Fordist organization has led to a decline in the importance of transfers to promotion and hence to social mobility, one would expect the cost to women of male career-building to have fallen. At the same time constrained migration associated with male unemployment appears to have risen. A rise in the proportion of female migrants who were single, rather than married, would help to explain evidence that migration is now more beneficial to women as a whole, but does not resolve our paradox in relation to migration by married women.

As far as rising labour market participation by married women is concerned, there are two distinct effects. Rising participation will, by definition, increase the numbers of women who are vulnerable in conventional career terms to tied migration. However, rising participation can limit the rate at which potential migration is translated into actual relocation, given the cross-sectional finding that rates of migration are lower – other things being equal – for dual- as against single-income households (Mincer 1978; Bruegel 1996). None the less, the relationship between women's earning power and their ability to avoid constrained migration turns out to be complex. The migration differential between dual- and single-income households has not changed since the 1970s, even though women employed full-time are contributing rather more to household income than they were in the 1970s (Bruegel 1996). Large-scale changes in attitudes of both men and women could have reduced migration detrimental to married women, whether they are employed or not (Green 1997). However, the pattern revealed in table 5.2 suggests that moves made by married couples retain a tendency to be detrimental to the immediate employment position of working wives. In summary, any analysis of gender and migration must be sensitive to differences in the degree of independence of the women, which largely still means differences in their effective marital status.

Measuring the social mobility effects of labour market migration

Another possible resolution of the paradox lies in the measure of social mobility, given the problem of comparing the social mobility of men and women arising from the very different structures of men's and women's jobs. Fielding's (1995) measure of upward and downward mobility, which compares positions on a five-point reduced socio-economic group (SEG) scale at two points in time may not be sensitive enough to compare social mobility for men and women (Crompton and Mann 1986). A further issue is the measure of social mobility for the large number of women who are not continuously in the labour market. Measures of social mobility restricted to those in the labour market at both points in time ignore the effects of career breaks and may well overstate the upward mobility of women compared with men.

Research that compares the social mobility of men and women, whether as migrants or non-migrants, is constrained by the measures of stratification available. Fielding uses very broad categories of managerial and professional jobs for his analysis, thus a move from a shop assistant to a window dressing job would constitute a move from a white collar to a professional job. On the other hand, the category of professional work is so broad that many of the typical negative effects of tied migration on women's jobs will be lost, as for example where the university lecturer takes a job as a librarian, or a top personal assistant takes a typing job. In terms of male pay, Fielding's category of service class jobs ranged from an average of £700 a week in 1995 for civil servants down to £317 for youth workers and £314 for building technicians (NES 1995). This covers much of the total range of occupational pay for men, the lowest recorded that year being £203 a week for male cleaners. Across Fielding's service class groups, women are typically concentrated in the lower end of that distribution.

There are always boundary problems in measuring social mobility, quite apart from the broader question of what we mean by social mobility. Fielding's five-point classification of SEGs, which classes some intermediate non-manual SEGs as professionals and others as white-collar groups is as legitimate as any other, but does have important consequences for comparing male and female mobility because of the very different distributions of men and women in these jobs. Between 1981 and 1991, according to Fielding (1995), more women managers became professionals (16.5 per cent) than comparable men (14 per cent); similarly, twice as many of the female petty bourgeoisie as the male entered professional jobs. Less than ten years after they left full-time education, 16 per cent of women were in professional jobs in 1991, as against 14 per cent of men.

To ascertain how far Fielding's use of a five-category scale may have overestimated the degree of female social mobility associated with migration, a continuous scale – the Cambridge score – was computed for each of the occupations of LS members for both 1981 and 1991. The Cambridge score was available for both 1981 and 1991 on the LS. It was derived as a measure of the 'social distance' between people in different occupations, based on their patterns of friendship (Stewart, Prandy and Blackburn 1980; Prandy 1990). It correlates reasonably well with the 36-point Hope Goldthorpe scale, a measure of the prestige of different jobs to produce a ranking of different occupations on an ordinal scale. The Cambridge score does not tackle the problem of comparing male and female social mobility, since it was derived from an analysis of men and their friendship patterns, but it gives some measure of the 'distance' people have moved in a given period. Table 5.3 provides the results of this analysis for all inter-regional migrants and for migrants to the South East and London.

On the Cambridge score measure, across the social spectrum as a whole, women in employment in 1981 and 1991 were indeed slightly more upwardly mobile than men, but the difference is far smaller than that implied by Fielding's results. Married women are, as expected, less upwardly mobile, both as migrants and as non-migrants. Indeed social mobility is much less strongly associated with geographical mobility for married women than for other groups. For women as a whole, nevertheless, migration to London and the South East was associated with a higher than average increase in Cambridge score than for men, in line with Fielding and Halford's (1993) findings for 1971–81. Over the ten-year period, even married women moving to a large extent as constrained migrants appear to have benefited from a move to the South East and especially to London.

The Cambridge score analysis in table 5.3 only covers people in jobs in both periods and therefore does not allow for the second problem of comparing male and female social mobility on migration. Differential 'drop-out rates' of men and women can bias the comparison of male and female social mobility over time. Fielding overstates the proportion of women moving from non-service to service class jobs compared with men. While almost all the 1981 base of non-service male employees were in the labour market in 1991, this is not true of women. In excluding from the 1981 base those women who left the

Table 5.3 Average change in social status (Cambridge score) 1981–91, by sex and migration

	All	*All men*	*All women*	*Married women*
All LS members	2.16	2.14	2.19	1.7
Inter-regional migrants	3.74	3.91	3.49	2.18
Migrants to South East	6.14	6.06	6.28	3.69
Migrants to London	7.75	7.2	8.67	4.66

Source: LS, Crown Copyright.

labour market by 1991, and including entrants to the market, the proportionate shift of women from non-service to service jobs will appear greater than it was. Around 14 per cent of female LS members who were in managerial jobs in 1981 were not in the labour force in 1991. Making an allowance for this reduces the proportion of 1981 women managers who became professionals in 1991 from Fielding's figure of 16.5 per cent to 14.5 per cent. Again, looking at migrants to the South East between 1981 and 1991, excluding women leavers from the base, suggests that 18 per cent of female migrants from outside service-level jobs moved into such jobs. Making allowance for women who left the labour market between 1981 and 1991, this proportion falls to 12 per cent of all those who were not in service jobs in 1981.

It is, of course, difficult to measure the occupational mobility of those who are not currently employed. The criticism being made here is not of a failure to crack the nut of ascribing a social status to full-time housewives and mothers in analysing the effects of labour market migration. Rather it is an example of the problem of comparing male and female attributes on one measure – change in occupational status between 1981 and 1991 – without taking into account the wider differences in gender roles, which will affect the comparisons.

The selectivity of migration

A third possible explanation for the paradox lies with the social composition of migrants. The relatively beneficial effects of migration on women may reflect the selectivity of the migration process, rather than migration *per se* (see also Cooke and Bailey 1996). As table 5.1 showed, marriage is a major block to upward social mobility for women. Rates of social mobility were particularly low for those women who married between 1981 and 1991. This reflects the occupational downgrading women experience after any break for childbearing (Joshi 1987). The apparent boost migration gives to social mobility could, at least in part, stem from differences in the social composition of migrants and non-migrants, since migration is heavily weighted towards the young and single, the more socially mobile groups. Higher-grade jobs, particularly in London, are weighted towards younger women in full-time employment, precisely the groups most over-represented in migration flows (Bruegel, Lyons and Perrons 1996).

The selectivity effect is likely to be particularly strong when inter-regional migration, rather than just job-related migration is considered, since inter-regional migration by people of working age is very varied. For many women, inter-regional migration is related to marriage and setting up a home, as Grundy (1987) showed, and cannot be thought of as labour market migration. Analysis of the LS from 1981–91 shows that women who married between these dates account for 30 per cent of all married female migrants in the period, although they were no more than 17 per cent of all married women in 1991.

Using the LS for the whole twenty-year period 1971–91 enables analysis of the relationship between inter-generational social mobility and geographical migration for both men and women. The results are shown in table 5.4. The measure of inter-generational mobility was crude, especially for women, since only people aged 10 to 15 in 1971 who were living with their fathers at that time were included, and the measure was a simple one of the proportion of those whose fathers had had manual jobs in 1971 who themselves had professional or managerial status by 1991. In line with our other findings, inter-generational social mobility was found to be particularly high for single women who moved from other parts of the country to London, and higher for married women migrants to London than for those who remained in one region, whether in London or elsewhere (not shown). Comparing migrants from a working class background with non-migrants, a degree of self-selection was evident. Working class migrants were much more likely to have had parents who were owner-occupiers and to have lived in relatively prosperous neighbourhoods in 1971 than were people from a similar background who stayed in their region of origin (table 5.4).

Part of the social mobility effect of migration to the South East arises from the tendency of women who migrate across regions for marriage or housing reasons to be of a higher social class than non-migrants. Where these women are not themselves in employment before migrating, they are highly likely to appear socially mobile over a ten-year period when they move from education or (more likely) childbearing through to employment. Even those in employment before migrating could be expected to rise in social status if they remained in employment. Coming as they do from higher social status groups, they are likely to have moved from more junior to more senior positions, as they gain experience, quite independently of the residential move. Even amongst married women, higher social mobility amongst migrants could be a selectivity, rather than a migration effect. Inter-generational analysis showed that men from working class backgrounds were far more likely than women to be in high-level jobs, even after discounting women no longer in the labour force.

Table 5.4 Inter-generational social mobility by migration and sex

	People from a manual background 1971		*Housing background of women with manual worker fathers*	
	% with professional and managerial status 1991		*% in owner occupation*	*% in 'best' localities*
Migration history 1971–91	*Male*	*Female*	*1971*	*1971*
Left London	31.3	10.1	34.8	21.5
Stayed in London	20.7	7.8	29.9	13.6
Moved to London	40.2	19.7	50.6	21.3
Stayed outside London	19.5	7.2	33.0	15.6

Source: LS, Crown Copyright.

Migrants are younger, more often single and more educated, but what stands out, particularly in relation to migration to the South East is the high proportion who were still in education in the year before the move. Some 12 per cent of female inter-regional migrants between 1981 and 1991 were in full-time education in 1981, compared to only 4 per cent of all women. A high proportion of women's inter-regional migration as defined by Fielding will be migration associated with higher education, especially in the form of migration to the South East. In this case, the escalator lies not in the labour market but in the tradition of people coming 'up to' the South East for post-school education and training. This does not account for the whole difference. The numbers coming from education are not sufficiently large to account for all the extra upward mobility enjoyed by female migrants to the South East. Fielding (1995) found, indeed, that a social mobility effect of women's migration to the South East remained after excluding those who came from education on to the labour market. Analysis of the LS and LFS shows, nevertheless, that social mobility might be overestimated when entrants from education are included amongst the socially mobile. At least some of our paradox is resolved by the fact that the socially mobile female migrants are not, generally, the 'same' women as those vulnerable to negative job effects from tied migration. This point suggests that analysis of female migration is more helpful when marital status is included as a variable (Robinson 1993).

The case of married women moving to the South East

If we confine analysis to married women who were in employment both before and after a move to the South East, to discount the effects both of marital status and entry from education, upward social mobility still appears to be greater for migrant than for non-migrant women. Migrants to the South East are also apparently more socially mobile than women who lived in the South East throughout. Table 5.3 showed that, compared with an average increase of 1.7 points, married women moving to the South East between 1981 and 1991 experienced a 3.7 point increase in their status on the Cambridge scale; for those moving to London it was greater at 4.7 points.

In an attempt to sift out the effects of the profile of female migrants on social mobility from any direct effects of migration, a regression analysis of the change in Cambridge score was run on the LS data covering all women in employment in both 1981 and 1991; a total of 191,000 cases, of which 9,000 (2.1 per cent) moved to the South East between 1981 and 1991. The model selected is shown in table 5.5. This shows that even after occupational group, qualifications and life-cycle stage are allowed for, women who lived in Inner London in 1991 experienced higher rates of social mobility than women living in other regions. Allowing for region of residence, the regression also shows that social mobility was enhanced by migration to the South East, and reduced by migration from the South East. Life-cycle stage matters. In particular, social mobility for

Table 5.5 Regression analysis of change in Cambridge score 1981–91, women members of the LS, living in the South East 1981 and/or 1991*

	Coefficient	*Standard error*
Constant	–4.03	.64
1991 characteristics		
SEG 1,2	10.2	.29
SEG 4	7.87	.204
SEG 5	13.9	1.2
SEG 6	16.4	.488
SEG 7	7.63	.167
SEG 11	–7.05	.483
SEG 12	–3.63	.269
SEG 13	–5.04	.157
SEG 14	–8.78	.176
SEG 18	–7.94	.774
Under 30, no child	1.53	.209
30–49, no child	–1.08	.151
Over 50	–2.21	.151
Degree	2.62	.643
Postgraduate qualification	7.57	.62
Resident Inner London	1.12	.62
Migration history		
Moved to South East 1981–91	1.78	.35
Moved from South East 1981–91	–1.47	.27
	$R2 = 0.13$	$F = 586.3$

Source: LS, Crown Copyright.

Note
* This includes women not in the labour market, on the basis of the score in their last job

women is associated with youth and lack of child-care responsibilities. Being under thirty years of age without a child and living in Inner London does rather more for women's occupational status than moving to the South East, but a South East migration effect, which needs explanation, is still evident.

Given the strong negative effects of age on social mobility, much of the migration-related social mobility would have been amongst young women in their first or early jobs in 1981. Those that remained in the labour market benefited from a move to London, possibly because higher-level jobs in the London economy appear to be rather more open to women, and to young women in particular, than similar jobs in other parts of the country (Bruegel 1999). It is not yet clear why this should be, although it appears to be a feature of metropolitan cities more generally.

The scale of London's labour market and the relatively good public transport system in the city might also be thought to reduce the negative effects of tied migration. This would be very much more of a 'London effect', shared by other metropolitan centres, than an effect of migration to the wider South East. Women involved in moves initiated by their partners, can be expected to do 'better' where they have access to a range of jobs and can retain and develop a degree of specialization, as against the situation described by Jordan, Redley and James (1994), where women find themselves moving to a small, relatively isolated labour market. In such circumstances, with a relatively limited set of jobs available, many women will accept a degree of occupational downgrading, often without complaint, as Bonney and Love (1991) point out. How far the growth of second car ownership has increased the range of choice for women in such circumstances, remains to be investigated.

Conclusion

Women moving to London in the 1980s, whether married or single, experienced a high degree of social mobility, albeit from a low base. For the most part that social mobility was achieved by young women without children, and stemmed to an extent from the role of London and the South East as centres of further and higher education for the country as a whole.

The degree of social mobility of women was shown to be overestimated where no account was taken of (largely temporary) exits from the labour market. None the less, whichever measure of social mobility was used, a higher rate of mobility was associated with migration, for women as well as men, and even for married women who in most cases were tied migrants.

Tied migration was, however, shown to have immediate costs for many women, in the form of a higher rate of unemployment and a fall in the proportion in high-status jobs, compared with non-migration and women-centred migration. The difference between the short- and long-term impact of migration for women, which is revealed in the contrasting results of the LFS analysis and that of the LS, is in line with some United States findings (for example, Yu et al. 1993). The ability of married women migrants to London and the South East to make good the immediate negative effects of migration suggests that this is a labour market effect depending on the pattern of jobs available to women in London, rather than a direct outcome of migration. On the other hand, there are stark contrasts between women local to London and inward migrants, suggesting an important selectivity effect, that migration filters in potentially mobile women as much as, if not more than, men.

These varied strands can help untangle our seeming paradox. First, single women dominate migration streams; hence, the theorized negative effects apply much less stringently, particularly once cohabiting single women are treated as married. Second, one in ten of household moves were classed as those made for the woman's job alone and reflect some increase in the numbers of cross-class households in Britain and the increasing importance of women's earning power. To that extent, the theorizations may be outdated. The rise in the two-car household will also tend to reduce the negative effects of household mobility on women's job chances. A paradox remains for the majority of married women, who were shown to do badly out of moves dominated by the interests of a partner's job and yet who also experienced upward social mobility when they moved to London and the South East. A small part of the discrepancy was shown to come from the method of measuring women's social mobility, using categorizations developed for a much more male workforce, and from not taking into account women who had left the labour market after migration. However, the main explanation of this remaining paradox lies with the selectivity of the migration process and its carry-over between partners in a household. Women living with male partners who are able and willing to migrate would appear to have higher aspirations and greater ability to move to new types of work. The relative buoyancy of the professional and managerial labour market in London, at least until the late 1980s, will have helped women realize those aspirations, but this is by no means the whole story, since migrants have tended to do better than London and the South East's longer-term residents.

Acknowledgements

Research for this chapter formed part of an ESRC-funded research project on 'Life cycles, life chances and migration in London 1971–1993' (reference R000221691). I am very grateful to Simon Gleave of the SSRU for extracting the LS data, to ONS for allowing me to use the LS and the LFS data, and to Joanna Brown for all her work on the data.

Notes

1 Snaith's (1990) study of childless married graduates of 1965 and 1972, found that almost 40 per cent felt that migration in pursuit of the husband's career had damaged their own. Johnson, Salt and Wood (1974), looking at migrants to four British towns in the 1960s, found that 40 per cent of the wives of migrants had given up their job following the move.

References

Bartholomew, K. (1991) 'Women migrants in mind: leaving Wales in the nineteenth and twentieth centuries', in C. Pooley and I. Whyte (eds) *Migrants, Emigrants and Immigrants: a Social History of Migration*, London: Routledge, pp. 174–87.

Bonney, N. and Love, J. (1991) 'Gender and migration: geographical mobility and the wife's sacrifice', *Sociological Review* 39: 335–48.

Boyle, P. and Halfacree, K. (1995) 'Service class migration in England and Wales, 1980–1981: identifying gender-specific mobility patterns', *Regional Studies* 29: 43–57.

Bruegel, I. (1996) 'The trailing wife: a declining breed? Careers, geographical mobility and household conflict in Britain, 1970–89', in R. Crompton, D. Gallie and K. Purcell (eds) *Changing Forms of Employment*, London: Routledge, pp. 235–58.

Bruegel, I. (1999) 'Gender inequality and the global city', in A. Hegewisch, R. Sales and J. Gregory (eds) *Women, Work and Inequality in a Deregulated Market*, London: Macmillan.

Bruegel, I., Lyons, M. and Perrons, D. (1996) 'Polarisation, professionalisation and feminisation in London, 1971–1993', Paper presented to the Institute of British Geographers Annual Conference, Strathclyde University, Glasgow, January.

Cooke, T. and Bailey, A. (1996) 'Family migration and the employment of married women and men', *Economic Geography* 72: 38–48.

Crompton, R. and Mann, M. (eds) (1986) *Gender and Stratification*, Oxford: Polity Press.

Fielding, A. (1992) 'Migration and social mobility: South East England as an "escalator" region', *Regional Studies* 26: 1–15.

Fielding, A. (1995) 'Interregional migration and intra-generational social class mobility', in M. Savage and T. Butler (eds) *Social Change and the Middle Classes*, London: UCL Press.

Fielding, A. and Halford, S. (1993) 'Geographies of opportunity: a regional analysis of gender specific social and spatial mobilities in England and Wales 1971–81', *Environment and Planning A*: 25: 1421–40.

Green, A. (1997) 'A question of compromise? Case study evidence on the location and mobility strategies of dual career households', *Regional Studies* 31: 641–57.

Grundy, E. M. (1987) *Women's Migration: Marriage, Fertility and Divorce*, London: Office of Population Censuses and Surveys, Longitudinal Study 4.

Halfacree, K. (1995) 'Household migration and the structuration of patriarchy: evidence from the U. S. A.', *Progress in Human Geography* 19: 159–82.

Jarvis, H. (1997) 'Housing, labour markets and household structure: questioning the role of secondary data in sustaining the polarization debate', *Regional Studies* 31: 521–32.

Johnson, J., Salt, J. and Wood, P. (1974) *Housing and the Migration of Labour in England and Wales*, Farnborough: Saxon House.

Jordan, B., Redley, M. and James, S. (1994) *Putting the Family First: Identity, Decisions, Citizenship*, London: UCL Press.

Joshi, H. (1987) 'The cost of caring', in C. Glendenning and J. Millar (eds) *Women and Poverty in Britain,* Brighton: Wheatsheaf, pp. 112–33.

Mincer, J. (1978) 'Family migration decisions', *Journal of Political Economy* 86: 749–73. NES (1995) New Earnings Survey, London: HMSO.

Prandy, K. (1990) 'The revised Cambridge scale of occupations', *Sociology* 24: 629–55.

Robinson, V. (1993) '"Race", gender and internal migration within England and Wales', *Environment and Planning A* 25: 1453–65.

Rose, M. and Fielder, S. (1988) 'The principle of equity and the labour market behaviour of dual earner couples', Economic and Social Research Council SCELI Programme, Working Paper 3, Nuffield College, Oxford.

Savage, M. (1988) 'The missing link? The relationship between spatial and social mobility', *British Journal of Sociology* 39: 554–77.

Snaith, J. (1990) 'Migration and dual career households', in J. Johnson and J. Salt (eds) *Labour Migration*, London: David Fulton, pp. 155–71.

Stewart, A., Prandy, K. and Blackburn, R. (1980) *Social Stratification and Occupations*, Basingstoke: Macmillan.

Wilson, F. and Tienda, M. (1989) 'Employment returns to migration', *Urban Geography* 10: 540–61.

Yu, L. C., Wang, M. Q., Kaltreider, L. and Chien, Y. (1993) 'The impact of family migration and family life cycle on the employment status of married, college-educated women', *Work and Occupations* 20: 233–46.

6 The effect of family migration, migration history, and self-selection on married women's labour market achievement

Thomas Cooke and Adrian Bailey

Introduction

The consensus within the migration research field is that, generally speaking, family migration has a negative effect on married women's labour market achievement. Theoretically, the human capital model of family migration predicts negative consequences of family migration on married women's labour market achievement (Sandell 1977; Mincer 1978), a hypothesis which is supported by the available empirical evidence (for example, Sandell 1977; Spitze 1984; Lichter 1980, 1983; Shihadeh 1991). Yet, Cooke and Bailey (1996), in response to methodological limitations of previous empirical research, find a 9 per cent *increase* in the probability of employment due to migration among a sample of married mothers living in the Midwest in 1980. The validity of their finding hinges on their arguments regarding the role of self-selection bias in cross-sectional models of the effect of migration on labour market achievement.

With cross-sectional data, the most frequent approach is to estimate a model of labour market achievement (ai) as a function of individual characteristics (xi) and migrant status (mi), where $mi = 1$ if the individual migrated during the recorded migration interval and $mi = 0$ if they did not:

$$\mathrm{E}\,(ai \mid xi) = \beta xi + \beta\, mi + \omega i \qquad (1)$$

Equation (1), therefore, attempts to measure the effect of migration (mi) on labour market achievement (ai) by estimating the difference in labour market achievement between migrants and non-migrants (βm), while controlling for individual characteristics.

Although underdeveloped in the literature, the human capital model of migration posits a sorting mechanism among individuals, such that individuals who are more likely to be successful by migrating are more likely to migrate and individuals who are more likely to be successful by not migrating are more likely to stay. This means that labour market achievement (ai)

and migrant status (mi) among migrants are positively correlated, pushing the observed distribution of ai to the right. Conversely, among non-migrants, labour market achievement and migrant status are negatively correlated, pushing the observed distribution of ai to the left. Together, this self-propelled sorting mechanism leads to upwardly biased estimates of the effect of migration on labour market achievement (βm) (Greene 1993). This is a classic case of sample selection bias. Following Heckman (1979) and Greene (1993), Cooke and Bailey (1996) identify the correct sample regression function for the mean of ai as:

$$E\,(ai \mid xi) = \beta\, xi + \beta m mi + \beta_{\lambda} \lambda i + \omega i \qquad (2)$$

where:

$$\lambda i = \frac{\phi\,[m_i]}{\Phi\,[m_i]}, \text{ for migrants}$$

$$\lambda i = \frac{\phi\,[m_i]}{1 - \Phi\,[m_i]}, \text{ for non-migrants} \qquad (3)$$

Equation (2) is easily estimated by first estimating a properly identified probit model of the probability of migration, then calculating λi from the predicted values of the probit model (following equation (3)), and including λi in the model of labour market achievement (estimate equation (2)).

Research hypotheses

In applying this method – known as a *treatment effects model* – Cooke and Bailey (1996) found that family migration actually had a *positive* effect on the labour market achievement of white, non-Hispanic, married mothers who lived in the Midwest in 1980. While their results clearly contradict both the human capital model of family migration and the available empirical evidence, it is not entirely clear to what degree their results are specific to the characteristics of individuals in the sample, the geographic and historic context of the sample, and the methods used. Given that the analysis demonstrates the need both to rethink how family migration decisions are made and to move beyond simple explanations based upon the calculus of human capital theory (see Halfacree 1995), it is important to revisit their data and methods to address some of the more important limitations of their analysis. In particular, this research addresses the following three questions.

(1) Does the positive effect of family migration on women's employment hold true for other measures of labour market achievement? Cooke and Bailey (1996) presented many possible explanations for their results:

- the labour market achievement of trailing spouses may be indirectly boosted by migration away from tight labour markets and toward buoyant labour markets;
- trailing spouses may invest in careers which are not adversely affected by frequent moving (for example, nursing);
- trailing spouses may enter the labour market following a move both to recoup the cost of moving and to make new friends;
- the intense economic restructuring in the industrial Midwest between 1975 and 1980 may have caused families to engage in return migration while, at the same time, women entered the growing service sector to supplement family income.

While not all of these hypotheses can be directly addressed by the data used in this chapter, it is clear that an analysis based just on employment has limited explanatory power. Therefore, this research investigates how family migration affects women's labour force participation, employment, and hours worked. (Other measures of labour market achievement, such as earnings, are not available for place of residence in 1980.)

(2) How does the effect of family migration on women's labour market achievement depend upon previous migration experiences? As noted in the previous paragraph, Cooke and Bailey (1996) hypothesize that their results could be related to return migration. Indeed, Cooke and Bailey (1998) find distinct differences in the effect of migration history on the labour market achievement of men and women using data drawn from the National Longitudinal Survey of Youth (NLSY). They find that initial and onward migration generally has a positive effect on men's labour market achievement, but return migration is associated with a decline in labour market achievement. In contrast, initial and onward migration has a negative effect on women's labour market achievement, but return migration is associated with an increase in labour market achievement. While the structure of the data used in this analysis prevents measurement of initial, return, and onward migration as precise as can be done with the NLSY, it is possible to estimate how the effect of family migration differs according to a simple measure of migration history.

(3) Do migration self-selection processes operate differently for married women and men? One limitation of the model presented by Cooke and Bailey (1996) is that the effect of migration self-selection was not differentiated according to gender. In contrast, the human capital model of family migration suggests that self-selection bias should have a more important effect on the labour market achievement of men because the labour market achievement of women is largely ignored in the family migration decision-making process. Although many researchers have taken issue with the human capital model of family migration (see Halfacree 1995), if married couples migrate together there is a case for suggesting that, in the aggregate, men and women should be similarly selected for migration. Therefore, this research attempts

to identify how differences in migration self-selection bias between men and women influence labour market achievement.

Data and methods

These research questions are addressed using a sample similar in most respects to that used by Cooke and Bailey (1996). Briefly, the data for the analysis are drawn from the Public Use Microdata Sample A (PUMS) of the 1980[1] US Census. The 1980 PUMS is used to select individuals who were out of school in both 1975 and 1980, married in both 1975 and 1980, had pre-school children in both 1975 and 1980, and lived in the Midwestern states of Illinois, Indiana, Michigan, Ohio and Wisconsin in both 1975 and 1980. The sample is limited further to white non-Hispanic men and women who had been married at least five years, never divorced, between the ages of 21 and 70, living with their own children, and with some labour force participation between 1975 and 1980 in the non-agricultural civilian labour force. Given these selection criteria, it is important to note that this sample is very specific in that it consists of spouses in traditional white Midwestern nuclear family households in which the husbands are the primary wage earners and the wives only work part-time. Table 6.1 lists variable names, definitions, and sample means for all variables used in the analysis according to gender and previous migration experience (= 1 if 1975 state of residence is different from state of birth). It immediately shows the greater involvement of men in the waged labour force.

The first step in the analysis is to estimate a model of the probability of migration. Migration is defined with respect to changes in residence between 1975 and 1980:

- from a metropolitan area to another metropolitan area;
- from a metropolitan area to a non-metropolitan county;
- from a non-metropolitan county to a metropolitan area;
- from a non-metropolitan county to another non-metropolitan county.

Following Cooke and Bailey's (1996) specification, independent variables include years of labour market experience, the square of years of labour market experience, years of education, employed in 1975, metropolitan residence in 1975, and professional occupation. In order to estimate how the effects of self-selection on labour market achievement differ by gender, and to estimate how previous migration experience mediates the effects of migration on labour market achievement, four separate models of migration are estimated (men with previous migration experience, men without previous migration experience, women with previous migration experience, and women without previous migration experience).

The second step of the analysis is to calculate λi from the probit model of migration based on the predicted probability of migration (see equation 3),

Table 6.1 Variable names, definitions and means

Variable name	*Definition*	*Men*	*Women*	*Men*	*Women*
		Previous migration history	*No previous migration history*	*Previous migration history*	*No previous migration history*
Migrant	=1 if migrant (defined in the text)	23.5%	16.8%	28.1%	17.8%
Labour force participation	=1 if participating in the labour force last week	98.1%	98.2%	63.0%	64.3%
Employment	=1 if employed last week	93.0%	92.5%	58.6%	59.8%
Hours worked last week	41.5	41.2	17.2		16.8
λ_i	self-selection bias control variable	0	0	0	0
Years of labour market experience	(age)-(years of education)–5	15.6	14.4	13.2	12.7
(Years of labour market experience)^2	279.9	230.5	193.9		175.9
Years of education	13.2	13.0	12.8		12.5
Employed in 1975	=1 if worked in 1975	97.1%	97.4%	41.0%	46.4%
Metropolitan residence in 1975	=1 if living in an MSA in 1975	75.9%	68.0%	71.9%	67.0%
Other family income	earned family income less earned personal income in 1979	$5,404	$5,094	$21,592	$19,890
Metropolitan residence	=1 if living in an MSA	75.1%	68.0%	71.9%	66.0%
Employed in 1979	=1 if worked in 1979	98.5%	98.9%	74.6%	76.6%
Work disability	=1 if reporting a work disability	4.6%	4.9%	1.5%	2.5%
Professional occupation	=1 if reporting a professional or managerial occupation	29.4%	21.7%	21.6%	14.8%
Unemployment rate	county group unemployment rate	7.6%	7.9%	7.6%	7.9%
N =		*1,890*	*5,047*	*1,197*	*3,323*

and to include it in the models of labour market achievement. Labour market achievement is measured by three variables: labour force participation (=1 if in the labour force), employment (= 1 if employed), and the natural logarithm of hours worked in the last week. Each dependent variable is estimated as a function of migrant status (yes = 1), λ, years of labour market experience, the square of years of labour market experience, years of education, other family income, metropolitan residence, employed in 1979, work disability, professional occupation, and county group unemployment rate. Consistent with the models of migration, separate models of labour market achievement are estimated according to previous migration experience and gender.

Results and discussion

Table 6.2 lists the parameter estimates of the probit model of migration. Results are generally similar across all four populations. The probability of migration is positively and significantly related to years of education and professional occupation (except for women with no previous migration experience), and is negatively related to metropolitan residence in 1975 (except for women with previous migration experience). Tables 6.3–6.5 show the parameter estimates for the models of labour market achievement. For each migration history/gender category, two sets of parameter estimates are listed. The biased estimates are models that do not include λi and therefore suffer from self-selection bias. The unbiased estimates include λi and are, therefore, unbiased. Neglecting (for the moment) the parameters associated with migrant status and λi, all of the models conform to expectations as to how labour market experience, education, family income, local unemployment rates, and other individual and contextual factors influence labour market achievement.

Several themes emerge with respect to the parameters associated with migrant status and λi. For men, migrant status is not significant in any of the biased models. Likewise, neither migrant status nor λi are significant in any of the unbiased models. This indicates that migration has no effect on the labour market achievement of men (in this sample) and that the family migration decision-making process does not sort men into self-selected migrant and non-migrant samples. Of course, this is in conflict with the human capital model of family migration which holds that the primary wage earner in a family drives the family migration decision-making process, such that the primary wage earner's resulting (family) migration behaviour is similar to the migration behaviour of a non-married migrant. This is clearly not the case because there is no evidence of self-selection among the men in the sample.

For women with a previous migration history, the biased models indicate no effect of family migration on any measure of labour market achievement, while the unbiased models show a strong negative effect of family migration on employment and hours worked last week. In both cases, the parameter estimates for λi are positive and significant. Similarly, for women with no previous migration history the biased models indicate a small, but signifi-

Table 6.2 Probit model of migration (1 = yes)

Variable	*Men*		*Women*	
	Previous migration experience	*No previous migration experience*	*Previous migration experience*	*No previous migration experience*
Intercept	–0.9446[a]	–1.2049	–0.9333	–1.8165
Years of labour market experience	0.0129	–0.0052	–0.0451	0.0357
(Years of labour market experience)^2	–0.0005	0.0002	0.0009	–0.0012
Years of education	0.0357	0.0431	0.0803	0.0737
Employed in 1975	–0.0503	–0.1367	–0.0377	–0.0941
Metropolitan residence in 1975	–0.5106	0.3175	–0.3888	–0.4255
Professional occupation	0.3227	0.2067	0.0368	0.1379

Note
a Underlines variables are significant at 0.05.

cant, negative effect of migration on all measures of labour market achievement. The unbiased models (with the exception of the model of employment) show a stronger negative effect of migrant status on labour force participation and hours worked last week. Again, in both cases the parameter estimates for λ_i are positive and significant. These results generally indicate that family migration has a strong negative effect on women's labour force participation and hours worked, but no effect on their employment.

These results qualify Cooke and Bailey's (1996) research. In this analysis the effect of migration on women's employment is statistically zero rather than positive. This is probably due to the fact that the models are estimated separately by gender and that measured self-selection effects are also estimated with respect to gender. None the less, this analysis demonstrates that self-selection is an important element of the migration behaviour of married women and, like Cooke and Bailey's previous analysis, suggests that the human capital model of family migration does not fully capture how family migration decisions are made. The positive parameter estimates associated with λ_i among women indicate a positive self-selection bias. This means that women who are more likely to migrate are more likely to be employed if they migrate. Women who are more likely to stay are more likely to be employed if they stay. In contrast, there is no evidence of self-selection among the men in the sample. Apparently, families are giving stronger weight to the wives' labour market achievement than is considered by the human capital model of family migration.

Table 6.3 Probit model of labour force participation (1 = yes)

Variable	*Men*				*Women*			
	Previous migration history		*No Previous migration history*		*Previous migration history*		*No Previous migration history*	
	Biased	*Unbiased*	*Biased*	*Unbiased*	*Biased*	*Unbiased*	*Biased*	*Unbiased*
Intercept	0.3338	–0.1522	–0.3710	–0.1070	–1.8816	–1.6175	–1.5735	–1.5988
Migrant	–0.0813	2.6010	0.0477	–2.2717	–0.1732	–1.6320	–0.1798	–1.3530
λ_i		–1.4820		1.2808		0.8768		0.6615
Years of labour market experience	0.0801	0.0773	0.0067	0.0015	0.1009	0.0760	0.0065	0.0122
(Years of labour market experience)^2	–0.0020	–0.0017	0.0002	0.0003	–0.0017	–0.0011	0.0003	0.0015
Years of education	0.0566	0.0352	0.0941	0.1145	0.0492	0.0864	0.0640	0.0842
Other family income	–2.0e–05	–1.9e–05	–9.6e–07	–9.8e–07	–1.5e–05	–1.5e–05	–1.1e–05	–1.1e–05
Metropolitan residence	–0.0100	0.2768	0.0355	–0.1005	–0.1841	–0.3267	–0.0211	–0.1198
Employed in 1979	1.1699	1.1652	1.8940	1.8642	1.5875	1.5870	1.6040	1.6051
Work disability	–1.1300	–1.1443	–0.9914	–0.9913	0.5478	0.5468	–0.2164	–0.2109
Professional occupation	0.5240	0.2820	0.1343	0.2742	0.0449	0.0681	0.0426	0.0934
Unemployment rate	–0.0516	–0.0508	–0.0528	–0.0536	–0.0125	–0.0135	0.0073	0.0073

Table 6.4 Probit model of employment (1 = yes)

Variable	*Men*				*Women*			
	Previous migration history		*No previous migration history*		*Previous migration history*		*No previous migration history*	
	Biased	*Unbiased*	*Biased*	*Unbiased*	*Biased*	*Unbiased*	*Biased*	*Unbiased*
Intercept	–1.4999	–1.4599	–1.9242	–1.8891	–2.1828	–1.8648	–1.8875	–1.9040
Migrant	–0.1464	–0.3506	0.0917	–0.2330	–0.1471	–2.0869	–0.1928	–0.9456
λ_i		0.1165		0.1769		1.1652		0.4243
Years of labour market experience	0.0870	0.0871	0.0355	0.0349	0.1402	0.1073	0.0085	0.0122
(Years of labour market experience)^2	–0.0017	–0.0017	0.0000	0.0000	–0.0028	–0.0021	0.0004	0.0002
Years of education	0.1194	0.1211	0.1489	0.1517	0.0463	0.0954	0.0842	0.0971
Other family income	–1.6e–05	–1.6e–05	–9.4e–07	–9.5e–07	–1.4e–05	–1.5e–04	–8.4e–07	–8.4e–07
Metropolitan residence	0.0050	–0.0187	–0.0107	–0.0295	–0.1494	–0.3395	–0.0569	–0.1204
Employed in 1979	1.3361	1.3374	1.7493	1.7453	1.5571	1.5576	1.5692	1.5692
Work disability	–0.6697	–0.6696	–0.4008	–0.4002	–0.0736	–0.0798	–0.3507	–0.3473
Professional occupation	0.6571	0.6779	0.5582	0.5763	0.1487	0.1792	0.0700	0.1028
Unemployment rate	–0.0707	–0.0709	–0.0798	–0.0799	–0.0310	–0.0321	–0.0068	–0.0067

Table 6.5 Linear model of λ_i (hours) worked last week

Variable	*Men*				*Women*			
	Previous migration history		*No previous migration history*		*Previous migration history*		*No previous migration history*	
	Biased	*Unbiased*	*Biased*	*Unbiased*	*Biased*	*Unbiased*	*Biased*	*Unbiased*
Intercept	0.9270	0.9154	0.6988	0.7124	–0.4820	0.0204	–0.1856	–0.2118
Migrant	–0.0009	0.0567	0.0705	–0.0642	–0.1426	–3.0498	–0.2273	–1.4217
λ_i		–0.0337		0.0749		1.7407		0.6749
Years of labour market experience	0.0293	0.0283	0.0112	0.0109	0.1538	0.1057	0.0053	0.0110
(Years of labour market experience)^2	–0.0005	–0.0005	0.0000	0.0001	–0.0029	–0.0018	0.0006	0.0004
Years of education	0.0633	0.0628	0.0657	0.0670	0.0289	0.1029	0.0902	0.1111
Other family income	–7.3e–06	7.3e–06	–3.1e–05	–3.1e–06	–2.0e–04	–2.1e–05	–1.8e–05	–1.8e–05
Metropolitan residence	–0.0029	0.0042	–0.0480	–0.0562	–0.1813	–0.4635	–0.1388	–0.2402
Employed in 1979	1.7445	1.7447	2.1337	2.1316	1.7024	1.7000	1.7403	1.7406
Work disability	–0.8230	–0.8230	–0.4803	–0.4801	–0.2760	–0.2834	–0.5247	–0.5195
Professional occupation	0.1181	0.1122	0.1096	0.1174	0.0708	0.1123	0.0015	0.0519
Unemployment rate	–0.0364	–0.0364	–0.0470	–0.0470	–0.0252	–0.0264	–0.0123	–0.0125

Conclusions

This research has attempted to qualify an analysis by Cooke and Bailey (1996) which found, contrary to both existing theory and empirical evidence, that family migration has a positive effect on the employment of married women. While an improved specification of the model failed to support Cooke and Bailey's previous findings (and also failed to find any effect of previous migration experience), the results suggest that families give greater weight to women's labour market achievement than assumed by the human capital model of family migration. Family migration has a strong negative impact on women's labour force participation and hours worked, but families decide to migrate (or to stay) in large part after considering the wife's skills and aptitudes. One way to interpret these results is to suggest that families with women who have strong local attachments (economically as well as socially) are less likely to migrate, not only because the initial act of migration will damage their labour market achievement but also because it may cause a loss of identification with their current locale. In contrast, families with women who have weaker local attachments are more likely to migrate because, although the initial act of migration damages their labour market achievement, the loss of identification with the current locale is not considered a significant problem. This suggests that the human capital model of family migration, in reducing all migration decisions to a simple economic calculus, fails to consider both how locally-constructed identities shape the willingness to migrate (especially for married women) and how married women are active shapers of family migration decisions. Finally, since there is not much research on how families actually make migration decisions, intensive qualitative research is needed in order to develop a more accurate portrayal of family migration causes and consequences (Halfacree and Boyle 1993).

Note

1 The 1980 PUMS is used rather than the 1990 PUMS because the latter does not include enough information on the individuals in the sample in 1985 to control for interdependence between labour market achievement, migration, and life-course events.

References

Cooke, T. J. and Bailey, A. J. (1996) 'Family migration and the employment of married women and men', *Economic Geography* 72: 38–48.

Cooke, T. J. and Bailey, A. J. (1998) 'Family migration and employment: the importance of migration history and gender', *International Regional Science Review* 21(21): 99–118.

Greene, W. H. (1993) *Econometric Analysis*, New Jersey: Macmillan.

Halfacree, K. H. (1995) 'Household migration and the structuring of patriarchy: evidence from the USA', *Progress in Human Geography* 19: 159–82.

Halfacree, K. H. and Boyle, P. J. (1993) 'The challenge facing migration research: the case for a biographical approach', *Progress in Human Geography* 17: 333–48.

Heckman, J. J. (1979) 'Sample selection bias as a specification error', *Econometrica* 47: 153–61.

Lichter, D. T. (1980) 'Household migration and the labor market position of married women', *Social Science Research* 9: 83–97.

Lichter, D. T. (1983) 'Socioeconomic returns to migration among married women', *Social Forces* 62: 487–503.

Mincer, J. (1978) 'Family migration decisions', *Journal of Political Economy* 86: 749–73.

Sandell, S. H. (1977) 'Women and the economics of family migration', *Review of Economics and Statistics* 59: 406–14.

Shihadeh, E. S. (1991) 'The prevalence of husband-centered migration: employment consequences for married mothers', *Journal of Marriage and the Family* 53: 432–44.

Spitze, G. (1984) 'The effect of family migration on wives' employment: how long does it last?', *Social Science Quarterly* 65: 21–36.

7 Family migration and female participation in the labour market

Moving beyond individual-level analyses

Paul Boyle, Keith Halfacree and Darren Smith

Introduction

There is a dominant consensus in the labour migration literature that the employment characteristics and quality of life of female partners often suffer as a result of long-distance family migration (McCollum 1990). Female partners have therefore been referred to as 'tied migrants' (Mincer 1978; Bielby and Bielby 1992) in a literature which usually adopts the human capital hypothesis (Spitz 1984; Mohlo 1986) as a starting point (Cooke and Bailey 1996). In short, the human capital hypothesis posits that family migration is motivated by the search for higher *household* (or family) incomes and may entail the female partner undertaking employment sacrifices in order for the family unit to reap the post-migration benefits obtained by the male 'breadwinner' (Halfacree 1995). It is assumed, according to the thesis of rational economic behaviour, that the potential income increment of the male partner will outweigh the loss of the female partner's income (Blau and Duncan 1967; Shihadeh 1991). While this simple conceptualization of labour migration underpinned many of the early empirical studies (e.g. Sandell 1977; Mincer 1978), the view that long-distance migration gives rise to constrained and detrimental employment opportunities for female partners has been challenged (Finch 1983; Bonney and Love 1991). The most recent departure has been provided by Cooke and Bailey (1996), in their study in the Midwestern United States, who concluded that family migration *increases* the probability of married women finding employment by 9 per cent. This led Cooke and Bailey to question strongly the limitations of the human capital hypothesis and call for the reconceptualization of labour migration. Importantly, Cooke and Bailey stress the need to ascertain how the effect of migration on married women's labour force participation, or quality of employment, is influenced by household structure – an issue considered in this chapter.

Similar claims for the reconceptualization of labour migration using a household structure perspective have been made, particularly in studies of dual-career couples where female partners are often less constrained in the labour market following migration (Dudleston et al. 1995; Hardill et al. 1997; Snaith 1990). Indeed, Green (1995) suggests that female partners are becoming increasingly less willing to hold supportive, and play secondary, roles to their male partners. This feature is linked to attitudinal and cultural shifts in contemporary society (McRae and Daniel 1991), which have facilitated greater employment opportunities for women and transformed their aspirations. Drawing upon the work of Kiernan (1992), Green outlines three intra-household cooperation strategies. These strategies are linked to particular types of partnership, termed 'traditional', 'middle' and 'egalitarian', in which different values are attached to the careers of the female ('her career') and male ('his career'). First, Green suggests that in 'traditional' partnerships the needs of 'his career' will dominate to such an extent that the female partner will 'drop out' of the labour market. This relation facilitates the 'female homemaker/male breadwinner' type of household. Second, in the 'middle' partnerships the female partner is still expected to fulfil homemaker and/or childrearer duties but there is a greater recognition of the needs of 'her career'. However, the needs of 'his career' still take precedence during intra-household decision making and female aspirations are often manifest in part-time employment and/or spatially ubiquitous occupations. Crucially, Green points to the rising significance of households where these relations are reversed (see also Jarvis's (1997) discussion of 'non-traditional' households). Third, Green identifies 'egalitarian' partnerships where the decision making recognizes the needs of 'his career', 'her career' and 'their career'. Such career cooperation is usually underpinned by the willingness of both partners to share household tasks and childrearing duties (Morris 1990; Gregson and Lowe 1993). This final type of partnership Green equates with 'dual-career' couples, especially those comprising young adults. We are interested in the impact that all these intra-household relations have upon the post-migration participation of the female (and male) partner in the labour market.

Despite the growing literature suggesting that women may not fare as badly from household migration as once supposed, empirical studies show that the employment opportunities of female partners generally suffer following long-distance migration (Boyle and Halfacree 1996; Gordon 1995). The career of the male partner frequently takes precedence over that of the female partner (Bielby and Bielby 1992). Various reasons have been suggested for this including the societal and cultural practices that give rise to the 'glass ceiling', which frequently constrains the career progression of women (Davidson and Cooper 1992; Hakim 1996; Hanson and Pratt 1995). As Bruegel (1996: 252) asserts, the tied migrant female partner

> has neither been killed off by organizational change nor by rising career aspirations amongst women. At most the breadwinner model may have been modified, rather than transcended.

We would argue that there is some validity on both sides of the argument but that the quantitative studies that have addressed the 'tied migrant' issue have not dealt adequately with intra-household factors, focusing too much on individual-level measures. Qualitative studies of labour migration have consistently shown that the decision-making process of labour migrants is not merely specific to individuals and involves many complex trade-offs and compromises between individuals within family units (e.g. Evetts 1996). Despite this, few studies have addressed why and how decisions are affected by the relational attributes of partners within different family units. As Bruegel (1966: 235) comments:

> decisions about household migration and location involve conflict between the interests of the different household members. Commonplace though this observation may be, it has largely escaped social science concerns with labour market change, migration behaviour and household relations in Britain.

Thus, the variables employed rarely offer insights into the effects of the relational characteristics of partners. By ignoring these factors, models based on individual-level characteristics alone may be mis-specified and we attempt to rectify this here (see also Long 1974; Markham and Pleck 1986).

More specifically, the use of individual-level data means that most previous studies have explored the effects of long-distance migration on female employment characteristics by simply comparing those individuals that did or did not move long distances (e.g. Boyle and Halfacree 1996; Cooke and Bailey 1996). There is no attempt in such models to identify partners who have moved long distances *together*. This problem is also addressed here. Moreover, the results will demonstrate that previous studies have understated the negative impact of long-distance migration on the employment characteristics of female partners.

Of course, this is not to suggest that migration studies in general have ignored family relations, just that studies in this particular research stream appear to have been quite narrowly focused. Recently, some have paid far more attention to household structure in migration research (e.g. Hayes et al. 1995). Green's (1995, 1997) discussion of 'dual-career' households in Britain deconstructs household types using data from the 1 per cent household file of the 1991 British Census Sample of Anonymized Records (SAR). This attempt to deal with households, rather than individuals, is to be commended but, unfortunately, Green adopts a 'household-led' focus rather than a 'family-led' focus, apparently justifying this approach by the falling number of nuclear families and the increase of childless and dual-career couples (see Duncan 1991). Essentially, Green's analysis considered household heads and their partners and excluded 'single-person' and 'multi-person households not containing two adults living as a couple' from her discussion. The remaining household types were 'households containing two adults living as a couple, with or without other household members', and this fails to distinguish between households with one couple and multi-couple households. Ignoring the partners (other than the household head and spouse) in multi-couple households appears to be for ease of data manipulation, rather than theoretically valid reasons.

Jarvis (1997) provides a more detailed breakdown of household types. Although it is not made clear explicitly, Jarvis does appear to include couples in multi-family households. Her sample includes a subgroup of 'nuclear family'-type households from the SAR which comprise married or cohabiting couples, under pensionable age, living with one or more dependent children. The strength of Jarvis's analysis is, however, what she terms 'a typology of household employment structures'. Based on a mixture of employment and occupational status, the typology differentiates between 'traditional', 'non-traditional', 'flexible', 'dual-earning' and 'dual-career' households and a similar typology is adopted here. 'Traditional' and 'non-traditional' households include couples where the male/female partner is in full-time employment and female/male partner is economically inactive, respectively. The 'flexible' category is assigned to couples where the male partner is in full-time employment and the female partner is working part-time. The 'dual-career' category is associated with couples where both partners are in full-time employment and employed in professional or managerial occupations (socio-economic group (SEG) I or II). Finally, 'dual-earner' couples are defined where both partners are in full-time employment but only one or neither partner is employed in SEG I or II occupations. Although Jarvis should be commended for scrutinizing the data and constructing these household types, there are a number of gaps which she does not address and these are discussed later.

In sum, the approaches adopted by both Green and Jarvis are useful, but limited in different ways. Green's method to identify 'dual-career' couples underestimates the true number. Jarvis's tightly defined household boundaries would appear not to consider multi-nuclear family households or households containing the co-existence of a nuclear family with another family type. Building upon this type of work, this study takes care to make sure that couples are not excluded simply because they live in multi-couple households, and it is the first large-scale quantitative analysis of the 'tied migration' question that achieves this. It then compares the migration of long-distance migrant couples with other couples to assess the effects of this mobility on women's employment characteristics.

Data and sample

This study utilized the 1 per cent linked household file of the SAR. The first stage involved identifying all partnered adults aged between 16 and 64 (males) and 16 and 59 (females) in the 1 per cent household file. This was achieved using a combination of a family number variable (for distinguishing families within multi-family households) and a derived household variable from the SAR that distinguishes between single and partnered adults (Holdsworth 1995). Once all partners had been linked in households, the second stage further refined the subsample by excluding couples where one or both partners were a student, permanently sick, retired, a member of the armed forces or a migrant from outside the UK. As a result of this second stage, 83,562 individuals were eliminated from the subsample of 264,342 individuals created at stage 1.

The sample at stage 2 therefore comprised 180,780 partnered individuals, 9.8 per cent of whom were migrants. These were divided into two groups based on the distance they had migrated. Long-distance migrants were individuals who had migrated 50 km or more (1.3 per cent of the total sample). Short-distance migrants were individuals who had migrated less than 50 km (8.5 per cent of the total sample) and these were grouped with non-migrants for the analysis. The distance cut-off between short- and long-distance migration was relatively arbitrary, as some long-distance migrants do not move for reasons of employment and vice versa, but it was expected that it would distinguish broadly between migrants moving for residential and labour purposes (Boyle 1995). Individuals who had not stated their place of origin prior to migration on the census form were subsumed within the short-distance category.

As stated earlier, quantitative studies of labour migration often identify long-distance migrants based on individual-level data. This simple error means that a number of these individuals will have moved alone, regardless of whether they were part of a couple living together at the time of enumeration. Here, we defined long-distance migrants both as those individuals who moved 50 km or more in the year prior to the census date, but also more rigorously by identifying those who moved 50 km or more *with their partner*. This has a number of important implications for studies of gendered labour migration, as borne out by table 7.1 which divides the sample by sex, based on both individual- and family-level definitions of the distance moved.

Table 7.1 Distance moved by individuals and those in family units (%)

	Male	*Female*	*Total*
Individual-level			
All long-distance migrants	1157 (1.28)	1228 (1.36)	2385 (1.32)
Other	89223 (98.72)	89172 (98.64)	178395 (98.68)
Family-level			
2x long-distance migrants moving together	958 (1.06)	958 (1.06)	1916 (1.06)
2x long-distance migrants joining	33 (0.04)	33 (0.04)	66 (0.04)
1x long-distance & 1x short-distance	206 (0.23)	206 (0.23)	412 (0.23)
1x long-distance & 1x non-migrant	197 (0.22)	197 (0.22)	394 (0.22)
Other	88996 (98.45)	88996 (98.45)	177992 (98.45)

According to the orthodox approach, in the upper part of the table, 1.36 per cent of females moved long distances, compared to a slightly smaller percentage of males (1.28 per cent). Alternatively, the lower part of table 7.1 disaggregates long-distance migrants by partnership types. Only 1.06 per cent of the couples (and therefore males and females) moved long distances together. There were 66 people in couples where both partners had moved *different* long distances, and 806 people in couples where only one individual moved a long distance and the other partner either moved a short distance or did not migrate. According to the family-based definition of long-distance migration, therefore, 469 of the long-distance migrants identified in individual-level analyses were ignored, as they did not move long distances with their partners.

Variables

The decision to migrate for employment opportunities and the probability of finding post-migration employment opportunities are influenced by many factors (Owen and Green 1992; Flowerdew and Halfacree 1994). Based on the most significant variables outlined in previous studies of gendered labour migration, eleven categorical variables were extracted from the 1 per cent household SAR (table 7.2). Two of the variables (car ownership and housing tenure) relate to the household level and the remaining nine relate to individual characteristics. Employment status was used as the *y* variable in the subsequent analysis, while the other ten were explanatory variables.

Relational variables were also constructed based on combinations of the individual-level characteristics of partners and it would appear that these types of variables have not been created previously in studies of 'tied migration' (table 7.3). First, a life-course variable was constructed to control for the impact of life-course events upon migration (Dex 1987; Joshi 1991). This was based on the presence/absence and age of dependent or non-dependent children. Five distinctions were made: couples with no child; couples with their youngest dependent child at pre-school age (0–4); couples with their youngest dependent child at primary school age (5–10); couples with their youngest dependent child at high school age (11+); and couples with non-dependent children. In line with the 1991 census, dependent children were defined as persons aged 0–15 or persons aged 16–18 who are in full-time education or are economically inactive and have never been married. Family units with no children dominate the sample but we should note that this category conflates non-family-forming units, family units at pre-family-forming and post-family-forming stages of their life course. Thus, it is not possible to identify in the SAR 'empty nest' couples, whose children have matured and vacated the family unit.

The second variable was an adapted version of Jarvis's (1997) household typology, described above. In contrast to Jarvis (and Hardill et al. 1997) the focus of analysis is the family unit, not the household. Moreover, the

Table 7.2 Individual-level explanatory variables extracted from the 1991 Sample of Anonymized Records

Variable	*Values*
Employment status[1]	Full-time
	Part-time
	Self-employed
	Unemployed/other inactive
Sex	Male
	Female
Migrant status	Non-migrant/short-distance migrant
	Long-distance migrant
Age	16–24
	25–34
	35–44
	45–54
	55–65 (male), 55–60 (female)
Socio-economic group	Service class
	Petite bourgeoisie
	White collar
	Blue collar
	Other socio-economic group
Qualifications	Non-graduate
	Graduate
Ethnic group	White
	Black
	Other ethnic group
Limiting long-term illness	No
	Yes
Marital status	Married
	Cohabiting
Tenure	Owner-occupied
	Public rented
	Private rented
Car ownership	No car
	One car
	Two or more cars

Note

1 Collapsed to employed/unemployed or economically active in the modelling analysis.

Table 7.3 Family-level explanatory variables derived from the 1991 Sample of Anonymized Records

Variables	*Values*
Life-course	No child(ren)
	Youngest dependent child at pre-school age
	Youngest dependent child at junior school age
	Youngest dependent child at high school age
	Non-dependent child(ren) only
Family type	Other
	Dual-career
	Dual-other
	Female-flexible
	Male-flexible
	Traditional
	Non-traditional
Relational occupational status	Other
	Male-dominated
	Female-dominated
Relational ethnic group	Partners belong to same ethnic group
	Partners belong to different ethnic groups
Relational age	Partners of same/one age group difference
	Partners of two age groups difference
	Partners of three age groups difference
	Partners of four age groups difference
Relational migrant status[1]	Two long-distance partners moving together
	Two long-distance partners moving joining
	Long-distance migrant/non-migrant
	Long-distance migrant/short-distance migrant
	Two short-distance migrants
	Short-distance migrant/non-migrant
	Two non-migrants
Relational limiting long-term illness	Neither partner
	Male partner only
	Female partner only
	Both
Relational qualifications	No graduates
	Male graduate only
	Female graduate only
	Both graduates

Note

Collapsed to two long-distance partners moving together/non- or short-distance migrants in the modelling analysis.

typology employed here adds greater relational detail, distinguishing between seven partnership types. In line with Jarvis's 'dual-career' couples were those where both partners were in full-time employment in professional and managerial occupations (SEG I or II). 'Dual-other' is where both partners are in full-time employment, but only one or neither partner was employed in a professional or managerial occupation, and this is preferred to Jarvis's label of 'dual-earner' since 'dual-career' couples are also 'dual-earners'. Jarvis's 'flexible' couples involve males working full-time and females working part-time, but this overlooks cases where the female partner is in full-time employment and the male partner is working part-time. Therefore, we differentiate between 'female flexible' couples, where the male partner is in full-time employment and the female partner is working part-time, and 'male flexible' couples, where the female partner is in full-time employment and the male partner is working part-time. Slightly amended versions of the 'traditional' and 'non-traditional' definitions used by Jarvis were also adopted. The former involves a male partner who is employed full-time and an unemployed or economically inactive female, while the latter is the reverse. The remainder of the sample was classified into an 'other' category comprising couples where both partners were either in part-time employment, unemployed, economically inactive or in a mixture of these circumstances. While this variable was used in the descriptive analysis that follows, it was not used in the modelling exercise because of the considerable overlap with the employment status dependent variable.

The third variable was devised to assess the occupational power relations within couples, based on Cambridge scores which are gender-specific occupational rankings allowing us to calculate the relative position of employed partners (see Stewart et al. 1980; Prandy 1990). The variable distinguishes between couples where the female had a higher Cambridge score, the male had a higher Cambridge score or where the Cambridge scores of the partners were equal (this category includes cases where neither partner had an occupation and could not be assigned a Cambridge score). It was assumed that the partner with the highest occupational status would probably influence the migration decision-making process most (Wheelcock 1990).

A fourth derived variable was also produced to take account of the partner's educational qualifications, as it was hypothesized that individual-level qualifications do not explain the entire relationship with employment status. Rather, we might speculate that non-graduate individuals may be less likely to work if they have a graduate partner, and this may vary with sex. Four categories were identified distinguishing between couples where: both partners were non-graduates, only the male was a graduate, only the female was a graduate, and both partners were graduates. Similarly, a four-way limiting long-term illness variable identified couples where neither partner was ill, only the (fe)male was ill, or both were ill, again because we anticipated that a partner's health status may influence an individual's employment status. The difference in age groups between partners in a couple was determined in

Table 7.4 Individual-level variables by distance moved (%)

	Individual-level			*Family-level*		
	All long-distance migrants		*Non- and short-distance migrants*	*Two long-distance migrants moving together*		*Other migrants and non-migrants*
	Male	*Female*		*Male*	*Female*	
	n = 1157	*n* = 1228	*n* = 178395	*n* = 958	*n* = 958	*n* = 178864
Age groups						
16–24	113 (9.8)	237 (19.3)	10985 (6.2)	69 (7.2)	130 (13.6)	11136 (6.2)
25–34	496 (42.9)	575 (46.8)	49915 (28.0)	407 (42.5)	448 (46.8)	50131 (28.0)
35–44	344 (29.7)	269 (21.9)	56203 (31.5)	301 (31.4)	246 (25.7)	56269 (31.5)
45–54	156 (13.5)	122 (9.9)	43770 (24.5)	137 (14.3)	110 (11.5)	43801 (24.5)
55+[1]	48 (4.1)	25 (2.0)	17522 (9.8)	44 (4.6)	24 (2.5)	17527 (9.8)
Socio-economic group						
Service class	690 (59.6)	493 (40.1)	49927 (28.0)	589 (61.5)	355 (37.1)	50166 (28.0)
Petite bourgeoisie	85 (7.3)	26 (2.1)	17893 (10.0)	71 (7.4)	22 (2.3)	17911 (10.0)
White collar	129 (11.1)	469 (38.2)	43213 (24.2)	106 (11.1)	381 (39.8)	43324 (24.2)
Blue collar	232 (20.1)	123 (10.0)	51376 (28.8)	178 (18.6)	95 (9.9)	51458 (28.8)
Other	21 (1.8)	117 (9.5)	15986 (9.0)	14 (1.5)	105 (11.0)	16055 (9.0)
Employment status						
Full-time	886 (76.6)	495 (40.3)	95022 (53.3)	750 (78.3)	310 (32.4)	95343 (53.3)
Part-time	19 (1.6)	177 (14.4)	28200 (15.8)	13 (1.4)	157 (16.4)	28226 (15.8)
Self-employed	120 (10.4)	45 (3.7)	20339 (11.4)	102 (10.6)	35 (3.7)	20367 (11.4)
Unemployed/ other inactive	132 (11.4)	511 (41.6)	34834 (19.5)	93 (9.7)	456 (47.6)	34928 (19.5)
Qualifications						
Non-graduate	674 (58.3)	804 (65.5)	147761 (82.8)	554 (57.8)	656 (68.5)	148029 (82.8)
Graduate	483 (41.7)	424 (34.5)	30364 (17.2)	404 (42.2)	302 (31.5)	30364 (17.2)
Marital status						
Married	862 (74.4)	907 (73.9)	158483 (88.8)	801 (83.6)	801 (83.6)	158650 (88.7)
Cohabiting	295 (25.5)	321 (26.1)	19912 (11.2)	157 (16.4)	157 (16.4)	20214 (11.3)

Note
1 Male (55–65), female (55–60).

the relational age variable and the final variable identified whether the partners came from the same or different ethnic groups.

Individual-level analysis

Table 7.4 presents some of the individual-level variables broken down by sex within the two definitions of long- and non-/short-distance migration, which differ most significantly between the two levels. Comparing the percentages in the first and second columns with those in the third column demonstrates which groups of individuals are most likely to migrate long distances and the results from both levels of analyses generally conform with previous studies of labour migration. As expected, males and females aged 16–34, those in service class occupations, graduates and cohabitees were more likely to be long-distance migrants. In addition, female white-collar workers, male full-time employees and unemployed or economically inactive females were also especially likely to migrate long distances and these gender differences are a first indication that migration involves different types of males and females; indeed this may be evidence of 'tied migrants' in the sample.

On the whole, five obvious differences are evident when comparing the results of the individual- (columns 1–3) and family-based (columns 4–6) definitions of long-distance migration. First, the individual-level analysis overemphasizes the incidence of young adults aged 16–24 among long-distance migrants. In contrast, based on the family-level definition, a greater proportion of long-distance partners was aged over 35. Second, the family-level definition suggests that female migrant partners were less likely to be members of the service class. Rather, a greater proportion of female partners belonged to the 'other' socio-economic group, which includes those with no occupational status. Third, according to the family-level definition, a lower proportion of females was employed full-time while the individual-level definition underemphasized the importance of female partners who were working part-time or were unemployed/other inactive following long-distance migration. These results suggest that studies that rely on the individual-level definition understate the negative impact of long-distance migration upon female participation in the labour market. Fourth, the family-based definition suggests that female partners were more likely to be non-graduates, when compared to the individual-level analysis and, finally, cohabiting couples were far less prevalent among the family-based definition of long-distance migrants.

Family-level analysis

Table 7.5 presents the most significant findings for some of the relational variables. Couples with no children, or with their youngest child at preschool age, 'dual-career' or 'traditional' couples and those in occupationally male-dominated relationships were most likely to migrate long distances.

Table 7.5 Family-level variables by distance moved (%)

	Individual-level			*Family-level*		
	All long-distance migrants		*Non- and short-distance migrants*	*Two long-distance migrants moving together*		*Other migrants and non-migrants*
	Male	*Female*		*Male*	*Female*	
	n=1157	*n*=1228	*n*=178395	*n*=958	*n*=958	*n*=178864
Life-course						
No child(ren)	535 (46.2)	626 (51.0)	55245 (31.0)	395 (41.2)	395 (41.2)	55616 (31.1)
Youngest dep. child at pre-school age	366 (31.6)	360 (29.3)	40890 (22.9)	337 (35.2)	337 (35.2)	40492 (22.9)
Youngest dep. child at junior school age	115 (9.9)	110 (9.0)	23927 (13.4)	106 (11.1)	106 (11.1)	23940 (13.4)
Youngest dep. child at high school age	95 (8.2)	88 (7.2)	31847 (17.9)	81 (8.5)	81 (8.5)	
Non-dependent child(ren) only	46 (4.0)	44 (3.6)	26846 (14.8)	39 (4.1)	39 (4.1)	26498 (14.8)
Family type						
Dual-career	199 (17.2)	219 (17.8)	14454 (8.1)	142 (14.8)	284 (14.8)	14588 (8.2)
Dual-other	215 (18.6)	251 (20.4)	40332 (22.6)	156 (16.3)	312 (16.3)	40486 (22.6)
Female flexible	163 (14.1)	174 (14.2)	55875 (31.3)	150 (15.7)	300 (15.7)	55912 (31.3)
Male flexible	12 (1.0)	7 (0.6)	831 (0.5)	6 (0.6)	12 (0.6)	838 (0.5)
Traditional	424 (36.6)	454 (37.0)	49140 (27.5)	400 (41.8)	800 (41.8)	49218 (27.5)
Non-traditional	43 (3.7)	30 (2.4)	3515 (2.0)	22 (2.3)	44 (2.3)	3544 (2.0)
Other	101 (8.7)	194 (7.6)	14248 (8.0)	82 (8.6)	164 (8.6)	14278 (8.0)
Relational occupational status						
Male-dominated	637 (55.1)	684 (55.7)	91141 (51.1)	555 (57.9)	1110 (57.9)	91352 (51.1)
Female-dominated	473 (40.9)	499 (40.6)	80896 (45.3)	368 (38.4)	736 (38.4)	81132 (45.4)
Other	47 (4.1)	45 (3.7)	6358 (3.6)	35 (3.7)	70 (3.7)	6380 (3.6)

The importance of having children as a constraining factor is especially noteworthy here (see Bradley 1989; Joshi and Hinde 1993; McRae 1993; Macran et al. 1996). The individual-level definition of long-distance migration (columns 1–3) overstates the presence of childless family units and, in comparison to the family-level definition (columns 4–6), underestimates the proportion of male and female partners with young children. 'Dual-career' and 'dual-other' families are also exaggerated and the emphasis attached to dual-career households in the labour migration literature perhaps needs further scrutiny. On the other hand, 'female flexible', 'traditional' and 'other' partnership types involving female partners are underestimated and in all these partnership types, the female partner is either employed in a part-time post, unemployed or economically inactive.

In sum the definition of migration adopted will influence dramatically the results of studies of this type. The traditional approach, which aggregates all the long-distance migrant individuals, appears to understate the detrimental impact of long-distance migration upon the employment opportunities of female partners.

Modelling aims and results

A descriptive analysis of the SAR data was provided in the previous section, but the cross-tabulation of pairs of variables ignores the potentially confounding effects of other correlated variables. Here we use logit models to estimate the log-odds that each individual in our sample will be unemployed or economically inactive, rather than employed (either full- or part-time). The remaining variables were used to explain these differences and of particular interest were sex and migrant status and the interaction between the two. Odds ratios were calculated for each parameter estimate, to assess whether they were significantly different from unity, and the results for three models are presented in table 7.6.

Two specific aims were investigated. First, we compared the two definitions of long-distance migration and the different modelling results are provided in Model A (individual-based definition) and Model B (family-based definition). The latter definition allows a more rigorous investigation of the effects of family migration on the female's employment characteristics and we aimed to test whether the conclusions from these two approaches would vary significantly.

Second, a number of variables constructed from the characteristics of both partners were considered. We argued above that little attention has been devoted to family-level factors that may influence migration, at least within the developed world literature (see chapter 1, above). Here, the aim was to test whether such family-level variables (excluding the family-type variable) were significant, controlling for individual-level variables that are known to influence migration behaviour. Indeed, we were interested to see the final results are presented in Model C (table 7.6).

Table 7.6 Logit models of unemployment and economic inactivity

Variable categories	*Odds ratios*		
Individual variable categories	*Model A*	*Model B*	*Model C*
25–34	1.11	1.08	0.86
35–44	0.57	0.56	0.76
45–54	0.47	0.45	1.03[2]
55+	0.68	0.66	1.69
Female	3.55	1.69	3.99
Graduate	0.77	3.45	0.49
Cohabiting	0.73	0.75	1.11
Public renting	2.26	2.27	2.05
Private renting	1.32	1.30	1.46
One car	0.55	0.55	0.53
Two or more cars	0.46	0.46	0.44
Black	0.80	0.81	0.71
Other ethnic group	1.37	1.37	1.13
Limiting long-term illness	2.10	2.10	2.19
Petite bourgeoisie	0.65	0.65	0.58
White collar	1.38	1.38	1.36
Blue collar	1.47	1.48	1.45
Other socio-economic group	86.46	87.09	86.73
Male long-distance migrants[1]	1.27	1.43	1.34
Female long-distance migrants[1]	4.01	10.60	12.09
Family variable categories			
Youngest dependent child at pre-school age			6.11
Youngest dependent child at junior school age			2.05
Youngest dependent child at high school age			1.18
Non-dependent child(ren) only			1.06[2]
Male-dominated			1.31
Female-dominated			1.02[2]
Male graduate			1.12
Female graduate			1.32
Both graduates			1.66
Male ill			1.28
Female ill			1.03[2]
Both ill			0.85[2]
Two age groups difference			1.08[2]
Three age groups difference			1.16[2]
Four age groups difference			3.36[2]
Different ethnic groups			1.17

Notes

1 The definition of long-distance migrants varies between Model A and Models B and C. In Model A, long-distance migrants are those individuals that moved 50 km or more. In Models B and C, long-distance migrants are those individuals that moved 50 km or more together with their partners.

2 Insignificant parameter estimates at the 95% level.

Redefining long-distance movers

All of the variables and the category-specific parameter estimates (expressed as odds ratios) were significant in Model A. Ignoring the migration and gender interaction terms for now, this model demonstrates that the unemployed and economically inactive in our sample were especially likely to be: in the 'other socio-economic group', female, living in public housing or suffering from limiting long-term illness. Those in the 'other socio-economic group' will include those who are unable to work because of illness, rather than self-reported limiting long-term illness, and those who have no occupation because they have never worked. This explains the extremely high odds ratio. Similarly, we would expect those suffering from self-reported limiting long-term illness to be less likely to be employed because of their health status. The role of public housing is more difficult to determine, as living in a particular type of property may, at the outset, seem unlikely to influence the probability of employment. In fact, there are various reasons why this parameter may be high, including the fact that the unemployed may have other characteristics that help them gain access to public housing, and even the cultural effects of living on a council housing estate among other unemployed individuals.

The least likely to be unemployed or economically inactive were those aged 35 or over, particularly those aged 45–54, those who owned one or more cars, and those in the petite bourgeoisie. The strong age effect demonstrated that unemployment and economic inactivity was more prevalent among young couples. Those families with one or more cars were less likely to be unemployed or economically inactive, as we would expect, as this has often been used as a crude surrogate for income. Those in the petite bourgeoisie are generally employed in occupations with low unemployment rates, and a significant number in this group are self-employed.

Here, we are particularly concerned with the effect for female migrants, and the incorporation of the interaction term between the sex and migration variables allows this to be investigated. Of the sample, 2,385 individuals were defined as long-distance movers. The odds ratio for non-migrant females was high (3.55), in line with the generally held assumption that women are more likely to drop out of, or never enter, the workforce than men. More surprisingly, at least in terms of neo-classical migration theory, was that the odds ratio for long-distance male migrants was significantly greater than unity (1.27). Those males that move long-distance were more likely to be economically inactive or unemployed than non-migrants or those that moved less than 50 km. This demonstrates that long-distance migration does not necessarily equate with improved labour market circumstances. Some of this group may be moving speculatively in search of work and have yet to find it, although the number of genuinely speculative moves that are made in the developed world has been argued to be small (Flowerdew 1992). On the other hand, a long-distance move for employment reasons could subsequently result in job loss, as those that have joined a company most recently may be most vulnerable in recession periods (Pissarides and Wadsworth 1989). Also, we should acknowledge that many individuals may make long-distance moves because they become unemployed, perhaps returning to their home areas where they may have more social support networks to help find work and cope with the period of unemployment more generally. In this sense long-distance migration is not as 'positive' as it is usually assumed to be. Unfortunately, it is not possible to determine how the characteristics of migrants changed between the origin and destination, as the census only provides information about the person at the date of enumeration, not before.

The interaction term between the sex and migration variables, which was significant, allows the calculation of the odds of women that are long-distance migrants being unemployed or economically inactive. The value (4.01) was far higher than that for migrant men, suggesting that women fared worse from family migration than men, in concordance with much of the literature on 'tied' migration. Note, however, that the value was only slightly larger than that for non-migrant, or short-distance migrant women.

The results from Model B were generally similar to those from Model A and all of the parameter estimates were again significant. The migration definition used in this case only considered those that moved with their partners as long-distance migrants, reducing the migrant sample by 469 to 1,916 individuals. However, the odds ratio for female long-distance migrants (10.6) was much greater than in Model A, despite the small change in the migrant definition. Women who move long distances *with* their partners are even more likely to drop out of the labour force than women that move long distances individually (as this accounts for the difference between the two definitions of long-distance migration used in Models A and B). This is an important finding. It suggests that previous studies, that failed to link partners who moved together (e.g. Boyle and Halfacree 1996; Cooke and Bailey 1996), underestimated the true impact of family migration on women's careers. It is also worth noting that the odds ratio was higher for male migrants who moved with their partners in Model B (1.43) than that recorded for all male long-distance movers in Model A (1.69).

Incorporating family-based variables

Model C (table 7.6) includes the family-based variables described above. First, we discuss whether the interpretation of the individual-level variables alters with the inclusion of these variables. Second, we examine the impact that these family variables have on the labour market characteristics of men and women. Third, we then test whether the migration and gender effects demonstrated in Model B are consistent, controlling for these family variables.

The ten individual-level variables remained significant, although one of the parameter estimates for specific categories now became insignificant. The odds ratio for those aged 35 and above became larger and less significant and the 45–54 category was not significantly different to the base category. It appears that family variables are correlated with age and that once these are accounted for age becomes less related to employment characteristics. Similarly, while cohabiting partners were significantly more likely to be employed than married partners in Models A and B, the odds ratio became slightly larger than 1.0 in Model C, although it was only just significant. Cohabiting status in itself may not influence the likelihood of employment, but it is correlated with other family circumstances that are now controlled for in the model. The male long-distance migrant variable also became slightly smaller than in Model B. The tenure variable retained its significance and both the public and private renting odds ratios were greater than unity, although the odds ratio for public renting was smaller than in the previous two models. As suggested above, much of the effect of tenure on employment is because the variable is a surrogate for other factors. Consequently, controlling for family-level variables reduces the importance of this explanatory variable, although it is still true that those in public housing are less likely to be employed than those in private renting or owner-occupied accommodation according to this model. The ethnic minority variable remained significant but the odds ratio fell for 'other ethnic minorities' and although it was greater than unity it was only marginally significant, controlling for these other individual- and family-level variables. This is a potentially important finding, suggesting that ethnic minority couples are not more likely to be unemployed than whites and this is in contrast to many commonly held views.

Five of the six family-level variables were significant in Model C (relational age was not), although some of the categories within the other variables were not significantly different to the base category. The life-course variable provided predictable odds values with a consistent gradient between those with a pre-school-age youngest child (6.11) to those with non-dependent children (1.06). Each of the categories associated with having children produced a higher odds ratio than the base category (1.0) which included couples without children. Undoubtedly, having children was a major factor increasing the likelihood that an individual was unemployed or economically inactive. The partner's qualifications also influenced the employment status of the individuals in this sample. Controlling for individual-level qualifications, there was an increased probability of unemployment or economic inactivity for those couples where one, or particularly both, partners were graduates. This seems surprising initially, although it may be related to the fact that graduate incomes are higher than those for non-graduates, making it more feasible for someone to drop out of the workforce if their partner is highly qualified. This ties in with the literature that demonstrates the growing importance of dual-earner households where both partners are employed in relatively poorly paid occupations (Martin and Roberts 1984). There may be less pressure to work in those families with at least one graduate whose salary may support the family adequately.

The primary focus in this chapter, though, is on the gender and migration effects. The odds of unemployment or economic inactivity for females increased slightly in Model C (3.99) compared with Models A and B (3.55 and 3.45 respectively). The odds ratio for long-distance female migrants rose to 12.09 in Model C, compared with only 4.01 in Model A and 10.6 in Model B. This more complex model, which accounts for a variety of both individual- and family-level variables also demonstrates that females that move long distances with their partners are much less likely to remain in employment after the move. Previous studies that ignored these family-level variables will have underestimated the significance of this effect.

Discussion

This research used some innovative strategies to investigate a question that has been attracting academic attention for some time. In summary, we have shown, first, that previous quantitative studies that failed to identify those individuals who moved with their partners were likely to underestimate the effects of long-distance family migration on women's employment. Second, we have demonstrated that family-level variables are important in explaining the likelihood of employment and studies that ignore these variables may be inadequate. Third, when these variables are included in the model, the negative impact of long-distance migration on women's employment status is demonstrated even more clearly.

Of course, despite these advances, there are inevitable drawbacks with this study, most of which result from the nature of the census data. The *y* variable distinguished simply between those that were employed and those that were unemployed or economically inactive. The census provides no socioeconomic information about individuals at the time they migrated and it is impossible to guarantee that female migrants were actually more likely to lose their jobs on migration; they may have been a self-selected group, although the work in chapter 6 (above) suggests this may be unlikely. This approach also ignores the specific type of job that each individual was employed in, compared to their jobs prior to migration. Thus, although many may be employed after moving, it is possible that they have been forced to take on a less career-orientated job. If this was more likely for women than men the models here underestimate the negative effects of long-distance family migration on women's careers because this is not accounted for in the analysis.

The definition of migration, which was obviously a key variable in the analysis, is also imperfect. Common to most censuses, the 1991 British Census simply asks whether the address at the time of enumeration was different from that of one year previously. If so, the individual is regarded as a migrant. However, some individuals will have moved more than once during this period, and these moves would be ignored, while a small number may even have moved back to their previous address so that they do not appear to have migrated at all (Boyle et al. 1998). Also, while the census provides information about 'wholly moving households' it provides no information about 'moving family units' which would have been especially useful here (Flowerdew 1998). Instead, we have been forced to use the distance moved by each individual to determine whether partners moved together or separately. Thirteen distance bands are defined in the SAR and, although unlikely, it is possible that a small number of partners will have moved long distances separately, even though they apparently moved the same distance.

While this study has considered a variety of original family-level variables that were related to individuals' employment status, there are other explanatory variables that have been ignored here. Most obviously, this analysis has been carried out for Britain as a whole and place differences have been ignored. In order to construct these family-based variables it was essential to have linked households but, unfortunately, the 1 per cent household SAR only provides a region identifier, rather than a SAR-area identifier as in the individual-level SAR. Even so, analysis is under way to investigate regional variations in these modelling results.

We also know nothing about the complex decision-making process that may precede the family's move. For example, a family may decide to follow the male's career (and the associated career-related moves) because they are anticipating having children and expect the female to leave her career for some time. It may be 'rational' for the family to take the decision to sacrifice the woman's career in such circumstances, even if her job was apparently better paid or more secure than the male's job. It is impossible to incorporate these factors into models based on census information.

Despite the drawbacks, this chapter presents some innovative analyses that suggest that women fare worse from family migration than men, at least in terms of their employment participation. Many of the problems identified above can only be investigated satisfactorily using qualitative methods, and this is the aim of the second stage of this research.

Acknowledgements

This research was funded by the ESRC (grant number R000237318). The census data are Crown Copyright and were bought by ESRC and JISC for use in the academic community. The ESRC Data Archive and Manchester Computing Centre made the data available to us, and Clare Holdsworth and Ian Turton provided valuable help with the data extraction. Tom Cooke and Robin Flowerdew made useful comments on a draft version of this chapter.

References

Bielby, W. T. and Bielby, D. D. (1992) 'I will follow him: family ties, gender-role beliefs and reluctance to relocate for a better job', *American Journal of Sociology* 97: 1241–67.

Blau, P. M. and Duncan, O. D. (1967) *The American Occupational Structure*, New York: Wiley.

Bonney, N. and Love, J. (1991) 'Gender and migration: geographical mobility and the wife's sacrifice', *Sociological Review* 39: 335–48.

Boyle, P. J. (1995) 'Public housing as a barrier to long-distance migration', *International Journal of Population Geography* 1: 147–64.

Boyle, P. J. and Halfacree, K. H. (1996) 'Gender issues in the migration of single, service class adults in Britain: an urban perspective', *RGS-IBG Annual Conference*, Strathclyde University, Glasgow, 3–6 January.

Boyle, P. J., Halfacree, K. H. and Robinson, V. (1998) *Exploring Contemporary Migration*, London: Longman.

Bradley, H. (1989) *Men's Work, Women's Work*, Oxford: Polity Press.

Bruegel, I. (1996) 'The trailing wife: a declining breed', in R. Crompton, D. Gallie and L. Purcell (eds) *Changing Forms of Employment*, London: Routledge, pp. 235–58.

Cooke, T. J. and Bailey, A. J. (1996) 'Family migration and the employment of married women and men', *Economic Geography* 72: 38–48.

Davidson, M. J. and Cooper, C. L. (1992) *Shattering the Glass Ceiling*, London: Paul Chapman.

Dex, S. (1987) *Women's Occupational Mobility: a Lifetime Perspective*, Basingstoke: Macmillan.

Dudleston, A. C., Hardill, I., Green, A. E. and Owen, D. W. (1995) 'Work rich households: case study evidence on decision making and career compromises amongst dual-career households in the East Midlands', *East Midlands Economic Review* 4: 15–32.

Duncan, S. (1991) 'The geography of gender divisions of labour in Britain', *Transactions, Institute of British Geographers* 16: 420–39.

Evetts, J. (1996) *Gender and Career in Science and Engineering*, London: Taylor and Francis.

Finch, J. (1983) *Married to the Job: Wives' Incorporation into Men's Work*, London: Allen and Unwin.

Flowerdew, R. (1992) 'Labour market operation and geographical mobility', in A. Champion and A. Fielding (eds) *Migration Processes and Patterns. Volume 1. Research Progress and Prospects*, London: Belhaven Press, pp. 135–47.

Flowerdew, R. (1998) 'The potential use of moving units in British migration analysis', in P. Rees (ed.) *Third Workshop. The 2001 Census Special Datasets: What do we Want?*, Working Paper 97/9, School of Geography, University of Leeds, pp. 81–2.

Flowerdew, R. and Halfacree, K. H. (1994) 'Logit modelling of migration propensity in 1980s Britain', *Migration Unit Research Paper* 7, University of Wales, Swansea.

Gordon, I. (1995) 'Migration in a segmented labour market', *Transactions, Institute of British Geographers* 20: 139–55.

Green, A. (1995) 'The geography of dual-career households: a research agenda and selected evidence from secondary data sources for Britain', *International Journal for Population Geography* 1: 29–50.

Green , A. (1997) 'A question of compromise? Case study evidence on the location and mobility strategies of dual-career households', *Regional Studies* 31: 641–57.

Gregson, N. and Lowe, M. (1993) 'Renegotiating the domestic division of labour? A study of dual-career households in North East and South East England', *Sociology* 28: 55–78.

Hakim, C. (1996) *Key Issues in Women's Work: Female Heterogeneity and the Polarisation of Women's Employment*, London: Athlone Press.

Halfacree, K. H. (1995) 'Household migration and the structuration of patriarchy: evidence from the USA', *Progress in Human Geography* 19: 159–82.

Hanson, S. and Pratt, G. (1995) *Gender, Work and Space*, London: Routledge.

Hardill, I., Green, A. E., Dudleston, A. C. and Owen, D. W. (1997) 'Who decides what? Decision making in dual-career households', *Work, Employment and Society* 22: 313–26.

Hayes, L., Al-Hamad, A. and Geddes, A. (1995) 'Marriage, divorce and residential change: evidence from the household Sample of Anonymised Records', *Migration, Kinship and Household Change Working Paper 3*, Department of Geography, Lancaster University.

Holdsworth, C. (1995) 'Minimal household units', *SARs Newsletter*, Number 5, Census Microdata Unit, University of Manchester.

Jarvis, H. (1997) 'Housing, labour markets and household structure: questioning the role of secondary data analysis in sustaining the polarization debate', *Regional Studies* 31: 521–31.

Joshi, H. E. (1991) 'Sex and motherhood as sources of women's economic disadvantage', in D. Groves and M. McClean (eds) *Women's Issues in Social Policy*, London: Routledge, pp. 179–93.

Joshi, H. E. and Hinde, P. R. A. (1993) 'Employment after childbearing in post-war Britain: cohort study evidence on contrasts within and across generations', *European Sociological Review* 9: 203–27.

Kiernan, K. (1992) 'The roles of men and women in tomorrow's Europe', *Employment Gazette* 100: 491–99.

Long, L. (1974) 'Women's labour force participation and the residential mobility of females', *Social Forces* 521: 342–48.

McCollum A. T. (1990) *The Trauma of Moving: Psychological Issues for Women*, London: Sage.

McRae, S. (1993) 'Returning to work after childbirth: opportunities and inequalities', *European Sociological Review* 9: 125–38.

McRae, S. and Daniel, W. W. (1991) *Maternity Rights: the Experience of Women and Employers*, London: PSI.

Macran, S., Joshi, H., and Dex, S. (1996) 'Employment after childbearing: a survival analysis', *Work, Employment and Society* 10: 297–318.

Markham, W. and Pleck, J. (1986) 'Sex and willingness to move for occupational advancement: some national sample results', *Sociological Quarterly* 27: 121–43.

Martin, J. and Roberts, C. (1984) *Women and Employment: a Lifetime Perspective*, London: HMSO.

Mincer, J. (1978) 'Family migration decisions', *Journal of Political Economy* 86: 749–73.

Molho, I. (1986) 'Theories of migration: a review', *Scottish Journal of Political Economy* 33: 396–419.

Morris, L. (1990) *The Workings of the Household*, Oxford: Polity Press.

Owen, D. and Green, A. (1992) 'Migration patterns and trends', in T. Champion and T. Fielding (eds) *Migration Processes and Patterns, Volume 1: Research Progress and Prospects*, London: Belhaven, pp. 17–38.

Pissarides, C. and Wadsworth, J. (1989) 'Unemployment and the inter-regional mobility of labour, *Economic Journal* 99: 739–55.

Prandy, K. (1990) 'The revised Cambridge scale of occupations', *Sociology* 24: 629–55.

Sandell, S. H. (1977) 'Women and the economics of family migration, *Review of Economics and Statistics* 59: 406–14.

Shihadeh, E. S. (1991) 'The prevalence of husband-centred migration: employment consequences for married women', *Journal of Marriage and the Family* 53: 432–44.

Snaith, J. (1990) 'Migration and dual-career households', in J. Johnson and J. Salt (eds) *Labour Migration*, London: Fulton, pp. 155–71.

Spitz, G. (1984) 'The effect of family migration on wives' employment: how long does it last?', *Social Science Quarterly* 62: 21–36.

Stewart, A., Prandy, K. and Blackburn, R. (1980) *Social Stratification and Occupations*, Basingstoke: Macmillan.

Wheelcock, J. (1990) *Husbands at Home: The Domestic Economy in a Post-Industrial Society*, New York: Routledge.

8 Migration, marriage and the life course

Commitment and residential mobility

Norman Bonney, Alison McCleery and Emma Forster

> Men ordinarily settle down to a career in a limited field and do not change jobs and careers with the alacrity of the proverbial economic man under changing economic conditions.
>
> (Becker 1960: 33)

Putting aside the sexist assumptions of this quotation, there is much of value to be gleaned from utilizing the concept of commitment, as articulated by Howard S. Becker, for the analysis of migration and residential mobility. In this chapter evidence is presented supporting this theoretical approach, which is subsequently applied to the analysis of age, employment, gender and marital differences in migration patterns. It is further argued that changes in gender patterns of labour force involvement are not as profound as frequently claimed and that they have had correspondingly less of an influence in changing overall patterns of migration. In this chapter attention is given to both inter-regional migration and more local changes of residence.

Commitment, migration and the life course

Becker used the concept of commitment as an explanatory tool for understanding how actors come to follow consistent lines of activity. Becker refers to a process through which several kinds of interest become bound up with carrying out certain lines of behaviour to which they seem formally extraneous.

> What happens is that the individual, as a consequence of actions he has taken in the past or the operation of various institutional routines, finds he must adhere to certain lines of behaviour, because many other activities than the one he is immediately engaged in will be adversely affected if he does not.
>
> (Becker 1963: 27)

He suggests that 'commitment' helps to explain consistency in social behaviour.

> One way of looking at the process of becoming an adult is to view it as a process of gradually acquiring . . . a variety of commitments which

> constrain one to follow a consistent pattern of behaviour in many areas of life. Choosing an occupation, getting a job, starting a family – all these may be seen as events which produce lasting commitments and constrain the person's behaviour.
>
> (Becker 1964: 51)

Migration and residential mobility are important aspects of an individual's – or a household's – behaviour and Becker's perspective helps to understand how inertia rather than movement has been, and still is, the fundamental characteristic of most people's residential mobility and migration behaviour. As long ago as 1963 it was demonstrated that just over two-thirds of people over age 15 had lived at the same address for five years or more (GSS 1966: 8). More recently, the 1991 Census of Population, adjusted to take account of under-enumeration, demonstrated that only 10 per cent of the total 1991 base population were migrants (defined as residents with a different address one year before), and in terms of households, as few as 7 per cent were wholly moving. Similarly, Buck (1996), using data from the British Household Panel Survey (BHPS), confirms that 10 per cent of the national sample moved house in any one year in the early 1990s, and that the longer people live at an address, the less likely are they to leave it.

In addition to the tendency to inertia, another feature of people's residential moves is that when they do occur they tend overwhelmingly to be short-distance ones which minimize other changes in their lives and enable them to preserve existing employment, kinship and social ties. This residential mobility, of which Rossi conducted a seminal study in the 1950s (Rossi

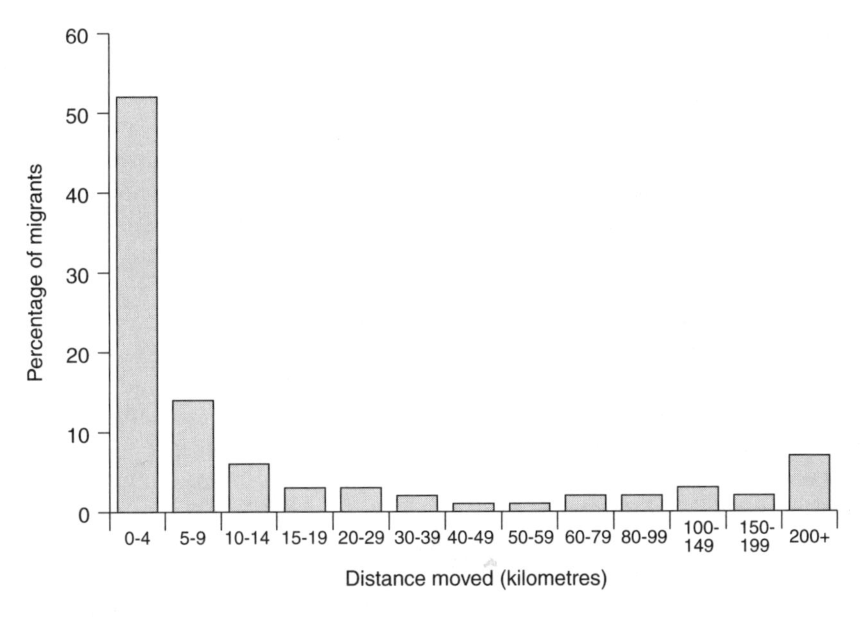

Figure 8.1 Distance moved by British migrants, 1990–91

1955), is distinct from longer-distance migration, a type of move which is far less frequent and which may be argued to be quite unusual or deviant in nature. Looking at migration in terms of actual distance moved, calculated from the 1991 Census, figure 8.1 shows that more than half of all residential moves within Great Britain (52 per cent) were between addresses within five kilometres of each other, and over two-thirds (69 per cent) were between addresses within ten kilometres (OPCS and GRO(S) 1994).

Similar results emerge from the BHPS, so that for the early 1990s in Great Britain, 63 per cent of all house moves were in the same local authority district whilst only 15 per cent constituted inter-regional moves (Buck 1996). Not only are most residential moves short-distance but there tends to be a self-perpetuating character to residential stability or movement. Past non-movers tend to continue in residence; those with a history of moving are more likely to move in the future. Buck found that the best predictor of mobility was, in fact, mobility in the previous year. According to this analysis of the BHPS, of those people who moved between years 1 and 2, 26 per cent were likely to move again between years 2 and 3, compared with only 9 per cent of those who had not moved in the first period. Much mobility is therefore concentrated among 'chronic movers' (Buck 1996: 2). This contrasting behaviour between stayers and movers suggests differing patterns of commitment to the local area, with stayers developing residential or area-based social networks, and movers developing contrary commitments, such as to an occupation or identity entailing occasional or frequent moves.

Further insights into the influences leading to residential mobility come from the same data source. Among the full respondents for the first five waves of the BHPS, when asked whether, if they could choose, they would stay in their present home or prefer to move somewhere else, 60 per cent did not express a desire to move, while some 40 per cent said they would prefer to move house. However, it must be stressed that the question relates only to expressed preferences as opposed to actual moves. Of the approximately 40 per cent of the respondents who said they preferred to move house when questioned in wave 1, less than half of these had actually moved by wave 2. Those who preferred to move but did not may be defined as latent movers who could move if ideal conditions should arise. The 60 per cent who say they prefer to stay constitute the majority and express no desire to move. Yet through unforeseen circumstances even they may be forced into moving.

Commitment to a locality will often have its origins in birth and upbringing in a certain location. For example, in some rural areas of northern Scotland the connection between place of origin and personal identity is so strong that incomers can sometimes never be fully accepted into the community. Only those born and/or brought up there are perceived to belong. Such a territorial basis of personal identity is not restricted to rural localities. Zimmer (1970) has demonstrated how length of residence is related to social integration into the community. The longer the period of residence in an area, the greater the commitment to the locality. Youth and early adulthood are the age phases most at risk of residential mobility – because commitment to the local community is less well developed at this phase of the life course. Failing the occurrence of migration at these times, the pressures derived from continued residential and employment commitment make migration later in the life course less and less probable (see also McGinnis 1968 on 'cumulative inertia').

Age has consistently been shown to be a major influence on migration (Lewis 1982). Just over half of those in the age group 25–29 years had moved addresses twice or more in the previous ten years, compared to just over a fifth of the whole sample in the previously mentioned 1960s survey (GSS 1966). Using the 1991 census Sample of Anonymized Records (SAR), figure 8.2 shows that rates of inter-regional migration are much higher among the under-40s and peak in the age group 25–29.

This evidence fits in with the argument that adolescents and young adults possess a relatively weak commitment to a particular community. Younger people are typically engaged in a process of extricating themselves from their family of origin and are engaged in a search for an adult identity and a social location for themselves, often through a lengthy process of trial and error. Frequently, they are excluded from, or placed low in priority for, access to positions of local social privilege, such as in employment or housing. Relatively free of commitments to a house, a job and dependants, they are more prone to experiment and rove geographically and socially for more satisfactory social and economic arrangements. Geographical mobility is thus just one form that this searching for a role takes in the phase of youth or early adulthood. Inevitably, there are social variations in the length and character of this interstitial phase of geographical and social exploration. It may be rel-

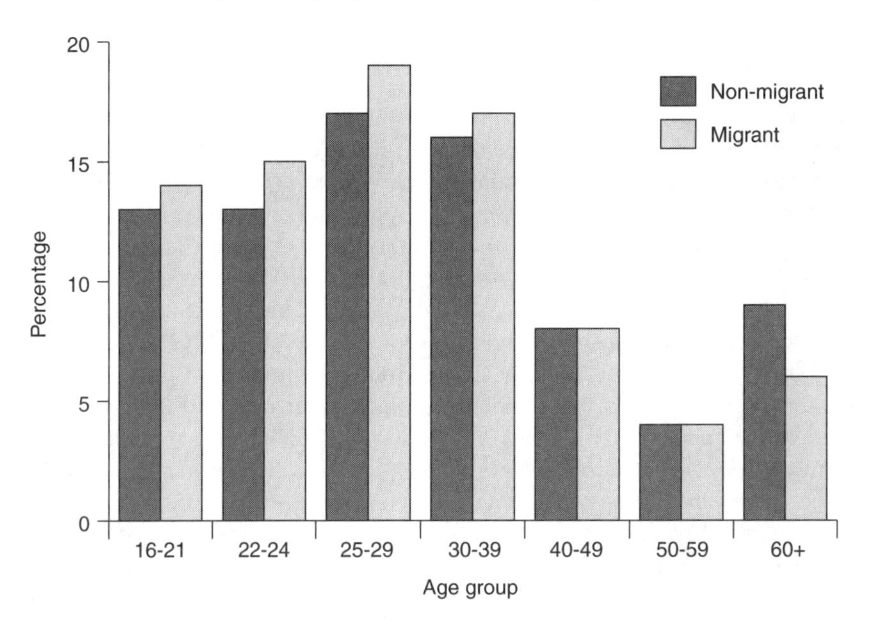

Figure 8.2 Age distribution of non-migrants and migrants between SAR regions

atively short for working class people. Early childbearing and obtaining a new home or occupation may lead to limited social exploration and an early commitment to roles and locations very similar to the parental ones. Residential commitments under these conditions may emerge as an unintended consequence of entanglements in other social roles. The net effect, however, is to minimize the likelihood of migration. For others, especially for the middle classes and those involved in extended post-secondary education, this interstitial phase is much longer and involves a greater degree of social and locational exploration. Commitments to roles, relationships and communities among such groups are much more tentative and provisional. A lesser degree of attachment to any particular local social system is thus likely to be accompanied by greater social and locational mobility and exposure to a variety of different opportunity structures.

In time, even the more geographically mobile build up commitments to a particular social and geographical location. For instance: marital or otherwise stable household arrangements emerge; permanent jobs are obtained; homes are furnished and modified to meet personal and household requirements; and networks of friendship and social support develop. This sociological perspective has affinity with migration decision-making models formulated by geographers. Thus, Forbes (1989) talks of 'plugging in' to local social and service networks which anchor households to their local area. Earlier, Wolpert (1965, 1966) introduced and developed the concept of 'place utility', whereby the current residential location has a utility to the resident, which has either to be undermined by changes in the current environment or superseded by changes in the external environment for migration to occur.

Employment and migration

Patterns of involvement or non-involvement in paid employment are major factors influencing the likelihood of migration. The 1991 Census of Population demonstrates that households with an economically active head (full- and part-time workers, those self-employed or on a government scheme, and those who are unemployed) have higher rates of migration than households with an economically inactive head (students, permanently sick, retired and other non-specified inactive groups). This contrast applies both to wholly moving households and to migrants generally (OPCS and GRO(S) 1994). Within the economically active group, the migration rate is highest for the unemployed (160 per thousand residents for men and 200 per thousand for women) (1991 Census of Population). Halfacree, Flowerdew and Johnson (1992) also found that recent migrants are disproportionately likely to be unemployed. The unemployed are both more willing to consider leaving an area than the employed (GSS 1966) and are actually more mobile (1991 Census of Population; Buck 1996), but they do tend to move lesser distances than other migrants. Examination of the Scottish House Condition Survey (1991) reveals that in terms of expectation to move, the unemployed have the highest, whilst the economically inactive categories of the retired, long-term sick and those looking after the home have the least (Munro, Keoghan and Littlewood 1995). The Scottish House Condition Survey also shows that this aspiration is matched by the reality i.e. in the main those expecting to move did so. There are also gender differences to be seen in the lowest rates of migration by occupation, with the self-employed group having the lowest rate of migration amongst men (85 per thousand), whilst among women, part-time employees had the lowest rate of all at 62 per thousand (OPCS and GRO(S) 1994: 8). In summary, patterns of labour force involvement heavily influence geographical mobility. Apart from the unemployed, lack of engagement in paid work generally leads to relative immobility.

From a life-course perspective, if entry is made to the local labour market soon after leaving the educational system the employment ties that develop will contribute to local residential commitment. Long-term knowledge of a particular job generates familiarity and ease of performance and knowledge of a wide range of contacts and procedures for undertaking it. What may have begun as temporary commitments become more permanent. A constellation of interests may thus combine to create commitment to a job and a reluctance to leave it even when it does not appear satisfying in other respects (Palmer and Parnes 1962). In turn, commitment to a job creates commitment to a locality. However employment can act not only as an anchor but sometimes as a catalyst to movement. This is because not only does commitment to a job encourage stability, but it can also, in some occupations, encourage mobility motivated by the aspiration for career advancement. Thus, professionals – including those with families – constitute part of a mobile elite. This is to be explained partly by professional orientation to job advancement and also by employer behaviour, with employers increasingly subsidizing the moves of those staff transferred around the country (Coleman and Salt 1992). Consequently, two more mobile groups, young adults and professionals, demonstrate similar characteristics of lower levels of commitment to localities, the one through the process of youthful experimentation and development, the other through processes of occupational motivation and control.

The influence of employment is particularly powerful in inter-regional migration as opposed to local residential mobility. Lack of employment and desire for employment advancement both function to promote the former. The 1991 census SAR shows that unemployment rates are slightly higher among inter-regional migrants than non-migrants. Among the employed, job aspirations play a major part in explaining inter-regional migration. In 1966, work reasons were the major motive for inter-regional and long-distance moves and the greater the distance moved, the greater the proportional significance of work reasons in the move (GSS 1966: 16–18). Three decades later the BHPS demonstrates that job-related reasons are twice as important in inter-regional moves as they are in inter-district moves of address, accounting for about 42 per cent of the former type of moves and 22 per cent of the latter. Broadly similar findings are also evident from the Migration and Housing Choice Survey (1991), with employment being of most importance in moves over 50 kilometres whereas housing is the major influence over shorter-distance moves.

As previously stated, most people's social circles are local and most changes of address take place in their immediate geographical and social environs as they acquire or change partners, and alter family or housing status. It takes strong economic factors such as employment to override continued local residence. Even though the unemployed may have higher overall rates of inter-regional migration, there is evidence that this is less the case among the more unskilled and in some strong occupational communities faced with major employment losses (Jones 1992; Markland 1975). Given the general tendency for residential stability to emerge in middle age, continuing movement in the latter years of the working life may be due to the relatively greater significance of employment pressures. Mann (1973) observed, for instance, that concern for job security was a more important motive for moving among older workers than younger ones in the sample. Amongst the latter, housing improvement was an important consideration. Given that middle-aged and older workers are more likely to have secured desirable housing accommodation and have developed a greater general level of commitment to a local area, it may reasonably be hypothesized that both employment inducements and constraints play a more significant role in their migrations. The 1963 official survey of labour mobility (GSS 1966) demonstrated that work reasons were a proportionately greater motive for moving house among the 45–59-year-olds than they were among the 20–44-year-olds while housing improvements were a considerably more important motive among the under-45s than among those over that age. Housing improvement-related moves are expressed predominantly as very short-distance ones. Such moves are disproportionately undertaken by people in the second quarter of an 80-year life span, as they aspire to housing which better matches their changing domestic circumstances. The 1966 GSS does not offer different findings in these respects from those of the 1996 wave of the BHPS. If groups of reasons are considered in respect not only of age but also of stage of the life course then similar results are evident. Local residential mobility to upgrade accommodation and neighbourhood is a major means by which people achieve an improved quality of life and represents historically and currently an important factor in the rise of the social and spatial phenomenon of suburbanization. The desire for a larger house, for a change of house type and for home ownership were the major reasons cited for house moves among short-distance movers in the previously cited Scottish study (McCleery, Forbes and Forster 1996) and similar influences are found to be near-universal migration goals throughout the developed world (for example, Thorns 1980: 56).

Gender, marriage and migration

This general application of the concept of commitment to the analysis of residential mobility can now be extended to gender differentials and the impact of marriage and partnership. Until recently, research on labour markets and migration has paid scant attention to gender differentials, although one of Ravenstein's (1885) famous laws of migration proposed over a century ago that females are more migratory than males. Lewis (1982: 84) observes that this has generally been shown to be false in modern western industrial societies. Evidence from the 1991 Census of Population in Great Britain and the BHPS, to be considered in this section, points to the complexity of the issues involved.

Work undertaken a generation ago yielded interesting findings on gender differentials that are consistent with the commitment perspective. The labour mobility survey (GSS 1966: 24) found that single female workers were more willing to consider a migration move than married female workers but that women workers as a whole were considerably less likely to be willing to move than male workers. Women workers, however, recognized a greater dependency on other members of their household with regard to the possibility of migration. They were much more likely than men to say that they would only move if the spouse or another member of the household was moving. The nature of employment ties and domestic relationships was a relevant factor in interpreting these findings since full-time and unemployed women workers were much more willing to consider a move than part-time workers (29 per cent to 15 per cent). The survey explained this lower willingness of part-time

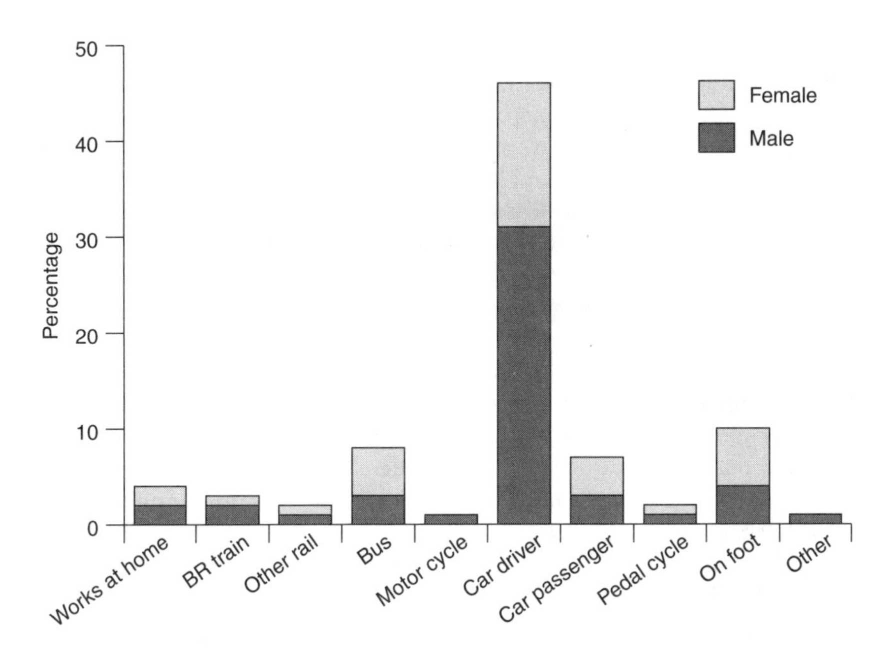

Figure 8.3 Mode of transport to work, by sex

working women to move as 'probably because they are married' and presumably playing a secondary role to a partner in the labour force.

MacKay et al. (1971), in their study of local labour markets, found that female employees were less likely than males to change residence when they took up new employment – they were more likely than men to base a job change on their current residence. Conversely, male job changes were more likely to be associated with a change of residence. Women travelled shorter distances to work and had smaller commuting zones and areas of job search. They appeared, as a general rule, more closely tied to the home than male workers. Correspondingly, employers in search of female labour had more geographically restricted job markets. These differential patterns of local mobility and employment opportunity between males and females were explained by MacKay et al. in terms of the household's dependence on a principal male earner and the 'flexible' nature of female employment. Women's jobs, they argued, tend to be more flexible in that they were low-status, lowly paid and not as secure as men's.

More contemporary data are consistent with this older picture and suggest, contrary to much opinion, that there have not been fundamental changes. Women are still more likely to play secondary labour market roles in partnerships with men (Hakim 1996) and generally have inferior means of mobility and less access to economic opportunities even in their own local labour markets. Women are, for instance, half as likely to drive a car to work as men. Figure 8.3, based on the total working and non-working census SAR sample, shows that only 15 per cent of all women use this mode of travel to work, compared with 31 per cent of men. In contrast, 5 per cent of women travel on buses and 6 per cent walk to work compared with 3 and 4 per cent of men, respectively.

BHPS data also show that women spend less time on travel to work and cover lesser distances to work than men. These patterns can plausibly be interpreted as being a result of most women's secondary employment status, not having priority access to the family car and having to use other means of travel to work. Women's greater attachment to the home base is also evident in BHPS data on house moving preferences and is evident in inter-regional as well as local geographical mobility. The question on whether people would like to move shows that women are more disposed to stay in their current residence while men are more evenly split about the possibility. Inclination to move is also related to marital status. The balance of preference among married people is in favour of staying, while among the never-married it is to move.

Gender and marriage are thus a major influence upon people's employment and residential patterns and a considerable debate has emerged about its relationship to migration and women's employment opportunities. The general debate has largely revolved around the extent to which the migration of couples is male-led and its effects on female partner employment careers (Finch 1983; Bonney and Love 1991). Clarity in analysing these issues would be helped by recognizing that the deemed negative consequences of marriage and motherhood relate only to a proportion of migrants. Data from the 1991

Table 8.1 Employment status of female inter-regional migrants and non-migrants by age: 1991 Census SAR

	Employment status									
	Full-time employment		*Part-time employment*		*Unemployed*		*Inactive*		*Others (e.g. sick, student, self-employed)*	
Age (years)	*M**	*N**	*M*	*N*	*M*	*N*	*M*	*N*	*M*	*N*
16–21	45	42	6	5	14	11	17	7	17	34
22–24	60	53	7	8	8	8	17	21	8	8
25–29	53	43	10	16	7	6	23	29	5	5
30–34	37	31	16	26	7	4	32	32	5	6
35–39	34	31	19	31	7	4	31	26	6	6
40–44	38	36	20	31	7	3	24	20	7	6
45–49	38	35	18	30	7	3	23	20	6	6
50–54	28	30	19	29	8	4	26	24	6	5
55–59	18	22	16	24	7	3	29	30	4	3

Source: Individual SAR, 1991 Census, Crown Copyright.

Note
* M = migrants, N = non-migrants.

census SAR show that single females without children constitute 12 per cent of the non-migrant female population under age 46 but comprise 32 per cent of the equivalent inter-regionally migrant population. Many women thus engage in inter-regional migration as single persons prior to marriage and motherhood. Nor is it by any means self-evident that migration is actually disadvantageous occupationally for women. According to the same data source, inter-regionally migrant women have a higher occupational status than non-migrant women: 27 per cent of the former, compared with 22 per cent of the latter were in top-level professional, managerial and technical jobs. Table 8.1 displays 1991 census SAR data on the employment status of inter-regionally migrant women compared with non-migrant age peers.

In conformity with earlier observations, in all age groups from 16 to 49 years, female migrants are more likely to be in full-time employment than non-migrants. They are less likely to be in part-time employment and are more likely to be unemployed or inactive. Given that these data are based on one year's inter-regional migration, the fact that a higher proportion of inter-regionally migrant women are in full-time employment compared with non-migrants is a remarkable testimony to their overall position of labour market *strength*. Their lower level of involvement in part-time employment may also reflect the lack of part-time options in higher-level occupations.

The 1991 census SAR data show that, in accordance with earlier observations, there is a general tendency for men to be more inter-regionally mobile than women, although the gross differences are small. Generally, in line with the commitment perspective, single people without partners or dependants are also much more likely to be geographically mobile. Male inter-regional migrants are also much more likely to be single than female migrants. Since women marry younger than men, the latter are less likely to be attached and thus are more likely to be available for migration. An analysis of inter-regional migration in the 1991 census SAR undertaken among the age group mostly likely to move house and to be exposed to the risk of having young children – those aged under 45 years – among whom the great majority of inter-regional migrants are located, found that single males constitute 42 per cent of migrants compared with the 18 per cent they constitute of the non-migrants. Similarly single women constitute 32 per cent of female migrants compared with 12 per cent of female non-migrants. If attachment to a partner contributes to relative immobility, so does having children. Some 26 per cent of male migrants under age 45 years have children co-resident, compared with 43 per cent of non-mobile men. The figures for female were 34 and 57 per cent, respectively. Inter-regional migrants are, then, less likely to be married or to have children than non-migrants. A thorough analysis of female and male employment and its relationship to migration thus has to take account of the life-course perspective and the relative duration of single-person status before marriage or partnership and the effects of having children on migration patterns. The dissolution of marital bonds leads to a reversion to single person status and partnership reconstitution that have the effect of increasing geographical mobility rates to a level higher than for married people (Devis 1983).

Boyle and Halfacree (1996), in examining whether migration is detrimental to married women's waged work, sought to explore the perception of women as tied movers, (mere) followers of the main breadwinner, a perspective with obvious connections to theories of patriarchy and capitalism. They conclude that migration does seem to be detrimental to married women's working. Such conclusions, while true, do not take into account that women's career strategies must balance short-, medium- and long-term goals; the short-term picture may be strongly connected with childrearing. Boyle and Halfacree restrict their concern to current employment status and do not concern themselves with the long-term perspective. Green (1996) has explored the bargaining between partners that can occur in such situations. An interesting observation which can be made from this research, and the earlier quoted finding about the employment status of inter-regionally migrant women, is that, for them, the reverse of the human capital model applies. There is – at least in the short term – a greater likelihood of unemployment after migration and not vice versa as expected. This is strongly connected with research findings on the gender differences in dual-career households, with women tending not to move to enhance their own career but instead tending to get part-time work or any sort of job nearby to fit in with family commitments – to use a partner's move to advance other aspirations and achieve childraising goals. Thus, women balance family commitments with career ones (Green 1994, 1995) and at certain times family goals may take precedence over career. Bruegel (1996) also investigated these issues. She found that women give little importance to employment moves and, if women do give employment as a reason, it may be because of their partner's job move.

The balance of this research suggests that marriage, partnership and parenthood, in so far as they relate to inter-regional mobility, thus act to reduce women's independence as individuals pursuing their personal employment careers. Instead, it places them within the context of a longer-term household strategy of family formation and career development which may only be detrimental to their own individual interests in the short term and not necessarily disadvantageous from a shared household perspective in the longer term. The argument so far, then, sustains a view that accords with the 'conventional' model of spousal relations, with a husband specializing in paid employment and a wife with a more domestic-focused set of social roles and a more marginal position in the labour market. This model has come under considerable recent challenge, ideologically as a normative model for behaviour, and substantively as an accurate account of actually prevailing patterns (see, for instance, the debate over the work of Catherine Hakim in the *British Journal of Sociology* 1996). It is widely believed that labour market changes such as rising male unemployment and rising female labour market participation have contributed to the erosion of customary gendered patterns of labour market participation. In turn, it is sometimes asserted that the increased incidence of dual-career partnerships among couples makes it far more difficult for attached individuals to undertake migration (for example, Abercrombie and Urry 1983). However, the actual pattern of change is much more complex than has generally been considered when such propositions have been put forward. Indeed, Hakim (1996) has questioned the widely accepted view that there has actually been a secular increase in women's labour force participation in the twentieth century.

A convincing interpretation of the historical evidence that Hakim presents is that the relatively static levels of women's overall labour force participation levels in Great Britain during this century (with the possible exception of trends in the most recent years) are the result of conflicting tendencies. On the one hand, there has been a lengthening of full-time education and increased participation in further and higher education which has reduced labour force involvement in the teens and early 1920s. This deferment of the entry to employment has involved a loss of years that were previously in full-time paid employment prior to marriage or childbearing. On the other hand, women have been working to an older age in paid work in full-time jobs before having children and then have been increasingly returning to work while they have young children. The net effect of this has been, again contrary to much popular and academic perception, a shift to part-time employment and increased intermittency, not continuity, in women's paid employment patterns (Buck et al. 1994; Hakim 1996).

This argument has various implications for gender differences in migration. Extended educational careers lead to later entry into paid work, a deferral, and in increasing proportions an abandonment, of marriage and childbearing – what Champion (1992) has called the second demographic transition with low fertility levels (see also Stockman, Bonney and Sheng 1995). If increasing proportions of younger women are pursuing full-time working lives as single or childless women following upon the completion of initial further and higher education, this should increase their potentiality for migration as independent individuals or equal partners with men. The net effects of labour market and demographic change do not then inevitably imply that there is any necessary increasing tendency for rising levels of female labour market participation to lead to increased levels of conflict between the employment priorities of dual-career couples. Nor should it be forgotten that nearly 60 per cent of both female and male inter-regional migrants are below the age of 30 and one-third below 25. With the increasing deferment of marriage and childbirth the population at risk of such potential handicaps is not as large as often assumed. Nor are dual-career couples the norm. After the birth of the first child a majority of women still abandon full-time employment. Only 17 per cent of the mothers of children below the age of 5 were in full-time paid employment in 1996, 31 per cent were in part-time employment and over one-half were unemployed or inactive (ONS 1997). Those partnered women in part-time employment or non-employment would find it easier than those in full-time employment to move geographically along with their spouses, since part-time work is more flexibly available to them in different local labour markets (Bonney and Love 1991). Perhaps, then, only small proportions of mothers of young children, and a smaller proportion of women in general, are open to the risk of career disruption through husbands' geographical mobility.

Conclusion

The idea of commitment is a useful concept in understanding people's patterns of residential mobility and inter-regional migration. The basic pattern of inertia is founded upon commitments to family, local community and employment but is subject to systematic modification by a range of countervailing influences that produce typical patterns of residential mobility. The forces that produce inertia also lead to most residential moves being of a primarily short-distance nature when they do occur. Life-course influences such as the transition to adulthood and family formation are major factors in local residential mobility; employment interests figure more highly in longer-distance moves both for the unemployed and middle class professionals. Commitments to other household members greatly influence local employment opportunities and longer-distance migration. Locally, family and child-care roles influence patterns of mobility and employment. Inter-regionally, the younger and the single are more prone to migration. Marriage and children inhibit longer-distance moves for both women and men. Those not economically active and those with lesser occupational commitments, such as the part-time workers, have less employment incentive to engage in longer-distance mobility.

Because of the high relative frequency of younger and single people in migration flows and the rising age of marriage and age at first birth especially among the middle classes, dual-career partnerships are less of an impediment to women's employment careers than is sometimes believed. Migrant couples are not different from couples in general in experiencing a temporary scaling down of the wife's involvement in paid work while the children are young.

Acknowledgements

The SAR are crown copyright and are provided through the Census Microdata Unit of the University of Manchester, with the support of the ESRC/JISC/DENI.

References

Abercrombie, N. and Urry, J. (1983) *Capital, Labour and the Middle Class*, London: Allen and Unwin.

Becker, H. (1960) 'Notes on the concept of commitment', *American Journal of Sociology* 66: 32–40.

Becker, H. (1963) *Outsiders: Studies in the Sociology of Deviance*, London: Collier-Macmillan.

Becker, H. (1964) 'Personal change in adult life', *Sociometry* 27: 40–53.

Bonney, N. and Love, J. (1991) 'Gender and migration – geographical mobility and the wife's sacrifice', *Sociological Review* 39: 335–48.

Boyle, P. and Halfacree, K. (1996) 'The employment impact of migration on married women: evidence from south-east England', Paper presented at the British Society for Population Studies Annual Conference, University of St Andrews, September.

Bruegel, I. (1996) 'Taking the stairs rather than the escalator? Married women and migration to London and the South East, 1970–1991', Paper presented at the British Society for Population Studies Annual Conference, University of St Andrews, September.

Buck, N. (1996) 'Using panel surveys to study migration and residential mobility', *British Household Panel Study News* Spring, ESRC Centre for Micro-social Change, University of Essex.

Buck, N., Gershuny, J., Rose, D. and Scott, J. (eds) (1994) *Changing Households. The British Household Panel Survey 1990–1992*, University of Essex: Economic and Social Research Council Research Centre on Micro-social Change.

Champion, A. G. (1992) 'Urban and regional demographic trends in the developed world', *Urban Studies* 29: 461- 82.

Coleman, D. A. and Salt, J. (1992) *The British Population: Patterns, Trends and Processes*, Oxford: Oxford University Press.

Devis, T. (1983) 'People changing address: 1971 and 1981', *Population Trends* 32: 15–20.

Finch, J. (1983) *Married to the Job*, London: Allen and Unwin.

Forbes, J. (1989) 'Migration monitoring and strategic planning', in P. Congdon and P. Batey (eds) *Advances in Regional Demography*, London: Belhaven Press, pp. 41–57.

Green, A. E. (1994) 'The geography of changing female economic activity rates', *Regional Studies* 28: 633–39.

Green, A. E. (1995) 'The geography of dual career households: a research agenda and selected evidence from secondary data sources for Britain', *International Journal of Population Geography* 1: 29–50.

Green, A. E. (1996) 'A question of compromise?', Paper presented at the International Conference on Population Geography, University of Dundee, September.

GSS [Government Social Survey] (1966) *Labour Mobility in Great Britain 1953–1963*, London: Ministry of Labour and National Service.

Hakim, C. (1996) *Key Issues in Women's Work: Female Heterogeneity and the Polarisation of Women's Employment*, London: Athlone.

Halfacree, K., Flowerdew, R. and Johnson, J. (1992) 'The characteristics of British migrants in the 1990s: evidence from a new survey', *Geographical Journal* 158: 157–69.

Jones, H. (1992) 'Migration trends for Scotland: central losses and peripheral gains', in J. Stillwell, P. Rees and P. Boden (eds) *Migration Processes and Patterns, Volume 2: Population Redistribution in the United Kingdom*, London: Belhaven, pp. 100–14.

Lewis, G. J. (1982) *Human Migration: a Geographical Perspective*, London: Croom Helm.

McCleery, A., Forbes, J. and Forster, E. (1996) 'Deciding to move house: a preliminary analysis of household behaviour in Scotland', *Scottish Geographical Magazine* 112: 158–68.

McGinnis, R. (1968) 'A stochastic model of social mobility', *American Sociological Review* 23: 712–22.

MacKay, D., Boddy, D., Brack, J., Diack, J. and Jones, N. (1971) *Labour Markets Under Different Employment Conditions*, Glasgow: University of Glasgow Press.

Mann, M. (1973) *Workers on the Move. The Sociology of Industrial Relocation*, London: Cambridge University Press.

Markland, J.A. (1975) 'Some theoretical and empirical considerations of Scottish population migration 1961–1971', PhD thesis, University of Dundee.

Migration and Housing Choice Survey (1991) http://datalib.ed.ac.uk/EUDL/ surveys/migration/

Munro, M., Keoghan, M. and Littlewood, A. (1995) *Mobility and Housing Aspirations: Analysis of the Scottish House Condition Survey (1991)*, Social Report No. 4, Research and Innovation Services, Scottish Homes.

ONS [Office of National Statistics] (1997) *Social Trends 1997*, London: ONS.

OPCS [Office of Population Censuses and Surveys] and GRO(S) [General Register Office (Scotland)] (1994) *1991 Census Topic Monitor: Migration, Great Britain,* London: OPCS.

Palmer, G.L. and Parnes, H.S. (1962) *The Reluctant Job Changer: Studies in Work Attachments and Aspirations*, Philadelphia: University of Pennsylvania Press.

Ravenstein, E. G. (1885) 'The laws of migration', *Journal of the Royal Statistical Society* 48: 167–235.

Rossi, P. H. (1955) *Why Families Move*, Glencoe: Free Press.

Stockman, N., Bonney, N. and Sheng, X. (1995) *Women's Work in East and West*, London: UCL Press.

Thorns, D. C. (1980) 'Constraints versus choices in the analysis of housing allocation and residential mobility', in C. Ungerson and V. Karn (eds) *The Consumer Experience of Housing: Cross National Perspectives*, Westmead: Gower, pp. 50–68.

Wolpert, J. (1965) 'Behavioural aspects of the decision to migrate', *Papers of the Regional Science Association* 15: 159–69.

Wolpert, J. (1966) 'Migration as an adjustment to environmental stress', *Journal of Social Issues* 22: 92–102.

Zimmer, B. (1970) 'Participation of migrants in urban structures', in C. Jansen (ed.) *Readings in the Sociology of Migration*, Oxford: Pergamon, pp. 71–83.

9 Residential relocation of couples

The joint decision-making process considered

Jenny Seavers

Introduction

Residential migration can be a highly disruptive process for all of those involved, particularly in moves involving considerable distance. It disrupts and fragments a household's social space, forcing changes in patterns of work and social life. This disruption also involves housing considerations, with the majority of migration in the United Kingdom seeing households moving into, within or out of home ownership. This is because the United Kingdom housing market is characterized by a very high proportion of home owners, around 70 per cent of households. Both residential relocation decisions and housing search are intrinsically tied to the household, how it functions as a decision-making unit and the weighting of power relations within it. The process of migration and the buying and selling of a property can be described as one of the most stressful events in a household's life experience. Place this within the context of the recent slump in house prices – the United Kingdom housing market is highly cyclical in nature – and this stress can increase tenfold. Not only does migration now involve the expenditure of considerable amounts of money, the majority of which is usually borrowed, but it is the exchange of private space. It is a construction of bricks and mortar that has been imbued with a sense of place that is intensely personal and has become ever-increasingly private as households have moved away from public renting. As such, moving house, particularly for home owners, is an immensely important operation involving serious amounts of debate and discussion for most couples before eventually reaching a decision to move and to buy a given property. This chapter therefore argues that to understand the decision-making processes involved in residential relocation and housing search it is necessary to consider the roles played and the power relations that make up the household unit. It presents some of the findings from a detailed study of joint decision making in housing migration, based on the findings from a small sample of couples who have recently moved house in two rural districts of lowland England.

Joint decision making and migration

In the study of migration decision making the focus has, in general, been on the household unit and the procedures adopted by the head of household in deciding whether to move or not. In this way, the head 'was simplistically thought to act either out of benign self-interest or as an impartial arbiter of conflicting demands within the household unit' (Robinson 1993: 1,453). However, within the behavioural literature there has been considerable debate about the validity of using only one partner in such investigations (for example, Davis 1970, 1971; Bokemeier and Monroe 1983; Monroe et al. 1985; Burns and Hopper 1986). General consensus appears to be that interviewing both spouses is crucial, with a number of studies explicitly investigating the roles which partners adopt in the decision-making process (Kenkel and Hoffman 1956; Davis 1970; Davis and Rigaux 1974; Haas 1980). Indeed, Monroe et al. (1985) observed that a quarter of the couples they interviewed would have been misrepresented had only one spouse been questioned about their decision-making roles. Nevertheless, although it has been observed that roles in joint decision making cannot be generalized without reference to the product being purchased (Davis 1970), Haas (1980) identifies a series of roles into which couples traditionally segregate, illustrated in table 9.1.

It is important to identify these various roles that are taken up by couples, because the way in which a number of them are perceived by the couple will reflect the way in which they make housing migration decisions. Thus, for example, the way they view whose responsibility it is to provide the family income is likely to determine whether a household moves as a result of one partner's job, even though it may be disadvantageous to the other partner's job prospects. Equally, child care or domestic responsibilities that are shouldered by one partner, for example, may determine the location of a new home near a

Table 9.1 Traditional segregated roles of couples

Role	*Description*
Breadwinner	Responsibility for earning the family income
Domestic	Responsibility for undertaking housekeeping chores
Handyman	Traditionally masculine tasks such as repairs
Kinship	Responsibility of meeting kinship obligations (for example, letters and gifts)
Child-care	Responsibility for doing routine child-care tasks
Major/minor decision-maker	Influence on major and minor decisions (often major decisions are delegated to men)

Source: (Haas 1980).

school, a crèche, shops, and so on. However, even though Haas (1980) identifies major and minor decision roles as an area that couples have to negotiate, she does not fully explore this aspect of the decision-making process.

Not only are the roles played by partners of significance, but the power that each partner holds is also important. It has been suggested that, in considering the process of decision making by couples, the focus should be on power within the relationship which can be simply defined as 'the potential ability of one partner to influence the other's behaviour' (Blood and Wolfe 1960: 11). Thus, the greater the partner's responsibility for decision making, the greater his or her power.

Migration within the owner-occupied sector involves a significant financial commitment and outlay. In a study of money and marriage, Pahl (1989) showed that husbands were more likely to be the dominant partner and that inequality outside the home in terms of earnings is linked to inequality within the household in terms of decision making and control of finances. Vogler (1994) has added to this by suggesting that equality in financial management within households depends not only on increased participation in the labour market, but through a change in the husband's traditional role as the breadwinner. However, in spite of these findings, O'Connor (1991) suggests that marital relations are usually viewed through a form of 'received wisdom': because marriage has become increasingly loving and compassionate, the power within marriage has therefore become more or less equally shared between partners.

The traditional picture of the role of partners in migration decision making was drawn by Pahl and Pahl (1971) in their study of middle class managers and their wives, where the role distinction was explicitly gendered and the wife and family followed the husband's career. The Pahls observed:

> Husbands more often look at a move in terms of their career, seeing it as a necessary preliminary to a better job; for them the domestic side of a move tends to take second place in their comments. Their wives on the other hand, will be thinking of a move *primarily in terms of friendship and kinship links, their own shopping patterns, their children's education and whether the old carpets and curtains will fit the new house.*
>
> (1971: 54–5, emphasis added)

They suggested that the characteristics of any house move were dependent on the stage that the couple have reached in their life cycle. However, the study did, even in the early 1970s, identify the beginnings of change in the joint decision-making process. For example, it was suggested that the 'feminization' of young men, through coeducational higher education, made them more sensitive to the feelings of their wives, whilst the rise of women with similar academic backgrounds reduced their willingness to be slaves to their husbands' careers.

Following on from the work by Pahl and Pahl there has been a whole series of papers written around the issue of dual-career households and migration decision making (for example, Poloma, Pendleton and Garland 1982; Hunt and Hunt 1982; Bird and Bird 1985; Bielby and Bielby 1992). Indeed, it has been suggested that households should be seen as occupying different positions along a continuum from 'leader' to 'follower', with these positions possibly changing over a household's life cycle (Dudleston et al. 1995). Marital decisions themselves can also be seen to be on a continuum, with those important yet infrequent decisions at one end and decisions that are frequent but are perceived to be unimportant at the other end (Hardill et al. 1995). One can argue, therefore, that women tend to be responsible for the frequent so-called unimportant decisions ('implementation power'), whilst men tend to hold the 'orchestration power' to make the infrequent important decisions, which clearly would include those about migration.

In a study that specifically considered the joint decision-making processes in housing migration decisions, Park (1982) used a method of 'decision plan nets' (henceforth referred to as DPNs) to examine both the similarities between partners' decision strategies and any changes in these strategies over time. Using separate questionnaires and individually constructed DPNs at three stages in the process – before, during and after search – he argued that joint decision making was a 'muddling through process'. His findings suggested that each partner follows their own decision strategy and the couple reach a decision through a disjointed, unstructured strategy assisted by the use of 'conflict-avoiding heuristics' (common preference levels on salient objective dimensions, such as price, number of rooms; task specialization based on an individual's expertise and concessions on preference differences).

To summarize, the literature suggests a number of significant points. First, there is a need to interview both partners when investigating the decision-making roles in a given experience. Second, the degree of responsibility for decision making is related to the power held by each partner, which in turn appears to be affected by any disparity in the earnings of partners and the extent to which the traditional breadwinner role is adhered to. Finally, traditional roles may be changing in the light of the increased involvement of women in paid employment, which may in some households result in roles changing over the course of the couple's life cycle or in response to career choices.

Methods used

This chapter provides some of the results from a detailed study of sixteen couples, drawn from an in-depth investigation into joint decision making which considered the processes that lie behind migration and housing search and examined the ways of conceptualizing the joint decision-making process. The study focused on home owners moving into or within two rural areas of lowland England, North Dorset and the rural part of East Northamptonshire. Both of these areas have been the focus of significant flows of in-migrants over the last few decades. The two areas form a band or crescent immediately adjacent to the South East Region Planning Area (SERPLAN) that extends from Dorset and Hampshire through Wiltshire and Oxfordshire to Northamptonshire and then across to East Anglia (Lewis et al. 1991), and have witnessed significant levels of population growth since the 1960s.

Drawing from the observations made by studies mentioned above, particularly the evidence of spousal inconsistency and the need to avoid the misrepresentation of the decision-making roles, this study questioned both partners. During the interview each couple (whose names have been anonymized) answered together a joint questionnaire. Each partner was then asked to complete an individual questionnaire separately and simultaneously without conferring. Included on this questionnaire was a question about the relative influence of each partner at various stages of the search process. Finally, couples were requested to complete a joint decision-making exercise using a DPN as a way of reconstructing the process that the couples went through during the search for a house.

The period of time since a couple moved house is clearly of utmost importance for memory recall. In this study, two out of the 16 households were still in the process of searching and one was in the process of purchasing a property at the time of the interview. Of the 13 remaining couples, the time lapse between completion and interview dates ranged from 1 to 12 months. This time span seems reasonable for recalling information concerning such a key event involving considerable household upheaval.

The DPNs used in the observation exercise are 'branching structures using attribute and situational factors to predict acceptance or rejection of an alternative' (Bettman 1979: 231). These were constructed in the following way. First of all, couples were asked to name the main features that they considered when they were searching for a property. Next, they were asked to rank these in order of importance, with one as the most important criterion. These were noted by the interviewer and written on a record card. These criteria then formed the main or primary branch of the DPN. This type of DPN is one that is operationalized at an attribute level (the couples' intended strategies toward each one of the criteria specified). Such a DPN is 'believed to reflect a DM's (decision maker's) internally generated decision plan in response to an internally generated stimulus such as problem recognition' (Park et al. 1981: 34–5). These criteria were then listed in rank order (with the most important at the top of the page) on a large sheet of paper linked together with lines labelled with a 'Y' to denote 'yes, the criterion is present'. This is illustrated in figure 9.1. The letter 'C' means that the couple would consider the property if it had all the criteria specified in the primary branch. The couple were then asked: 'If the property does not have this feature would you reject the property or would you still consider it?' If they said that they would still consider the property they were then asked under what conditions they would still consider it. These questions were asked of each of the criteria in turn.

For example, in figure 9.1, the couple gave 'cost £80–100,000' as the most important criterion for the new property. When asked what they would do if it was not within this price range, they indicated that they would consider a property that was lower than £80,000. However, for a property over £100,000 they were prepared to go to a maximum of £120,000 providing all the other criteria specified were present. Otherwise they would reject the property in question. This type of decision has been termed a 'relative preference dimension' (RPD) (Park 1978, 1982; Park et al. 1981) because the couple specify a 'differential threshold' of acceptance for that particular criterion or dimension.

There are two other types of decisions or dimensions that can be identified on the DPN. The first is a 'rejection-inducing dimension' (RID); an example of this is the 'thatched cottage' criterion in figure 9.1. In this instance the couple have established a minimum acceptable threshold so that when a property does not satisfy the condition they reject the property even if it has some or all of the other specified features. The other dimension that can be identified from the DPN is a 'trade-off dimension' (TD). This differs from the RPD because the absence of a satisfactory TD criterion needs to be offset by an improvement on another feature for the property to be acceptable. In figure 9.1, 'not adjacent to the road' is an example of this dimension.

Clearly, the assumption made in using such a methodology as DPN is that the decision plan which couples construct would have existed *prior* to the interview. However, such an assumption may not be unreasonable because couples purchasing a property would be expected to have undertaken some form of 'pre-planning' for such a major financial decision. In addition, Park and Lutz (1982) found, in their study of decision making at the different stages of house purchase, that the original decision plan constructed at the 'before-search' stage was a good predictor of post-search choice. In other words, individuals do appear to carry with them a set of key criteria to which they attach a measure of relative importance and the most important of which they retain throughout the decision-making process.

What is perhaps potentially more problematic is the terminology used and the meanings attached to them that may have been interpreted differently by different couples or indeed partners within couples. Attempts were made to ensure that respondents defined the various criteria that they used, without imposing a set vocabulary upon them as some studies have done. Nevertheless, the conceptual frameworks that were adopted by the respondents and the interviewer may not be 'mutually intelligible' (Sayer 1984: 214). Although this is a problem endemic to social science, it does clearly need to be borne in mind when explanations and interpretations of the data are developed.

Another important consideration concerns the effect of the interviewer on the respondents. Sayer (1984: 212) describes the 'objects' investigated by social scientists as 'structured mess' which is susceptible to change. He argues that:

> people are self-interpreting beings who can learn from and change their interpretations . . . their causal powers and liabilities are considerably more diverse and changeable (even volatile) than those of non-human objects.
>
> (1984: 213)

As a result of this, the knowledge that is developed in social science can effectively alter the 'objects' or, in this instance, the behaviour being studied. A crucial concern, then, for a study of this nature is whether the gender of the interviewer actually influences the *type* and *volume* of information that is received from the respondents. Kenkel's (1961) study of the gender of the interviewer and the joint decision-making roles of couples takes this further to suggest that the presence of a female interviewer would remind wives of 'the emergent role of modern woman and her place *vis-à-vis* her husband, and subtly [to] suggest to them to act in accordance with their interpretation of the emergent role' (1961: 185). As a result of this, he observed that the women were more talkative, had more influence on the decision outcome and were less likely to confine their role to social-emotional leader than those interviewed by a male interviewer. Such concerns may be less prominent in the 1990s, when the role of 'modern' woman has moved on from simply being in its emergent stages. Nevertheless, this may still produce expectations that female partners may try to adopt during the interview. Clearly, this raises questions about the observations made by the interviewer in this study and emphasizes the point that behaviour studies can not be seen as objective observations by a detached interviewer.

Selected key findings

The following analysis is based on the responses of sixteen couples interviewed between August 1992 and May 1993. The sample covers the range of types of household and a broad age distribution. The analysis presented here compares the responses given to a self-reported question about migration and search processes with observed responses of partners. The self-reported question was answered by each partner separately and produces a single categorical response for each decision stage given. The observed responses, however, allow a much more detailed view of each decision and crucially the reasoning behind the decision. They are based on a triangulation of responses given in the DPN, the structured interview and the discussion which took place during the construction of the DPN. In this way, any weaknesses in the simple binary DPN method were lessened by the results of the interview and the discussion between partners during its construction.

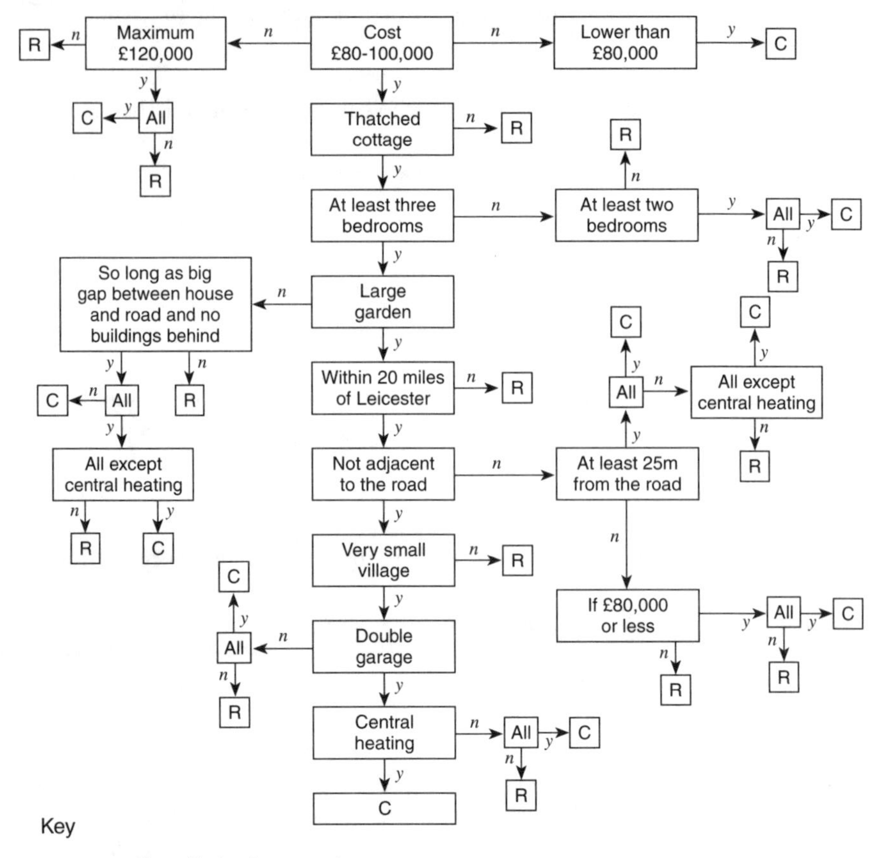

Key

y yes, the criterion is present

n no, the criterion is not present

C consider the property

R reject the property

Figure 9.1 An example of a decision plan net

Table 9.2 The housing migration decisions

A	Decision to search for a new house
B	The area searched
C	Type of settlement (villages, towns, cities)
D	Type of property considered
E	Size of property
F	Internal layout of the property
G	External appearance/style of the property

Table 9.3 A scale of relative influence in decisions

Value	*Meaning*
1	Male partner decides
2	Male partner more than female partner
3	Equal
4	Female partner more than male partner
5	Female partner decides

The individual questionnaire contained a self-reporting question that asked respondents *who* had made the various decisions regarding the house move. This was adapted from a question used by Munsinger, Weber and Hansen (1975), who used seven elements of the house purchase decision to analyse husband and wife decision roles.[1] The seven stages of the housing migration process that were included are listed in table 9.2. Respondents were asked to give a value on a five-point scale for each stage that reflected who had the most influence in each decision. Table 9.3 shows the five-point scale and its meaning.

From this scale, a dominance measure was calculated by taking the mean value of the responses given on the five-point scale by both male and female partners. A score that is less than three indicates *male-partner dominance*, a score of three suggests *syncretic* or *non-dominated* decision-making by the couple, and a score of more than three suggests *female-partner* dominance. Table 9.4 shows the resulting dominance values for each of the sixteen households in the study. Female-partner-dominant households form the majority in this small sample.

Such a measure provides an overall dominance score for the couple but tends, however, to obscure the complex reality of the joint decision-making process. For example, it was clear from the observations that in fact the dominance or relative influence of the partners changed at different stages in the migration process. As an illustration, figure 9.2a shows the dominance at each stage of the housing migration process for Mr and Mrs J. The single dominance score suggested female-partner dominance; however there appear to be clear distinctions between the various phases of the housing migration process that are dominated by one partner or the other. In terms of the property itself and the facilities of the settlement, Mrs J appears to have been the dominant party. It was very important to her that they bought a property

Table 9.4 Responses to dominance measure

Value	*Dominance measure*	*No. of households (%)*
Less than 3	Male-partner dominance	3 (18.8)
3	Syncretic/non-dominated	4 (25.0)
More than 3	Female-partner dominance	9 (56.2)

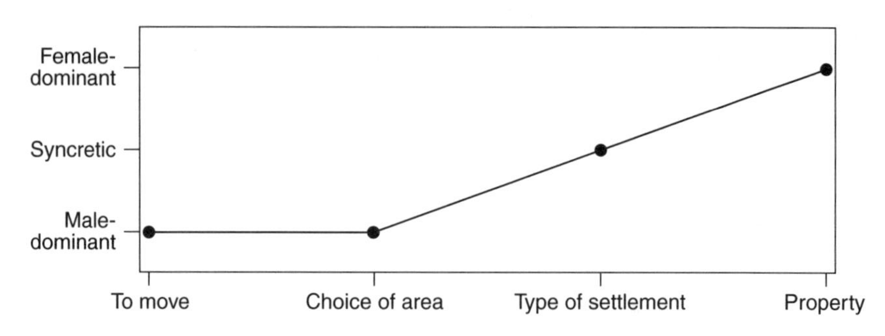

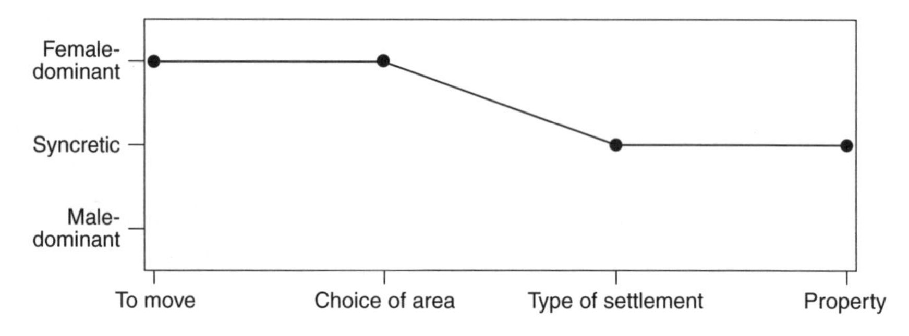

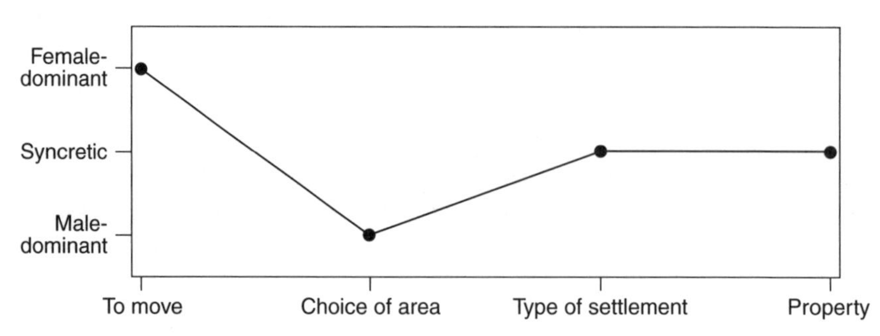

Figure 9.2 Dominance scores for different couples

that was modernized and did not require refurbishment. They both wanted to live in a place with some facilities, but the choice of area and the reason for moving were dictated by Mr J's job relocation.

Similarly figure 9.2b shows that although Mr and Mrs M apparently had a syncretic overall dominance score, Mrs M was the dominant force behind both the decision to move and the choice of area to move to. Her need to get out of the jointly owned house (with close relatives) prompted the decision to move and her desire to remain near to their daughter limited the choice of area. However, in terms of the property itself, the decision appeared to be syncretic, as did the desire for a village.

A final example of this changing dominance is shown in figure 9.2c. Mr and Mrs P had a single dominance score of male-partner-dominant. However, the decision to move was clearly female-partner-dominant. Mrs P wanted to buy a home that they had jointly chosen rather than live in the property that Mr P had had before they had got married. However, the area they searched in was clearly dictated by Mr P's employment that confined them to a specific area. Both decided on a village location and the majority of the characteristics of the property appeared to have been syncretic decisions.

In defining roles of influence or dominance within the joint decision-making process there are also a series of other points that need to be made. First, there is a need to distinguish between the *impetus* to move as stimulated by local or regional job moves and the *actual decision* to move. In other words, if the male partner's job is relocated within the region or if he decides to change his job intra-regionally it should not be assumed that this was a male-partner-dominant decision to move. Clearly, the couple can decide whether to move or stay because the relatively short distance may mean that migration is unnecessary. Indeed, circumstances may not permit such a move (for example, partner's job, elderly parents, key schooling time for the children, negative equity). Equally, the household may decide to move. The decision to move may well be syncretic – the impetus provided by the job change, but the decision to move jointly achieved. Table 9.5 provides an example of this.

In table 9.5(A) the assumption is made that the change of job for Mr F means that the decision to move is to be a male-partner-dominant decision, but as table 9.5(B) shows the reality was that although the change of job location provided the impetus for the move, Mr and Mrs F had already wanted to move house and this local job move enabled them to realize their

Table 9.5 Distinction between the impetus to move and the decision to move

A. *Seen as male-partner dominant decision*	B. *Seen as a syncretic decision*
Mr F.	Mr and Mrs F.
Job in Poole: 'on-call' work requires proximity to workplace.	Both desire to move back to the area near Little Hinton where Mr F came from originally.
Therefore Mr and Mrs F live in Poole.	But, Mr F's job is in Poole.
Mr F's Job is moved to Blandford.	'On-call' work requires proximity to workplace.
Near to Little Hinton, where Mr F came from originally.	Therefore they live in Poole.
Decision to move to Little Hinton.	Mr F's job moved to Blandford.
	Near to Little Hinton.
	Therefore enabling Mr and Mrs F to decide to move to Little Hinton.

Note: Little Hinton is a ficticious name.

aspirations. The decision was, therefore, in this instance, one jointly made. Clearly, this is not the case in every instance but it is important to be aware of this point, particularly when considering the relationship between employment opportunities and housing migration decisions. Distinction also needs to be made between the general area migrated to and the specific housing search space. The general locality may be determined by the female partner's job, and is therefore a female-partner-dominant decision, but the specific area agreed upon to be searched within may well be reached syncretically.

An example of this is Miss D and Mr E. Mr E's job was relocated to Weymouth where he moved into rented accommodation. Miss D then joined him after completing her postgraduate teaching certificate and found a job locally. They then jointly made the decision to buy a property rather than continuing to rent. They searched for a property that was within easy driving distance of their two workplaces. In this example, the general area that they searched in was determined by Mr E's job (a male-partner-dominant decision). However, the specific search space, once the decision to buy had been made, was decided by both of them on the basis of access to both workplaces (a syncretic decision).

Another important consideration that needs to be made with respect to the dominance or influence at different stages of the housing migration process is the differential weighting given to different dominances in the various stages of the housing migration process. This will clearly alter with each couple. Thus, different dominances may be distributed between different phases of the housing search process, yet dominance or relative influence by one partner over a particular issue may be held so strongly and fixedly that it supersedes all the other dominances. An example of this is Mr and Mrs F, who took the opportunity of a local job relocation to move to a bigger and better home near to Mr F's family, as illustrated in table 9.5. The decision to move was, as we have seen, syncretic. However, in terms of the choice of area this was a male-partner-dominant decision because Mr F was adamant that it had to be a move to one of two specific villages, preferably only the one (Little Hinton), as this was where he had grown up. In terms of the characteristics of the property, this was dominated by Mrs F, who wanted a big, old, detached house with lots of rooms and three bedrooms. They both agreed that they did not want a thatched property but a large garden was important. Ultimately they bought a *small, thatched house* with only a *few rooms* and a *large garden*, but it was in Little Hinton! In this example, Mr F's overwhelming determination to move back into one specific village dominated the decision about which property they were able to buy. In such a small village they had minimal choice and the result was a set of fairly fundamental compromises on the final property that they purchased.

As has already been noted, it has been suggested that women tend to be responsible for the frequent so-called unimportant decisions ('implementation power') whilst the men tend to hold the 'orchestration power' to make infrequent important decisions (Hardill et al. 1995). This grouping would clearly include housing migration decisions. However, it was clear from the findings that migration is not one homogeneous entity but has clear phases that show distinct differences between them in terms of who holds the dominance or power. In addition, if the distinction between 'orchestration power' and 'implementation power' is so clearly gendered then one would expect that the crucial decisions relating to the decision to move and the choice of a property to be male-partner-dominant decisions. Table 9.6 shows the results of the observed responses.

Two significant points emerge from this table. First, in general, women are not the most significant decision makers; rather joint decision making is either still male-partner-dominated or, where this has changed, women's roles form part of the decision in a joint capacity. Clearly this finding needs to be considered further in a larger sample of couples. Second, the decision to move, the choice of settlement type and the choice of property show no obvious tendency towards male-partner dominance. However, most noticeably the choice of area is gendered, being predominantly male-partner-dominated, and noticeably two-thirds of these male-partner-dominated decisions were job-related reasons for the choice of area.

Discussion

The findings presented in this chapter indicate that the process of joint decision making in housing migration often involves the dominance of one partner or the other, though this relative influence or dominance changes at different phases of the migration process. In addition, this dominance often has differential weighting at the various stages of the housing migration process, which affects the overall outcome of the decision-making process, often overriding other aspects of the housing migration process. Also, it was shown that there is a need to distinguish between the impetus to move stimulated by local or regional job moves and the decision to move. However, although there was variation, women do not form the highest proportion group in any of the housing migration decisions. This suggests that joint decision making is either still male-partner-dominated or, where this has changed, women's roles form part of the decision in a joint capacity in the majority of cases.

Earlier in the chapter it was noted that a 'leader–follower' continuum, with households occupying different points along it, might be a way of conceptualizing couples based on their decision making. However, it was clear from the analysis of the different stages of the housing migration process that the dominant partner or 'leader' can and does change at different stages of the process. The relative complexity of the results indicates that suggestions of a linear continuum based on two poles such as orientation power/implementation power, and leader/follower do not adequately explain the joint decision-making process of couples in the housing migration process.

Table 9.6 Observed dominance by stage of the housing migration process

Decisions	*Male-partner dominated*	*Syncretic*	*Female-partner dominated*
Decision to move	2	11	3
Choice of area	9	4	3
Type of settlement	1	13	2
Property characteristics	2	11	3

Perhaps a more appropriate way of conceptualizing this process can be adapted from Downs's (1970) conceptual schema for research into geographic space perception, shown in figure 9.3. Downs attempted to explain spatial behaviour patterns of geographic space perception. His schema is, as he acknowledged, 'a blatant over-simplification of a highly complex situation' (1970: 84). Nevertheless, it provides a clear, broad conceptual schema upon which to expand and develop some form of joint decision-making framework. Crucially, it is a *dynamic* model which allows for the changing nature of human behaviour as it interacts with society. In addition, Downs observes that the key objective of this schema is 'to view man [*sic*] as a decision maker' (1970: 85).

The starting point for the schema is the 'real world'. This provides the information to the person who then uses a system of perceptual receptors through which an interaction between the person's value system and their image of the real world provides meaning to the information. This meaning is then incorporated into the person's image. The information may mean that the person has to make an adjustment to align with the real world – a decision. This decision results in one of two responses from the person. Either the person searches the real world for more information or the decision produces a response in the form of behaviour. Lewis (1982) added to this

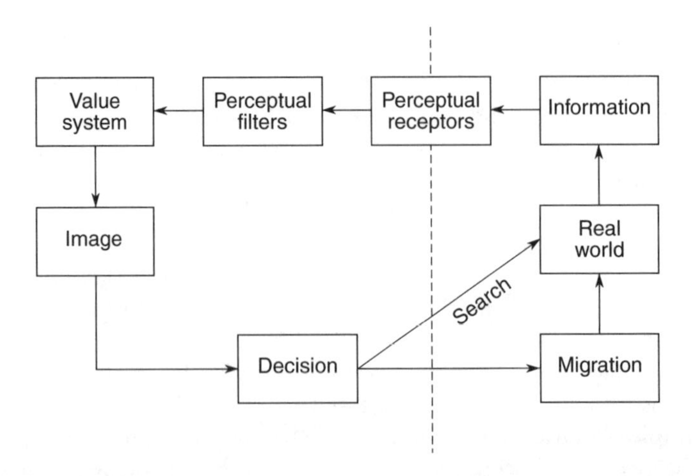

Figure 9.3 Down's conceptual schema for research into geographic space perception

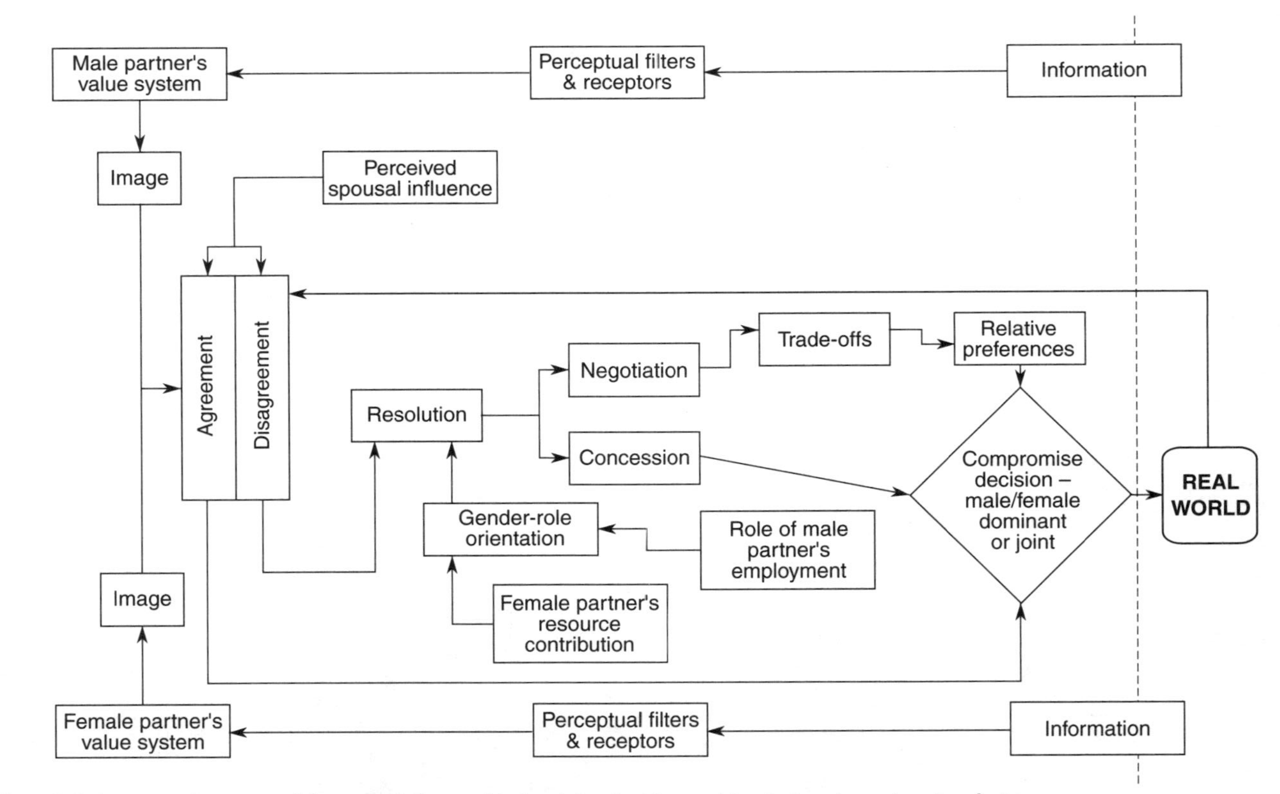

Figure 9.4 A proposed conceptual framework for considering joint decision making in housing migration decisions

schema perceptual filters and, more importantly, a migration component so that the decision based on information received results in the person either deciding to migrate, remain *in situ*, or continue to search for a property by gathering more information in the real world.

Using Downs's schema as a starting point, and incorporating the findings from previous work as well as this study, a simple model of the joint decision-making process is proposed in figure 9.4. This suggests that each partner enters the decision process with an image of their desired outcome of the decision as they perceive it. Within this there will be areas of both disagreement and agreement. The areas of disagreement will require the couple to discuss, negotiate and undertake a series of trade-offs to develop a relative preference or concede their preferences. These agreements and disagreements will be affected by perceived spousal influence, which is related to socialization of cultural norms. In addition, Qualls (1984) noted that the method used to resolve disagreement also appears to be related to the sex-role orientation, which Bird and Bird (1985) defined as the degree to which individuals adopt roles (on a continuum from traditional to egalitarian) that reflect normative prescriptions for husband–wife behaviour. In this way, traditionally orientated couples appeared to adopt essentially one-sided resolutions based on one partner giving in to the other's preferences, whilst modern-orientated couples tended to be characterized by mutual satisfaction of both partners' individual preferences. Burns and Hopper (1986) suggested that this in turn may be related to the wife's socio-economic contribution, so that the wife's relative influence over purchase decision-making is related to her personal contributions to the family's socio-economic status (as well as the product itself). Furthermore, Pahl's (1989) findings suggested that inequality outside the home in terms of earnings is linked to inequality within the household, and that this inequality is reflected in the decision-making roles and the control of finances. As a result of this, the interaction in the areas of disagreement will differ for each couple and be related to the relative and perceived influence or power that each partner holds within the relationship. The final decision will reflect the balance of power in the relationship. If there are indeed particular aspects of purchasing decisions that are female-partner-dominated or male-partner-dominated, then clearly the process of negotiation or concession and the relative influence or power of each partner will differ accordingly.

Table 9.7 shows an example of this model applied to data from this study. On the left are the various boxes that are depicted in the model, whilst on the right there is a case study of Mr and Mrs J. Essentially, Mr J's office was being closed down and he was requested to reapply for vacancies in other offices. Both Mr and Mrs J wanted to remain locally and not have to move house but Mr J was unable to obtain a position in any of the local offices and had to apply to some further afield. As a result of this, he was relocated to the Peterborough office and they had to move house. Both Mr and Mrs J had an image of the type of house and area that they would like, influenced by their experiences and the information that they had gathered over the years

Table 9.7 An example of the joint decision-making model applied to the housing migration process

• Information received	Mr J's office in Civil Service based in Matlock closed down and he has to apply for jobs in other offices. Attempt to remain local is unsuccessful. Job relocated to Peterborough office. Have to move house – too far to commute.
• Male partner's image	Mr J desires a small town or village, an old property near to the family and close to work.
• Female partner's image	Mrs J also wants a small town or village with facilities (shops, buses, schools), and an old yet modernized property near to the family.
• Agree	On an area to the north of Peterborough convenient for family and work; in a small town or village with facilities, and an old modernized property.
• 'Real world'	Lack of older property modernized in price range and area.
• Resolution	Based on the properties available on the market and in their price range.
• Role of male partner's employment	Mr J is the main income-earner of the household (breadwinner).
• Female partner's resource contribution	Mrs J's job provided an additional supplement to the household income (main roles: childcarer and domestic work).
• Trade-offs	Mrs J's complete refusal to purchase a property requiring refurbishment. Need for basic facilities, such as shops, schools and buses, locally. Mrs J does not drive, which therefore ruled out smaller villages.
• Relative preferences	They looked in a larger search area and at modern properties in larger villages and small towns but still near to Mr J's workplace.
• Compromise decision	Decision to buy a modern house at the right price by a good local builder renowned for quality modern houses in a small town near to both family and work.
• Final outcome	Decision to move – male-partner-dominant Choice of area – male-partner-dominant Type of settlement – syncretic Characteristics of property – female-partner-dominant

about their 'ideal home'. They discussed these views and reached an agreement. It is difficult to note what the effect of perceived spousal influence had in this stage of the process or the fact that Mrs J's income was subsidiary to Mr J's, except to state that these roles appear to be accepted as 'normal' and that Mr J's employment was seen to be the key determinant in migration decisions for them. Nevertheless, they agreed on a set of criteria which they then used to search for a property. However, although they had agreed on an older yet modernized property they found that there was a paucity of these properties in their price range on the market at that time. This meant that they had to re-evaluate their desires in the light of the available property and then agree on an adjusted set of criteria that suited the prevailing market conditions. This required a series of negotiations and trade-offs to occur and, in this instance, it was apparent that Mrs J was adamant that she would not settle for an old unmodernized property and that the location of the property had to be in a large village or small town that had some facilities because she could not drive. The final outcome clearly reflects the changing dominances throughout the housing migration process, and how the desires and images that individual partners' start off with are tempered to take into account both their partner's views and the type of properties available on the market at the time of migrating.

It is possible that this model is only representative of joint decision making in housing search behaviour. As noted earlier, Davis (1970) suggested that such roles cannot be generalized without reference to the product. Clearly, such a model would need to be applied and tested for other purchases before such a generalization might be made. In addition, different types of households from different social classes and ethnic groups may well display different decision-making patterns. In households that, for cultural or religious reasons, hold very strong patriarchal views the whole process may be simply reduced to the male partner deciding for the household with little or no discussion or negotiation. The model developed here is based on middle class movers and therefore represents a conceptualization of the type of process that these types of households work through in order to reach a decision. It tries to argue that within geographical studies of housing migration there is a need not only to look at gender, but to move away from the view of the household as a 'black box' (Pahl 1989) and consider the decision-making processes that occur *within* the household and the interplay of relations within it.

The effect of socialization on the joint decision-making process and the roles that couples play is extremely complex. Halfacree (1995) considered the various perspectives on migration and gender and makes a very useful distinction between what he terms 'internal' and 'external' perspectives. The 'internal' perspectives he defines as those whose primary focus is upon the migrant and her family, and the gender dimension of migration is couched in terms of net human capital gains or sex roles within the family unit. These focus on the internal workings of the household and roles played within the relationships. The 'external' perspectives provide a much broader interpretation of migration decisions within society's structures and the influential role of society upon the decisions. External perspectives are important because, when considering the joint decision-making processes of couples around housing migration decisions, it is important not to focus on the internal workings of the relationships to the exclusion of the external pressures of the society in which the couple live. These perspectives emphasize the need to locate a couple's decision making within society as a whole and, thereby, to take account of the 'structure–agency' interaction within which their roles are effectively operationalized.

Indeed, these internal and external perspectives can be broadly seen in terms of the spaces of reproduction and production, of home and work, which have historically been seen as separate sets of social relations occupying different spaces (Rose 1993). Rose argues that the two spheres need to be seen as part of a single process:

> the spatial division between reproduction and production is not universal, but is a consequence of specific historical-geographical changes; and that despite the ideological and spatial division of the two spheres, production and reproduction were intimately connected.
>
> (1993: 119)

She observes that reproduction and the home are not only explained by patriarchy and gender relations, and production and the workplace are not confined to explanations of class and capitalism, they intersect and overlap. Nevertheless, the existence of such a dualism may serve to emphasize the notion that external structures and agencies in the space of production influence a couple's roles, whilst the interpretation, negotiation and enactment of these structures is internally generated within spaces of reproduction and the home and will vary over time and by the couple's experiences, backgrounds, and so on. In this way, our understanding of the decision-making processes can enable us to explain how the decisions made help to reproduce these 'dual roles'. Thus, the separation of the migration decisions into different stages and the differential weighting of influence in different stages can either reinforce the dual roles or indicate any shift towards a merging of roles.

Conclusion

This chapter has shown that the process of joint decision making in housing migration often involves the relative influence of one partner more than the other. However, this relative influence of partners appears to change at different stages of the migration process, often with differential weighting influencing the overall purchase decision. The complexity of the joint-decision-making process prompted the need to rethink the way in which housing migration decision making is conceptualized, and a possible framework was proposed. Finally, it was argued that studies of couples and their migration decisions need to be cited in the context of their position within society, as well as with respect to the roles adopted in the home. The interaction of 'internal' and 'external' factors is likely to produce and reproduce roles of much greater complexity than a simple reproduction of patriarchy as it exists in the workplace or in society as a whole.

Note

1 Munsinger, Weber and Hansen used the following seven housing decision elements: decision to seek a new residence, to rent/to buy, floor plan, style, price, location and size. The current study placed greater emphasis on the geographical component.

References

Bettman, J. R. (1979) *An Information Processing Theory of Consumer Choice*, Reading, MA: Addison-Wesley.

Bielby, W. T. and Bielby, D. D. (1992) 'I will follow him: family ties, gender-role beliefs, and reluctance to relocate for a better job', *American Journal of Sociology* 97: 1241–67.

Bird, G. A. and Bird, G. W. (1985) 'Determinants of mobility in two-earner families: does the wife's income count?', *Journal of Marriage and the Family* 47: 753–58.

Blood, R. D. and Wolfe, N. M. (1960) *Husbands and Wives: the Dynamics of Married Living*. London: Collier-Macmillan.

Bokemeier, J. and Monroe, P. (1983) 'Continued reliance on one respondent in family decision making studies: a content analysis', *Journal of Marriage and the Family* 45: 645–52.

Burns, A. C. and Hopper, J. A. (1986) 'An analysis of the presence, stability, an antecedents of husband and wife purchase decision making influence assessment agreement and disagreement', *Advances in Consumer Research* 13: 175–80.

Davis, H. L. (1970) 'Dimensions of marital roles in consumer decision making', *Journal of Marketing Research* 7: 168–77.

Davis, H. L. (1971) 'Measurement of husband–wife influence in consumer purchase decisions', *Journal of Marketing Research* 8: 305–12.

Davis, H. L. and Rigaux, B. P. (1974) 'Perception of marital roles in decision processes', *Journal of Consumer Research* 1: 51–62.

Downs, R. M. (1970) 'Geographic space perception. Past approaches and future prospects', *Progress in Geography* 2: 65–108.

Dudleston, A. C., Hardill, I., Green, A. E. and Owen, D. W. (1995) 'Work rich households: case study evidence on decision making and career compromises amongst dual career households in the East Midlands', *East Midlands Economic Review* 4, 15–32.

Haas, L. (1980) 'Role-sharing couples: a study of egalitarian marriages', *Family Relations* 29: 289–96.

Halfacree, K. H. (1995) 'Household migration and the structuration of patriarchy: evidence from the U. S. A.', *Progress in Human Geography* 19: 159–82.

Hardill, I., Dudleston, A. C., Green, A. E. and Owen, D. W. (1995) 'Decision making in dual career households: theoretical perspectives and empirical evidence from case studies in the East Midlands', Paper presented at the 'Gender Perspectives on Household Issues' conference, University of Reading, April.

Hunt, J. G. and Hunt, L. L. (1982) 'Dual-career families: vanguard of the future or residue of the past?', in J. Aldous (ed.) *Two Paychecks. Life in Dual-Earner Families*, Beverly Hills, CA: Sage, pp. 41–59.

Kenkel, W. F. (1961) 'Husband-wife interaction in decision making and decision choices', *Journal of Social Psychology* 54: 255–62.

Kenkel, W. F. and Hoffman, D. K. (1956) 'Real and conceived roles in family decision making', *Marriage and Family Living* 18: 311–16.

Lewis, G. J. (1982) *Human Migration. A Geographical Perspective*, London: Croom Helm.

Lewis, G. J., McDermott, P. and Sherwood, K. B. (1991) 'The counterurbanization process: demographic restructuring and policy response in rural England', *Sociologia Ruralis* 31: 309–20.

Monroe, P. A., Bokemeier, J. L., Kotchen, J. M. and McKean, H. (1985) 'Spousal response consistency in decision-making research', *Journal of Marriage and the Family* 47: 733–38.

Munsinger, G. M., Weber, J. E. and Hansen, R. W. (1975) 'Joint home purchasing decisions by husbands and wives', *Journal of Consumer Research* 1: 60–6.

O'Connor, P. (1991) 'Women's experience of power within marriage: an inexplicable phenomenon?', *Sociological Review* 39: 823–42.

Pahl, J. (1989) *Money and Marriage*, London: Macmillan.

Pahl, J. M. and Pahl, R. E. (1971) *Managers and their Wives*, London: Allen Lane.

Park, C. W. (1978) 'A conflict resolution choice model', *Journal of Consumer Research* 5: 124–37.

Park, C. W. (1982) 'Joint decisions in home purchasing: a muddling-through process', *Journal of Consumer Research* 19: 151–62.

Park, C. W. and Lutz, R. J. (1982) 'Decision plans and consumer choice dynamics', *Journal of Marketing Research* 19: 108–19.

Park, C. W., Hughes, R. W., Thukral, V. and Friedmann, R. (1981) 'Consumers' decision plans and subsequent choice behaviour', *Journal of Marketing* 45: 33–47.

Poloma, M. M., Pendleton, B. F. and Garland, T. N. (1982) 'Reconsidering the dual-career marriage: a longitudinal approach', in J. Aldous (ed.) *Two Paychecks. Life in Dual-Earner Families*, Beverly Hills, CA: Sage, pp. 173–92.

Qualls, W. J. (1984) 'Sex roles, husband-wife influence, and family decision behaviour', *Advances in Consumer Research* 11: 270–75.

Robinson, V. (1993) '"Race", gender, and internal migration within England and Wales', *Environment and Planning A* 25: 1453–65.

Rose, G. (1993) *Feminism and Geography. The Limits of Geographical Knowledge*, Cambridge: Polity Press.

Sayer, A. (1984) *Method in Social Science. A Realist Approach*, London: Hutchinson University Library.

Vogler, C. (1994) 'Money in the household', in M. Anderson, F. Bechhofer and J. Gershuny (eds) *The Social and Political Economy of the Household*, Oxford: Oxford University Press, pp. 225–66.

10 To follow the chicken or not?

The role of women in the migration of Hong Kong professional couples

Lin Li and Allan Findlay

Patriarchy and international migration

Most models of migration behaviour have been developed on the premise that it is individuals who move. Even where groups of people, such as a household, migrate, it has been assumed that there is a primary decision maker within the household who is accompanied by dependants (typically a wife and children). This perspective favours interpreting marriage and family relations as an extra variable that produces deviations of migrant behaviour relative to mobility patterns of a single mover. Mincer (1978), for example, notes that so-called 'marriage ties' reduce mobility, while Clark (1986) offers several generalizations about the migration of North American couples, which include the dampening effect of marital status on migration, especially if both spouses work. According to Clark (1986: 71), 'the fact that working wives generally earn less than than their husbands makes it far more likely that females will be tied movers or stayers'.

Gender inequalities in access to and progression within education and the labour market have been widely documented. Since both education and professional experience are strongly associated with migration opportunities, it is not surprising that a gender bias is evident in both internal and international migration. If indeed gendered social structures favour male educational and career prospects, it is understandable that 'in the past most labour migrations . . . were male-dominated, and women were often dealt with under the category of family reunion' (Castles and Miller 1993: 9). Although, as highlighted by Castles and Miller, there has been a trend towards the feminization of international migration, this trend is not one which involves the migration of married couples. Married women caught up in the international trade in domestic servants and low-wage service workers usually leave behind not only their husbands but also their children. By contrast with the trend towards feminization of certain types of international flows, research concerned with recent movements of highly skilled labour suggest that it is still strongly male-dominated, characterized by a pattern of moves 'led' by the husband, with wives trailing behind in a passive role

(Beaverstock 1991; Johnson and Salt 1990). Something of the frustration for migrant wives of the resulting career breaks and increased family stress have been reported by Gordon and Jones (1994) and Ford (1992).

However, it is not simply gender inequalities in education and career progression in the labour market, which are significant in understanding the international migration behaviour of married women. What is just as important is the gendering of values and meanings that affect decision-making within the household. Gender models of migration and other behaviour suggest that within the household, just as much as within the external labour market, power relations operate to the material and ideological advantage of men (Ferree 1990). From this perspective, gendered power relations within the household also produce circumstances in which women become 'tied migrants' as a result of male dominance in household decision-making processes.

Studies of migration and gender therefore involve much more than simply inclusion of 'women' or 'marital status' as extra variables for analysis. Such a position has not only been dubbed 'gender-blind', but fails to recognize the dominance of patriarchal structures in producing gendered meanings to migration. Thus, 'we cannot obtain an adequate understanding of married women with respect to . . . migration without taking into account their location *within society*' (Halfacree 1995: 174). The situatedness of gendered value systems relative to socially constructed boundaries results in married women giving different meanings to migration from their husbands. Moreover, such meanings have shifted historically and geographically in a quite dramatic fashion. The emergence of ever-increasing numbers of dual-career households appears to support the view that, in western societies at least, there has been a shift towards a new model of household organization, albeit one biased towards the more educated and better-off elements of society and geographically focused in certain major urban labour markets where both partners can pursue their careers without the necessity for residential migration to facilitate upward occupational mobility (Snaith 1990; Green 1995). Gendered power relations within the household vary also between cultures, as has the response of married women to feminist ideas. Much work remains to be done on how changes in gendered power relations affect migration decisions, and how cultural specificity mediates to produce diverse gendered meanings for the migration act in different places around the globe.

Hong Kong provides an interesting example of the changes in gender relations that have occurred in the so-called 'newly industrializing countries' of southern and eastern Asia. The position of women in Hong Kong can be interpreted as one that is subject to both traditional and modern influences (Lilley 1994). On the one hand, the Confucian teaching of patriarchal hierarchy in which men dominated women heavily structured traditional Chinese interpersonal relationships (Chan and Leong 1994). On the other hand, modernization and globalization mean that western feminist ideas have infiltrated into Hong Kong and the promotion of the awareness and rights of women has gained momentum in recent years (Wu 1995). At the same time, the economic success of the territory is linked to better education and employment opportunities for both sexes and it is not uncommon for women to have successful careers. While the improved status of women has left some scholars (Bulbeck 1994) with an impression of more equal opportunities for the two sexes in Hong Kong, particularly in the commercial sector, others (Wu 1995; Yeung 1996) note that patriarchal ideologies are still extensive in the territory. With gendered power structures being caught between tradition and modernity, Hong Kong therefore offers an interesting site for the examination of how changes in patriarchal power relations impinge on migration, particularly on the role married women play in international moves.

Methodology

The results reported here are drawn from two separate but related large-scale studies conducted by the authors on migration to and from Hong Kong. Further details of other aspects of this research are reported in Findlay et al. (1994, 1996) and Li et al. (1995). The two studies included both questionnaire surveys and in-depth interviews with migrants and non-migrants. The discussion here is limited to interviews with twenty Hong Kong Chinese couples selected from the larger surveys, with the interviews undertaken in Cantonese in 1993 and 1995. The majority of the interviewees were immigrants to Britain or Canada. The age of the couples ranged from early thirties to mid-forties. The husbands had all been trained either as doctors or engineers. The interviews were therefore intentionally located within a very specific socio-demographic group and no attempt was made to represent the full spectrum of Hong Kong emigrants. At the time of the interviews, a quarter of the wives were not employed and the others were in a wide range of occupations. Rather than seeking to be representative in some statistical sense, the purpose of the in-depth interviews was to gain an insight into the understanding of migration held by the migrants, and from there to develop a more fruitful theorization of migration as a socially and culturally embedded process (Findlay and Li 1997).

Women as tied migrants?

In the literature review we suggested that because gendered social structures favour men in education and employment opportunities, for heterosexual married couples the geographical mobility of women is more likely to be linked to their husbands' occupations than to their own. However, the improved status of women in developed societies such as Hong Kong raises the issue to what extent family migration is still characterized by women being the 'tied' migrants.

Our research shows that in terms of official migration status, a significant proportion (a third) of the wives interviewed were actually the principal applicants for immigration visas to Britain or Canada. In other words, in these cases the wife's occupation, rather than the husband's, was instrumental in securing overseas residence status. However, this does not necessarily mean that the women were the 'leaders' in migration in terms of decision-making, nor that the decision was undertaken for the advancement of their careers. The gender dimension in the migration decision-making process will be examined more closely in a later section. As far as the reason for the migration is concerned, many interviewees clearly stated that their intention was to obtain foreign citizenship, largely because of the political changes in Hong Kong (Findlay et al. 1993). Thus, whether the husband's or wife's occupation was used in the visa application was mainly a strategic choice taken to maximize the success rate of the application. Their official migration status seemed to be largely determined by their destination country's immigration policies. An example is the severe restriction on entry to Canada of doctors who have obtained their medical qualifications in other countries. As a result of this restriction, the majority of male doctors in our study who had migrated to Canada had done so as dependants of their spouses or of their parents. In other words, immigrant status was often 'tied' to the wives' entry visas.

Nevertheless, the fact that some of the wives were the principal applicants not only reflects the educational and occupational achievements of the women in the sample, but also perhaps suggests a more egalitarian gender position within the family. Yet this position often had to be negotiated against the traditional patriarchal structure. For example:

> He found it hard to accept being my dependant in the application for British right of abode . . . He was very reluctant [to be my dependant] because this concerns his lifelong status.
>
> (Lam)

Lam was in a senior management position and her husband worked as a senior engineer. The couple's migration decision had been made only after much discussion between them. However, even in such an apparently egalitarian marital relationship, the dominance of the patriarchal system was evident. Although both Lam and her husband wanted to obtain foreign citizenship, the decision that the man should be the tied migrant, albeit a strategically sound one, challenged the male status of her husband. Lam described it as a 'lifelong' status, which suggests, first, that it was probably of primary importance to her husband; and second, that such a status was the result of 'lifelong' socialization of behaviour and attitudes which draw on and reproduce the institutional structures of male superiority. In our sample, the fact that a large number of women were officially the principal professional migrant does not therefore reflect a revolution in the traditional construction of gender roles and power relationship within these households; more a negotiated temporary status accepted to achieve a shared goal.

Power and negotiation in migration decision making

As official migration status was inadequate in revealing the influence of gender on migration, a close examination of the migration decision-making process was undertaken to find out how the female interviewees influenced the migration decisions. A multiplicity of experiences was reported, ranging from women who represented themselves as submissive to those who considered themselves as assertive. However, as illustrated below, categorization of the women into 'followers' and 'leaders' would be an oversimplification.

During the discussion of the part they played in the migration process, some women spontaneously referred to the traditional Chinese saying, 'when you are married to a chicken, you follow the chicken'. The saying depicted a very submissive role played by married women. In Cantonese, the term 'married to' is *ga*, an expression conventionally used only to refer to women, and it bears the connotation of a woman leaving her parental home and joining another paternal family. The saying therefore endorses the subordinate position of women in a patriarchal hierarchy. Although the expression refers to a general subordination of women to their husbands, the term 'follow' when taken in a geographical sense makes the saying particularly pertinent to the migration situation of the women as perceived by them. However, detailed analysis of the migration decision-making process reveals that even those women who applied this saying to their situation were not as passive and totally submissive as defined by the traditional cultural norms.

Consider the case of Fu, who was thirty years old, born and educated in Hong Kong up to post-secondary level. She has not been in salaried employment since the birth of her first child in 1992. She migrated to Canada as a dependant of her husband in 1993. She described herself as 'passive' and 'timid' and perceived the traditional saying as applicable to her relationship with her husband. One might thus assume that her husband dominated the migration decision-making process. However, from her husband's perspective, she had an important influence on their decision to migrate, even though the trigger for the final decision was politically based:

> After [the Tiananmen massacre in Beijing on] 4 June I felt very sad . . . As I woke up and heard the news about what had happened to the students, I felt–er–I actually cried . . . I suddenly felt that it seemed to be meaningless . . . It was not me who thought of emigrating but my wife who was very afraid. She wanted to leave and I said fine.
>
> (Fu's husband)

Interestingly, in Fu's own description, she attributed to herself a much less central role in the decision making:

> I think my views are very different from those of your other interviewees. They are more realistic in their considerations – politics and the

> future etc. As to me . . . I had always wanted to live abroad since the age of four. Basically I liked the environment . . . I considered immigrating here . . . when I was married because I liked this place when I was on holiday here. As to the fear of 1997, if my husband had decided to stay, I would have followed him . . . When I first got to know my husband, he said he would not emigrate. It was the 4 June incident which made him decide to emigrate. Of course I didn't object because I always wanted to – I didn't like living in Hong Kong. *(Researcher: Did 4 June have any influence on you?)* Not in any significant way.
>
> (Fu)

Rather than being the person who pushed for the move, as described by her husband, she seemed to suggest that her husband raised the option of emigration, by saying that 'I didn't object' and that if her husband had wanted to stay, she 'would have followed him'. Furthermore, her husband's understanding of her motivation to emigrate is slightly different from hers. While her husband attributed fear of political instability as a major factor for her intention to move, she herself played down the importance of this factor and stressed that her intention to emigrate had long predated the issue of 1997.

According to both Fu's own and her husband's accounts, Fu also played a crucial part in the choice of emigration destination. Before their emigration, the couple had obtained British citizenship through the British Nationality Selection Scheme (Jowett et al. 1995). In terms of employment prospects, it might have been more advantageous for the husband to migrate to Britain rather than Canada, since his professional qualifications were fully recognized in Britain but not in Canada. However, they decided to go to Canada mainly because of Fu's preference. So, although Fu appeared to adopt a subordinate role to her husband by endorsing the traditional saying and playing down her influence on her husband's decision, her views were crucial in shaping the decisions relating to the couple's migration.

Unlike Fu, other women were more overtly assertive in the part they played when negotiating with their husbands over the migration decision, and they also disputed the relevance of the traditional saying to their own positions. An example was Cheung, who was forty years old. At the time of the interview, she and her husband were residing in Hong Kong but had applied to immigrate to Canada. Since her husband had obtained his medical qualifications in Hong Kong, the visa application was lodged with Cheung, an experienced psychologist, as the principal applicant. During the process of making their migration decision, especially with regard to the destination, she had taken a very proactive role:

> I was very active in initiating the discussion [with my husband]. For example, when we were travelling abroad during last summer holiday, I suggested that we put aside some time to talk about the priority of the different migration destinations. My husband said it was not an appro-

> priate time to discuss this, but I said 'if we always say it is not an appropriate time, we'll never have the chance to talk about it'. So we sat down and discussed. Afterwards when we were back in Hong Kong, we took half a day off specially to discuss this, to analyse the pros and cons of different places . . . I feel I am more eager [to discuss] than my husband.
>
> (Cheung)

Cheung's insistence on a systematic analysis of the desirability of the various destinations clearly illustrates her leading role in the migration decision-making process – a far cry from the traditional Chinese saying. However, such active participation of the female interviewees in the decision-making process does not mean that they have replaced their husbands in dominating the migration decision. Rather, negotiation, 'give and take', and compromise were often emphasized. Cheung had actually made a compromise since it was her husband who had suggested emigrating and who was more eager to leave:

> My views change from time to time. When you interviewed my husband, we had just made the decision to go to Canada . . . But now I feel I actually don't want to go so much, but I believe the likelihood of leaving is quite high because my husband wants to go. He really wants to go . . . Mainly I do not like to leave my parents . . . I feel a bit reluctant to go, although I think I have tried not to be reluctant and go happily, but in actual fact I can't help it . . . If my husband is reluctant, I would not want to go. But I am quite strange . . . I don't like to go simply because you say go. I am often the devil's advocate. Like the process now, he says, 'should we go?' So I think about what leaving would mean or what staying would mean.
>
> (Cheung)

Although Cheung would not sit back and let decisions be made for her and insisted on negotiating with her husband, she did not attempt to dominate the migration decision either. The above quotation reveals that she felt somewhat ambivalent about the decision to emigrate and was more inclined not to leave Hong Kong. However, because of her husband's stronger preference to emigrate, she had acquiesced in proceeding with the application to migrate to Canada. Her demand for a rational analysis of the situation was in a way an attempt for her to reconcile her reluctance to leave with her husband's intention to emigrate. The discussion had not helped her to resolve fully her internal struggles and, despite her saying that 'I don't like to go simply because you say so', she reckoned that she would most likely emigrate with her husband because of his preference.

Cheung's remark above also illustrates the ambivalence the migrants sometimes experienced about their migration decision, and because of this their inclination to move or stay would fluctuate from time to time. The couples therefore had to negotiate and renegotiate their decisions. Part of the reason for the ambivalence and the need to negotiate is that international migration is a major decision that can potentially disrupt marital harmony and family dynamics and cohesion. Like Cheung, other female interviewees felt it was important to discuss with their husbands before the migration decision was made, but they were prepared to compromise, often so that marital harmony and family cohesion could be maintained. That some of the women perceived their situation as conforming to the traditional cultural norms of 'when you are married to a chicken, you follow the chicken' raises the issue of whether the greater goal of family harmony may not have helped to disguise the powerful influence of the patriarchal system. In short, despite their greater control over their lives and their more active participation in migration decision-making than might have been the case in the past, ultimately women were probably still more likely than their husbands to make compromises. Furthermore, when compromises were made, individual sacrifices were shaped by socio-cultural constructions of gender roles which further affected how the impact of migration was interpreted.

Migration and cultural construction of gender roles

When considering whether to migrate, the interviewees often emphasized the impact of migration on their career because Hong Kong was seen as economically more successful with much better employment prospects than Britain and CanadA. However, although all but one of the women were employed in Hong Kong before migration, and emigration would have an equal, if not greater, impact on their employment as on their husband's, the women seemed to be generally less concerned about this. There was an overwhelmingly consistent perception of career ambitions as being structured by gender, especially in relation to the concern for family and children. This was a view held even by women who had successful careers and who considered themselves as 'not traditional', such as Lam:

> My career [as a senior manager in Hong Kong] would certainly end [after emigrating]. My career is out of consideration . . . As for my husband, generally this is more of an obstacle for men. They look at it purely from a career point of view, whereas women adopt a viewpoint which you can say is not so pragmatic. Men want a sense of achievement whereas women generally want a better life, better value system, want to bring up their children with better values and not just for earning money . . . I am not a traditional woman who submits to her husband . . . [But] I'm always prepared for the possibility of accompanying my children [if they go abroad to study], whereas it would be more difficult to ask the man to do so.
>
> (Lam)

Numerous couples, as illustrated in table 10.1, reiterated such gendered construction of values and roles in relation to career and family.

Table 10.1 Examples of construction of gender roles and migration

Wives	*Husbands*
'I don't feel particularly strongly about returning to Hong Kong to work. My priority is given to my two sons . . . To my husband, work is the most important but to me the children are – I have to consider the children . . . The children are still young. It would be different if they were older. Say if they were already 13, 14 years old . . . then I would say, you can go back to work and I stay here to look after the children . . . Now the children are still young. I would not agree to him going back to Hong Kong to work and we see each other only during holidays because the children need their father.' (Siu, immigrant to Britain, freelance accountant)	I have career and family goals. To my wife, there is no career goal. So she focuses on the family. She looks after the children and wants the children to grow up here. The attraction of going back to Hong Kong is my career advancement but the children wouldn't be able to advance. My wife doesn't have a career and she focuses on the children . . . My wife doesn't like Hong Kong... But I feel that I have reached the stage when I can only get a promotion here if my boss retires.' (Shum, immigrant to Britain, engineer, Siu's husband)
'I think there are clear gender differences. Since my migration here, I've clearly noticed that work is very important to men . . . Since I came here, I have actually felt and experienced that unemployment – or if they cannot do what they want to do – is really devastating to men . . . On the other hand, women don't like changes . . . Women have a greater responsibility with children, they have to . . . help their children to adjust as well. Unlike men who can just leave because of their jobs, women have to consider more.' (Pong, immigrant to Canada, social worker)	'My wife and I didn't have any major disagreement. I think that – you can call it gender differences, or personality differences – women, especially those who are married, look at career in a different way from men. I don't mean that she is not ambitious but I feel that it is easier for her to be settled in her job but men think about many other things. They may compare with other things or they may not be contented to stay in the same post for their whole life.' (Poon, immigrant to Canada, engineer, Pong's husband)
'I think my husband's employment is more important than mine . . . As long as I have a job [it's fine]. I would consider his career to be the most important . . . I would [consider my own employment] only after his job and my son's school have been settled . . . Men put more emphasis on their career and women more on their family.' (Mui, immigrant to Canada, commercial buyer)	'I don't know whether the education system here will change. The training here seems to result in the younger generation having no ambition. People from places like where we have come from are more ambitious . . . I think it is better to have ambition . . . There are some differences between my wife and myself on this issue. She doesn't think this is important.' (Man, immigrant to Canada, engineer, Mui's husband)

The remarks shown in table 10.1 reflect a particular 'gender order' (Connell 1987) which shaped the interpretation of the meanings of migration. The 'gender order' which is unveiled here is perhaps not unexpected. What is rather surprising is that similar perceptions of gender differentiation in career ambition were expressed by both male and female voices, and by a large number of couples who varied in the relative dominance of each partner in the migration decision-making process. Women from diverse occupational backgrounds also made similar comments. The responses clearly demonstrate the widespread belief of gender differentiation in career ambition and how such a belief influences the identity of the interviewees. Men, particularly in the context of the intense competition and high social mobility in Hong Kong, were seen to be 'career minded'. They were perceived not only to have the responsibility to support the family financially but also to aspire to a sense of achievement through career advancement. Women, on the other hand, particularly those with young children, were portrayed as putting the welfare of the children and the whole family before their own career, and as lacking ambition in career development. The men and women in our study therefore evaluated their migration decisions according to different criteria. The men stressed the impact of the migration decision on their career development and were often seen to be less willing to emigrate from Hong Kong or more eager to return there for better career prospects. In contrast, the women were more ready to sacrifice their own employment for the family. They emphasized the impact of migration on their children and family life. Many considered the less polluted and crowded environment and the less examination-orientated educational system in the West as more beneficial to their children, and some appreciated the closer relationship within the nuclear family unit they experienced after migration.

Changing position of patriarchy

The above discussion seems to suggest that contradictions exist in how gender influences the migration of heterosexual couples from Hong Kong. On one hand, traditional cultural influences, in the form of the saying 'when you are married to a chicken . . .' or in the way social roles are gendered, still impinge on the interviewees' perceptions of the meanings of migration and of the part they played in migration decision making. On the other hand, the female interviewees also appeared not to be totally submissive, but exerted significant influence on the migration decision. While this contradiction can be explained partly by the multiplicity of experiences and subjectivities, it can also be understood in the context of the changing nature of patriarchy situated in the cultural specificity of Hong Kong. It is not our intention to provide a detailed analysis of the socio-cultural forces that have shaped the development of female identities in Hong Kong. Instead, we briefly examine the experiences of a 30-year-old female participant, Ng, in order to illustrate how both traditional and modern influences can operate at one and the same time.

Ng, an engineer, described herself as rather assertive in relation to her husband. She was one of the few female interviewees who had migrated before getting married, in order to obtain postgraduate professional qualifications. She was also atypical of the sample as a woman who displayed career ambitions. Yet examination of her biography reveals that, even in her case, rather than having been exposed to a revolutionized gender order, she had been subject to both traditional cultural constraints as well as more modern values. Her report of how her family reacted to her intention to study abroad clearly demonstrates these two different forces in operation:

> At first, I didn't consider further education because of my family's financial situation. Also I am female. My mother would prefer to support my elder brother first if he wanted to study, but my elder brother did not want to study and so I benefited. . . . My mother always encouraged us to study. She did not say that I should not study, but I think the priority went . . . my brother, if there was money enough only for one person to study abroad . . . If my brother wanted to [study abroad], I would not complain because firstly, he was male, and secondly, he was my elder brother. [When I considered going abroad], my eldest brother analysed the situation with my mother. At first, she was a bit worried, because . . . girls who studied too much might have difficulty getting married. Very typical. Also, she thought: when she returns from her study, she'll be too old. I would be 25 when I finished my study. To them at the time, if one did not get married by 25, one would be considered to be too old. She was worried that if I studied too much, boys might be afraid of approaching me . . . But my brother said: the world nowadays is different; let her go . . . So, she was also very supportive.
>
> (Ng)

The quotation shows how traditional patriarchal order influenced not only how Ng's migration was perceived by her mother, but also how Ng herself accepted such an order. Although Ng's mother, like many Hong Kong parents, considered education as desirable (Findlay and Li 1997), she also expressed the view that too much education would reduce the chance of marriage for women. Such a view reflects the traditional male superiority in heterosexual relationships, which is still prevalent in Hong Kong (Yeung 1996). Amongst the couples interviewed, none of the women had achieved higher academic or professional qualifications than their husbands, whereas some husbands had attained higher qualifications than their wives. This suggests that, even if opportunities for higher education are available to both sexes, patriarchal cultural beliefs can discourage women from accessing these opportunities.

Although Ng might not agree to her mother's analysis of the impact of further education and migration on her chance of getting married, she none the less embraced the intra-familial male-dominated hierarchy. Masculine authority was evident. As her mother was separated from her father, the most powerful male figure in the family was her eldest brother, whose influence was obvious in the way he was engaged as an intermediary by Ng in the attempt to change her mother's views. Furthermore, Ng explicitly stated her acceptance of the access to family resources for education being defined by gender and birth order.

Despite the influence of traditional values passed on from the previous generation, Ng was not bound by tradition and she would not hesitate to make use of the opportunity available to her for self-advancement. Her brother's remark that 'the world nowadays is different' further reflects how tradition has been challenged by new orientations. Such a challenge was mirrored in other social influences that impinged on Ng's identity, such as the media:

> When I was in secondary school, I wanted to be a 'strong woman' because there was this TV drama called *Strong Woman* . . . My goal then was to go into the commercial world, just like the woman in the drama and be successful and earn money.
>
> (Ng)

The image of career women portrayed in the media affected how Ng – and probably other women in Hong Kong, too – positioned themselves in society. Yet, as shown in the previous section by the still ubiquitous perception of the woman's place being primarily within the family, the influence of such images was perhaps limited and not permanent. These images may have helped to modify the patriarchal order, but have not led to a total erosion of patriarchy. One can also argue that the promotion of such images of 'strong women' is essentially an assertion of 'masculine' capitalist values and serves little to bring about structural changes for more egalitarian gender relationships. While it is possible for the better-educated women to have a successful career in Hong Kong, it is probably the economic environment, rather than a fundamental change in the cultural construction of gender roles and relations, which supports this possibility.

Conclusion

The interview materials reported above have illustrated that a bipolar categorization of 'leaders' versus 'followers' is inadequate to achieve an understanding of the complexity of how gender influences migration decision making. In our specific example of Hong Kong professional migrants, the 'complexity' is associated with the way in which the subjectivities of the women in our sample were caught between tradition and modernity.

On one hand, modernization, improved access to educational and employment opportunities and exposure to alternative ideologies as a result of globalization, facilitated by electronic media, empowered the women in the sample, so that they no longer identified themselves as being totally subordinate to their husbands. In terms of official migration status, or influence on the migration decision-making process, the women in our sample were certainly not just 'followers'. Yet, while recognizing both that women are active agents in the migration process, and that traditional gender power relationships do not have as strong a prescriptive hold on people as they did in the past, we have to acknowledge that the traditional Chinese 'gender order' is still operative amongst our interviewees. The fact that the saying 'when you are married to a chicken, you follow the chicken' is still regarded by some people as relevant to the gender relations of Hong Kong as a modernized, global city is perhaps somewhat surprising and demands explanation. The centrality of the family system, particularly in a society where public welfare is poorly provided for, has been suggested as one of the factors that sustain male dominance (Leung 1995). The importance of the family as a source of support as well as the social space in which patriarchal values are produced, reproduced and challenged, was evident in the example of Ng. For the heterosexual migrant couples in our study, international migration highlighted the centrality of the family. When the migration decision required compromises to be made for the sake of familial harmony, the prevalence of traditional sex-role stereotyping meant that women were still more likely than their husbands to compromise by sacrificing their career and personal development.

Our observation of the still widespread perception that men's career development is privileged over their wives' has two methodological implications. First, we found that such perceptions affected how the meaning of migration was interpreted even for couples whose immigration visa had been secured using the wife's occupation. This suggests that in order to understand the effects of patriarchy on migration, it is not enough just to enumerate how often moves are 'tied' to the husband's or the wife's employment. Understanding the effects of patriarchy requires in-depth examination of the underlying motives of migration and how couples negotiate the migration decision. Second, unlike other research on gender and migration which studies employment-related moves, our study has focused on couples whose moves were largely motivated by the intention to gain citizenship in another country. The perceived importance of male career development suggests that for migration that would advance the husband's career, patriarchy may be more powerful than in the kind of international citizenship-motivated moves we have studied.

We have argued that when studying the relationship between gender and international migration, it is necessary to analyse how patriarchal ideology contributes to the gendered meanings of migration (Halfacree 1995). This analysis should take into account the changing nature of patriarchy, with subjectivities being formed and re-formed, or hybridized (Li and Findlay 1996), from the forces of tradition and modernization. By using Hong Kong professional migrants as an example, we wish to emphasize the need to consider cultural- or place-specificity in such an analysis. We believe that migrants from other cultures or places have been influenced by similar tensions between tradition and modernity, but further research would be required to examine the forms that such tensions take in different milieux.

References

Beaverstock, J. (1991) 'Skilled international labour migration: an analysis of the geography of international secondments within large accountancy firms', *Environment and Planning A* 23: 1133–46.

Bulbeck, C. (1994) 'Sexual dangers: Chinese women's experiences in three cultures – Beijing, Taipei and Hong Kong', *Women's Studies International Forum* 17: 95–103.

Castles, S. and Miller, M. (1993) *The Age of Migration*, London: Macmillan.

Chan, S. and Leong, C. W. (1994) 'Chinese families in transition: cultural conflicts and adjustment problems', *Journal of Social Distress and the Homeless* 3: 263–81.

Clark, W. (1986) *Human Migration*, New York: Sage.

Connell, R. W. (1987) *Gender and Power: Society, the Person and Sexual Politics*, Cambridge: Polity Press.

Ferree, M. (1990) 'Beyond separate spheres: feminism and family research', *Journal of Marriage and the Family* 52: 866–84.

Findlay, A. M. and Li, F. L. N. (1997) 'An auto-biographical approach to understanding migration: the case of Hong Kong emigrants', *Area* 29: 34–44.

Findlay, A. M., Li, F. L. N., Jowett, A. J., Skeldon, R. and Brown, M. (1993) 'Emigration of Hong Kong doctors', *Applied Population Research Unit Discussion Paper* 93/3, Department of Geography, University of Glasgow.

Findlay, A. M., Li, F. L. N., Jowett, A. J., Skeldon, R. and Brown, M. (1994) 'Doctors diagnose their destination', *Environment and Planning A* 26: 1605–24.

Findlay, A. M., Li, F. L. N., Jowett, J. and Skeldon, R. (1996) 'Skilled international migration and global city', *Transactions of the Institute of British Geographers* 21: 54–65.

Ford, R. (1992) 'Migration and stress among corporate employees', Unpublished PhD thesis, University of London.

Gordon, E. and Jones, M. (1994) *Portable Roots*, The Hague Presses: Interuniversitaires Européennes.

Green, A. (1995) 'The geography of dual career households', *International Journal of Population Geography* 1: 29–50.

Halfacree, K. H. (1995) 'Household migration and the structuration of patriarchy: evidence from the USA', *Progress in Human Geography* 19: 159–82.

Johnson, J. and Salt, J. (eds) (1990) *Labour Migration*, London: David Fulton.

Jowett, A. J., Findlay, A. M., Li, F. L. N. and Skeldon, R. (1995) 'The British who are not British and the immigration policies that are not: the case of Hong Kong', *Applied Geography* 15: 245–65.

Leung, B. K. P. (1995) 'Women and social change: the impact of industrialization on women in Hong Kong', in V. Pearson and B. K. P. Leung (eds) *Women in Hong Kong*, Hong Kong: Oxford University Press, pp. 22–46.

Li, F. L. N. and Findlay, A. M. (1996) 'Placing identity: interviews with Hong Kong Chinese immigrants in Britain and Canada', *International Journal of Population Geography* 2: 361–77.

Li, F. L. N., Jowett, A. J., Findlay, A. M. and Skeldon, R. (1995) 'Discourse on migration and ethnic identity', *Transactions of the Institute of British Geographers* 20: 342–56.

Lilley, R. (1994) 'Chronicle of women', *Australian Journal of Anthropology* 5: 86–112.

Mincer, J. (1978) 'Family migration decisions', *Journal of Political Economy* 86: 749–73.

Snaith, J. (1990) 'Migration and dual career households', in J. Johnson and J. Salt (eds) *Labour Migration*, London: David Fulton, pp. 155–71.

Wu, R. (1995) 'Women', in S. Y. L. Cheung and S. M. H. Sze (eds) *The Other Hong Kong Report 1995*, Hong Kong: Chinese University of Hong Kong, pp. 121–56.

Yeung, L. (1996) 'Just the job for a jealous male', *South China Morning Post*, 20 May, p. 21.

11 Gender variations in the characteristics of migrants living alone in England and Wales 1991

Ray Hall, Philip Ogden and Catherine Hill

Introduction

This chapter is part of a wider research project, 'Household structures, household transitions ang geographical mobility', which aims to increase our understanding of the sources of change in the structure of households, in particular changes in non-pensioner, one-person households, that have taken place over recent decades. The role played by migration in household change, and the influence of gender on these changes in one particular theme of the research.

The focus of the chapter is the influence of gender on the migration characteristics of younger people (aged under 60) living alone in 1991, asking whether gender variations are more or less important than the differences between migrants living alone and all migrants. The relationship between changes in household status – termed here a household transition – and migration is discussed, particularly in terms of transitions between family and non-family household types. The characteristics of those who have migrated to Inner London between 1981 and 1991 are examined more specifically. We conclude with a brief discussion of possible gender differences in motivations for migration.

In particular, the chapter seeks to establish how far we can provide evidence for the hypothesis that individuals attracted by the employment opportunities of Inner London are more likely to decide to live alone and that the housing market, in turn, responds both with a greater degree of flexibility expressed particularly by a greater propensity to rent, as well as responding to particular lifestyle demands. To what extent are people living alone particularly well-suited to the labour market requirements of global cities, such as London, which require highly mobile populations able to move in and out of jobs relatively easily (Cadwallader 1992)? To what extent are there gender differences between the migrants living alone, or are they a distinctive subgroup with more similarities than differences? Are those migrating to Inner London distinctive from those migrating elsewhere in the country?

One-person households increased rapidly in number during the 1980s so that by 1991 they made up more than a quarter of all households in England and Wales, a trend shared by many other European countries and the United States (Kaufman 1994). Changing household structures are closely related to other social, economic and geographical processes – such as the professionalization of the labour force (Hamnett 1994a, 1994b), changes in the housing market, particularly gentrification (Smith 1996), together with less tangible processes which have been described as the rise of 'postmodern' individualism (Harvey 1989). Changing household structures and the tendency towards living alone are a particularly remarkable feature of large cities in the late twentieth century. At the same time, changes in the distribution of younger one-person households over the last decade also suggest that it is becoming a more diffuse geographical phenomenon than has previously been the case (Hall, Ogden and Hill 1997).

None the less, younger one-person households are not distributed evenly throughout the country – there is a particular concentration in London and other large urban centres. Inner London, in particular, has a much higher proportion of one-person households – 38 per cent – than any other region in the country; here, 17 per cent of the total population live alone compared with 10.8 per cent in England and Wales as a whole. A higher proportion of these are under retirement age than nationally – 78 per cent of males living alone are under retirement age, compared with the national figure of 69 per cent, as are nearly half of all women, compared with 29 per cent nationally. Thus, London may be characterized as an 'escalator region' attracting 'many upwardly-mobile young adults living in single-person households' and then encouraging 'their out-migration in nuclear family or empty-nest households to other regions in later middle-age or at, or close to, retirement' (Fielding 1993: 158).

Nationally, the majority of people living alone are older females but their dominance has declined over the last twenty years and by 1991 more younger (under retirement age) men and women were living alone. The overall proportion of those living alone who were under retirement age increased from 35 to 44 per cent between 1981 and 1991. By 1991, the age structure of one-person households was bimodal with a first, lower peak around the age of thirty years and a second peak at retirement. This distribution is most clearly seen among males living alone but is also evident for females. For example, for men only 3 per cent of the 25–29 year age group lived alone in 1971, by 1991 11 per cent were doing so. For women over the same period, the proportions of the 25–29 year age group living alone increased less dramatically – from 2 to 6.5 per cent.

Sources

The principal source used in this chapter is the Longitudinal Study (LS) for the period 1981 to 1991. This is a 1 per cent sample taken from the Census of Population for England and Wales, starting in 1971 and followed through subsequent censuses (Dale 1993), used increasingly in studies of geographical and social mobility (for example, Fielding 1989, 1993; Hamnett 1990). Using the LS, the same individuals can be followed from census to census so that changes in household or social status as well as residence can be tracked. Variables included in the migration tables, which give change of residence between the two censuses of 1981 and 1991, were age, gender, distance travelled, social class, tenure, marital status and region. Note that the groups selected for analysis were those aged under 50 years in 1981 and thus under 60 years in 1991.

There are a number of problems involved in using the LS, not least of which is the level of detail available. In general we have had to use quite large categories to ensure adequate sample size. Moreover, the LS can give only an imperfect picture of the relationship between migration and transitions to living alone. The data may tell us that both a change of residence and household status has occurred between the two dates but we have no indication of the timing of the two events. The assumption that the two events occurred concurrently, let alone that there might be a causal relationship between the two, may well be erroneous.

Some brief reference is also made here to findings from the British Household Panel Study (BHPS) (Buck et al. 1994). Here, we have longitudinal data that do show household changes and migration occurring concurrently, although it is a much smaller data set. The BHPS also includes a question on reasons for moving.

Household transitions and migration

A variety of changes in household living arrangements are likely to be accompanied by a geographical move: for example, a child leaving the parental home or a couple separating and at least one partner moving to establish a separate household. Geographical mobility and household transitions are closely linked but difficult to investigate so that there has been relatively limited investigation of this relationship or of its geographical impact (Grundy 1992). Grundy (1985) used 1971 LS data to show the relationship between marriage termination, remarriage and geographical mobility with remarried women having an 'excess' of moves around the time of remarriage. Life-course factors have an important influence on migration (Grundy and Fox 1985; Warnes 1992a) and it has been argued that age variations in migration are largely a reflection of life-cycle stages (Carter and Glick 1970). Consequently, mobility associated with older ages has been the focus of a number of recent studies (for example, Warnes 1992b, 1992c; Warnes and Ford 1995), with specific studies relating household change and geographical mobility among the old (for example, Grundy 1987; Speare and McNally 1992).

If we examine household transitions and geographical mobility for LS members aged less than 60 years in 1991 and also present in 1981 we can see that different transitions have different levels of mobility. Table 11.1 shows that 59 per cent of all household transitions in the decade were associated

Table 11.1 Proportion of movers by household transition category 1981–91, England and Wales

	% movers	*% moving 50 miles+*
Males		
All transitions	63.3	14.1
To living alone 1991	77.1	16.0
No change in household status[1]	50.7	13.2
Stayed living alone	52.8	16.1
Females		
All transitions	55.0	14.4
To living alone 1991	70.7	16.3
No change in household status[1]	52.0	13.2
Stayed living alone	49.4	14.0

Note
1 Excluding those who remained living alone.

with a geographical move, with changes of household status among males showing considerably higher mobility rates than among females (63 and 55 per cent respectively). Household transitions to live alone had the highest rates of mobility, but here the gender difference in rates was much less – 77 per cent for males and 71 per cent for females. Transitions to live alone also had slightly higher proportions of migrants moving 50 miles or more compared with all transitions. Those who remained living alone between 1981 and 1991 had very similar rates of mobility to the total population who remained in the same household type: all around 50 per cent. Slightly higher proportions of males who remained living alone moved 50 miles or more compared with all males who experienced no change in household composition or females who remained living alone.

Examining the characteristics of the household transitions in more detail, the highest rates of geographical mobility are associated with the move from a child in the household to either living alone or in a household with at least one unrelated other person. The significance of the move to live alone emerges clearly if we combine the categories into a broad family/non-family division, grouping those living alone with those living with at least one other unrelated person, as shown in table 11.2. Those transitions to live alone or with unrelated others have the highest rates of mobility with slightly higher rates for males (79 per cent) than females (74.5 per cent). The major difference in these mobility rates can be explained by the much higher rates of mobility of males changing from being a parent/spouse in 1981 to living alone in 1991. Child/others who changed their household status either to live alone or with unrelated others or to become a parent or spouse had the highest rates of geographical mobility – over 80 per cent – reflecting, of

Table 11.2 Proportion of migrants out of all males and females who changed their household status between 1981 and 1991, England and Wales

	Household status in 1991 (%)		
	Alone or with 1+ non-family	*Parent or spouse*	*Child or other in household*
	Proportion of male movers		
Alone or with 1+ non-family	74.6	84.5	75.3
Parent or spouse	71.4	48.7	55.8
Child or other in household	81.8	87.6	39.8
Total	78.8	68.5	41.2
	Proportion of female movers		
Alone or with 1+ non-family	74.0	86.0	74.1
Parent or spouse	59.2	34.9	42.0
Child or other in household	83.4	85.9	33.9
Total	*74.5*	*57.9*	*35.5*

course, the fact that such a change in household status almost inevitably entails leaving the family home.

Household transitions are clearly associated with increased rates of geographical mobility and both males and females with a transition to live alone have the highest rates of geographical mobility. These men and women who have migrated during the decade and are living alone in 1991 are the focus of the chapter. The majority of people living alone in 1991 were not living alone in 1981 (88.2 per cent), although it is worth noting that 42 per cent of those who were living alone in 1981 were still alone in 1991, demonstrating that living alone is not necessarily a transitory or short-term state.

Regional variations in in-migration

Table 11.3 shows that the regions with the highest proportions of migrants (meaning any residential move) between 1981 and 1991, within the total population aged 10–59 years in 1991, were East Anglia, the South West and the Rest of the South East. Both Inner and Outer London had less than the national percentage of migrants in their population. There were higher proportions of migrants among those living alone in all regions. For men, the region with the highest proportions was the Rest of the South East, followed by the South West and East Anglia. For women, the highest proportion was found in East Anglia, followed by the South West and the East Midlands. East Anglia, the South West and Rest of the South East were the regions experiencing the fastest rates of population growth between 1981 and 1991 and they also had the largest increases in the numbers of one-person households in the population (between 43 and 48 per cent). The East Midlands had a lower rate of increase but still above the national level.

Table 11.3 Percentage of migrants by region, England and Wales 1991

Region	*All migrants*						*In-migrants to region*			
	% of migrants within total population		*% of migrants among those living alone*		*% of migrants living alone out of all migrants*		*In-migrants as a percentage of total population*		*% of all in-migrant living alone*	
	M	*F*	*M*	*F*	*M*	*F*	*M*	*F*	*M*	*F*
Inner London	55.9	54.8	72.3	65.8	17.6	15.4	26.8	23.9	20.0	17.2
Outer London	57.5	56.5	71.3	67.4	9.0	7.5	21.3	21.1	10.9	9.0
Rest of South East	63.1	62.8	80.4	69.7	7.6	5.3	15.5	15.3	8.5	5.4
North	52.8	54.4	64.0	62.2	8.7	6.2	5.6	5.8	11.2	10.0
Yorkshire/Humberside	56.2	58.0	72.6	63.9	8.4	5.3	7.7	7.8	9.5	6.0
North West	54.0	53.6	68.3	64.1	8.9	6.0	6.3	5.9	10.8	7.8
East Midlands	57.5	57.9	72.4	71.5	7.3	5.2	13.0	12.5	8.5	6.4
West Midlands	54.2	55.8	69.9	66.9	7.7	5.2	7.4	7.3	10.1	7.1
East Anglia	65.1	66.6	77.9	75.2	6.7	4.9	17.2	18.7	7.3	5.8
South West	65.0	64.9	78.9	74.9	7.7	5.9	17.2	16.8	9.2	6.6
Wales	54.8	55.0	72.2	65.4	7.0	5.1	10.1	9.5	7.5	6.4
England and Wales	*58.2*	*58.5*	*73.2*	*67.7*	*8.3*	*6.0*	*12.6*	*12.4*	*10.1*	*7.5*

However, the region with the highest proportion of migrants living alone was Inner London: 17.6 per cent of all male and 15.4 per cent of all female migrants were living alone in 1991, compared with only 8.3 and 6.0 per cent nationally.

If we look at the proportion of total in-migrants to each region who were living alone in 1991, then the picture is rather different. Fewer women in-migrants were living alone in 1991 than male in-migrants in all regions. Although the gender differential remained for Inner London, the proportion of migrants of both sexes living alone was much higher than for any other region: 20 per cent of male and 17 per cent of female migrants. All other regions have figures within one or two percentage points of the national figure of 10.1 per cent in the case of males and 7.5 per cent in the case of females.

Distance travelled

From table 11.4, those regions attracting well above the national average of long-distance migrants (migrants over 50 miles) were the South West, East Anglia and Inner London, with negligible differences between male and female rates. Slightly higher proportions of migrants living alone in 1991 had moved 50 miles or more compared with all migrants both nationally (16.3 and 14.0 per cent respectively) and within every region. Gender differences were generally small.

Table 11.4 Long-distance migrants (50+ miles), England and Wales

Region	% *all migrants long distance (50+ miles)*		% *of migrants living alone who moved 50+ miles*		% *of all long-distance migrants living alone*	
	M	*F*	*M*	*F*	*M*	*F*
Inner London	21.4	19.7	22.6	20.2	18.7	15.9
Outer London	10.6	10.8	16.2	11.9	13.8	8.3
Rest of South East	14.3	14.5	15.7	15.8	8.4	5.8
North	8.9	10.0	9.4	14.6	9.3	9.1
Yorkshire/Humberside	10.9	10.7	13.6	12.2	10.4	6.0
North-west	8.7	7.9	10.6	10.3	11.3	7.8
East Midlands	14.1	14.4	16.8	17.6	8.7	6.4
West Midlands	10.1	9.9	12.4	12.2	9.5	6.4
East Anglia	21.1	22.1	21.7	23.9	6.9	5.3
South West	25.2	26.1	28.7	27.1	8.8	6.1
Wales	15.9	15.2	15.8	19.6	7.0	6.6
England and Wales	14.0	14.1	16.3	16.2	9.7	6.9

Looking at all migrants over 50 miles as a group, the picture is rather different. Nationally under 10 per cent of male and 7 per cent of female long-distance migrants were living alone. Inner London, however, again emerges as distinctive in having a much higher proportion of long-distance migrants living alone than any other region – nearly 19 per cent of males and 16 per cent of females who had migrated over 50 miles were living alone. For males, Outer London is also significant, with almost 14 per cent of long-distance migrants living alone, but no other region emerges as important as a receiver of long-distance male or female migrants to live alone.

Gender characteristics of migrants living alone in 1991

We have shown that household transitions to live alone have higher rates of mobility but with a smaller gender difference in rates than for all household transitions. Inner London emerges as a distinctive region, with the highest proportion of migrants living alone in 1991 and which was also attracting the highest proportion of long-distance migrants to live alone of any region in England and Wales. In this section the characteristics of males and females who migrated between 1981 and 1991 and were living alone in 1991 are examined and the national and Inner London figures compared.

Age structure

Table 11.5 demonstrates that the age structure for all migrants, both male and female, and for males living alone shows the highest proportions

Table 11.5 Age distribution of all migrants and migrants living alone in 1991, England and Wales

Age	*All migrants (%)*	*All migrants 50+ miles (%)*	*Living alone (%)*	*Migrants 50+ miles living alone in 1991 (%)*
Males				
20–9	30.5	33.5	34.1	45.0
30–9	33.1	31.9	31.2	32.0
40–9	21.6	22.2	20.6	15.0
50–9	14.8	12.4	14.1	8.0
Females				
20–9	33.9	37.3	32.9	45.0
30–9	30.9	30.5	23.0	24.9
40–9	22.0	20.7	19.1	14.1
50–9	13.2	11.5	25.0	16.0

between the ages 20–39 years, with decreasing proportions from ages 40–59 years. Female migrants living alone, however, have an older age structure with over 44 per cent aged 40–59 years and a quarter aged over 50 years. This is reflected further in table 11.6, which shows the proportions of migrants who were living alone in each age group. We can see clear gender differences in the age structure of male and female migrants living alone in 1991. The men showed a younger profile than the women, with modal categories of 20–29 years and 50–59 years, respectively.

All long-distance migrants have a similar age structure to all migrants. By contrast, long-distance migrants living alone in 1991 had a younger age structure than all migrants to live alone: 77 per cent under 40 years for males (compared with 65 per cent under 40 years for all male migrants to live alone) and 70 per cent under 40 years for females (compared with 56 per cent). For such migrants, the gender difference is much less evident: 45 per cent of both males and females migrating 50 miles or more and living alone were aged

Table 11.6 Proportion of migrants who were living alone in 1991 in each age group, England and Wales

Age	*Males*	*Females*
10 – 19	0.6	0.8
20 – 29	11.4	6.8
30 – 39	9.6	5.2
40 – 49	9.8	6.1
50 – 59	9.8	13.4
All	8.5	6.0

Table 11.7 Age distribution of all migrants to Inner London 1981–91: total population and those living alone 1991

	All males		*Males living alone*		*All females*		*Females living alone*	
Age	*Total %*	*50 miles+ (%)*	*Total %*	*50 miles+ (%)*	*Total %*	*50 miles+ (%)*	*Total %*	*50 miles+ (%)*
20–9	50.1	61.6	49.5	62.4	63.4	74.6	58.2	68.7
30–9	29.0	24.5	30.3	23.9	21.0	16.1	21.3	19.8
40–9	12.9	9.0	13.6	9.4	10.1	6.0	13.3	6.3
50–9	8.0	4.9	6.6	4.3	5.5	3.4	7.1	5.2

20–29 years, although there are still higher proportions of women aged 50–59 years in this category (16 per cent compared with 8 per cent of men).

The age distribution of migrants to Inner London is much younger than nationally, with female migrants even younger than males: 63 per cent of all female migrants and 58 per cent of female migrants living alone were aged 20–29 years compared with 50 per cent of both all males and males living alone; overall, 79 per cent of all male and 84 per cent of all female migrants to Inner London were aged 20–39 years (table 11.7). Female migrants living alone had a younger age structure than the men, but a slightly older age structure than all female migrants. For Inner London, therefore, the gender differences in age structure observed for migrants living alone in England and Wales disappear.

The age structure of long-distance migrants to Inner London was even younger than the overall picture: 86 per cent of all male and 91 per cent of all female migrants of 50 miles or more were aged under 40 years (six percentage points more than the figure for all male and female migrants to Inner London). For males and females migrating more than 50 miles and living alone the figures were almost the same: 86 per cent and 89 per cent respectively were aged under 40 years. Gender differences in the age structure for all migrants and migrants to live alone disappear in Inner London, attracting as it does a very specific group of young migrants, including those to live alone, a youthfulness which is particularly pronounced amongst the long-distance migrants.

Social class

A larger proportion of males and females living alone in 1991 had migrated in each social class category compared with the total population (apart from females without an assigned social class – which includes housewives, long-term unemployed and retired people). However, men living alone had consistently higher migration rates than females living alone. This is shown in table 11.8.

Table 11.8 Proportion of migrants 1981–91 by social class category 1991, England and Wales

	Professional/ Managerial/ Technical	*Skilled non-manual*	*Manual*	*Undefined*	*All*
All males	67.8	59.8	53.8	53.6	58.1
All females	68.9	60.4	53.3	53.6	58.4
Males living alone	81.9	77.1	67.3	61.2	73.0
Females living alone	76.2	68.9	61.6	51.5	67.5

Table 11.9 shows that a much higher proportion of males and females living alone in 1991 (each around 39 per cent) were in the professional, managerial and technical class than all migrants (31 and 23 per cent respectively), with the difference between the two groups of females being particularly pronounced. Thus, there is no gender difference for those living alone in proportions in the highest social class, but a difference of nearly ten percentage points between the proportions of all male and female migrants in this social class. By contrast, in both the skilled non-manual and manual social classes there were clear gender differences for both all migrants and migrants living alone, with women dominating skilled non-manual (around 30 per cent both for all women and those women living alone) and men dominating manual occupations (around 42 per cent compared with about 22 per cent for women). Fewer than 10 per cent of both male and female migrants living alone were in the undefined category, whereas much higher proportions of all migrants were so classified.

Some 57 per cent of male and 45 per cent of female migrants to Inner London were concentrated in the professional, technical and managerial social class, with even higher proportions of those migrants who were living alone in 1991: 62 per cent of males and 53 per cent of females. The differ-

Table 11.9 Social class distribution of all migrants and those living alone in 1991, England and Wales and Inner London

	All migrants		*Migrants living alone in 1991*		*All migrants to Inner London*		*Migrants living alone in Inner London*		*All migrants moving 50+ miles to Inner London*		*Migrants living alone moving 50+ miles to Inner London*	
	M	*F*	*M*	*F*	*M*	*F*	*M*	*F*	*M*	*F*	*M*	*F*
Professional/ managerial/ technical	31.3	22.3	39.2	38.7	57.1	44.7	62.2	52.9	55.6	46.8	54.6	57.7
Skilled non-manual	9.5	30.2	11.7	30.6	12.2	29.3	11.3	29.1	13.1	26.8	12.6	22.7
Manual	41.6	22.1	42.3	20.9	19.8	11.6	21.1	11.9	19.7	13.5	25.2	14.4
Undefined	17.6	25.5	6.8	9.8	10.9	14.4	5.5	6.2	11.6	12.9	7.6	5.2

Table 11.10 Social class change 1981–91 for all migrants and those living alone in 1991, England and Wales and Inner London[1]

	All migrants (%)		*Migrants living alone in 1991 (%)*		*All migrants to Inner London (%)*		*Migrants living alone in 1991 in Inner London (%)*	
	M	*F*	*M*	*F*	*M*	*F*	*M*	*F*
Stayed in same social class	79.5	79.8	73.3	71.0	71.6	72.8	70.3	67.4
Up a social class	14.5	13.1	18.7	20.7	22.2	21.1	20.3	27.6
Down a social class	5.9	7.1	8.0	8.3	6.2	6.1	9.4	5.1

ence in proportions between females living alone and all female migrants in this social class is greater than for males. Longer-distance female migrants living alone have even higher proportions in the professional, technical and managerial category (58 per cent), while male long-distance migrants living alone have somewhat smaller proportions in this category (55 per cent). For this group, therefore, gender differences are negligible.

One of the advantages of the LS is that it enables us to see, in very broad terms, how the social class of migrants changed between 1981 and 1991. The majority of migrants remained in the same social class, although there was a proportion who improved their social class, particularly amongst those living alone, with 19 per cent of males and 21 per cent of females going up a social class compared with around 14 per cent of all migrants (table 11.10). The move to live alone, especially for females, would appear to be associated with greater social class changes and perhaps, therefore, lifestyle changes than is total migration. There was little difference in proportions of the various groups of migrants who moved down a social class.

Migrant women living alone in Inner London are even more likely to have raised their social class than migrant men living alone: 28 per cent moved up a social class compared with 20 per cent of men (more of whom moved down a class). Once again, female migrants living alone in Inner London emerge as a particularly distinctive group – with a high proportion moving up a social class, while male migrants living alone are much more akin to all migrants to Inner London.

Housing tenure

Demographic events and processes are closely intertwined with mobility levels and patterns which, in turn, interact with the housing market: for example, American research has shown one-person households more likely to be renters and movers and to live in the central city (Moore and Clark 1990). Housing tenure is a significant variable in developing our understanding of the inter-relationship of migration, gender and household status.

Table 11.11 shows that, although all migrants were less likely to be in

Table 11.11 Housing tenure of total population, all migrants and those living alone in 1991, England and Wales

Tenure in 1991	*Total population (%)*		Total *living alone in 1991 (%)*		*All migrants (%)*		*Migrants living alone in 1991 (%)*		*All migrants 50+ miles (%)*		*Migrants 50+ miles and living alone in 1991 (%)*	
	M	*F*	*M*	*F*	*M*	*F*	*M*	*F*	*M*	*F*	*M*	*F*
Owner occupied	77.3	75.3	59.6	59.5	76.2	73.8	59.5	58.4	74.1	73.9	57.9	60.1
Local authority/ housing association	15.9	18.0	21.3	24.2	14.5	17.0	18.4	21.2	7.8	8.6	9.0	10.4
Rent furnished	4.3	4.4	7.2	8.1	5.3	5.6	7.5	9.3	8.0	8.9	7.8	10.1
Rent unfurnished	2.6	2.3	11.9	8.2	4.0	3.6	14.6	11.1	10.0	8.7	25.4	19.5

owner occupation than the total population, the major difference in tenure is between the total population (including migrants), with high rates of owner occupation, and the population living alone (including migrants), with much higher rates of renting. Gender differences in rates of owner occupation between both groups are insignificant. Migrants living alone, particularly those who had migrated 50 miles or more, were the most likely to be privately renting. Gender differences were small, although men were somewhat more likely to be renting unfurnished and women furnished property.

For in-migrants to Inner London, given in table 11.12, women were slightly less likely to be owner occupiers than men, with females moving 50 miles or more and living alone having the lowest rates of owner occupation and the highest proportion in local authority or housing association property. Male long-distance migrants had the lowest rates of privately renting. Overall, living alone is a much more significant factor in tenure than gender, although there are small gender differences in tenure in Inner London.

If we examine changes in tenure between 1981 and 1991, shown in table 11.13, again it is evident that gender differences are slight, with differences

Table 11.12 Housing tenure of all migrants and migrants living alone in 1991, Inner London

	All male migrants (%)		*Male migrants living alone (%)*		*All female migrants (%)*		*All female migrants living alone (%)*	
Tenure	*Total*	*Total 50+ miles*	*Total*	*Total 50+ miles*	*Total*	*Total 50+ miles*	*Total*	*Total 50+ miles*
Owner occupied	52.2	45.2	52.0	41.2	48.5	40.5	47.6	35.1
Local authority/ housing association	20.8	20.4	18.9	19.3	20.9	21.5	22.0	26.8
Rent furnished	6.5	6.8	6.6	8.4	8.2	7.0	10.6	9.3
Rent unfurnished	20.5	27.6	22.6	31.1	22.5	31.0	19.8	28.9

Table 11.13 Housing tenure change 1981–91, all migrants and migrants living alone in 1991, England and Wales

Tenure change 1981–91	*All migrants (%)*		*Migrants living alone in 1991 (%)*	
	M	*F*	*M*	*F*
No change in tenure	67.4	66.9	58.0	59.2
Owner occupied to local authority/housing association	3.4	4.1	5.4	6.6
Owner occupied to private renting	4.6	4.7	12.1	11.8
Local authority/housing association to owner occupation	13.8	13.5	9.2	7.2
Private renting to local authority/housing association	1.8	2.0	2.5	3.0
Private renting to owner occupation	6.5	6.4	6.6	7.1
Local authority/housing association to private renting	2.4	2.4	6.1	5.0

Source: LS Tables Crown Copyright.

between all migrants and migrants living alone being greater. Fewer migrants living alone remained in the same tenure at both dates. About 18 per cent of migrants living alone moved out of owner occupation to renting, compared with about 8 per cent of all male and female migrants; and around 15 per cent moved into owner occupation compared with 20 per cent of all migrants. These differences, partly at least, reflect household status changes – dependent and non-dependent children moving away from the parental home, for example, may well be moving from owner occupation to renting.

Marital status

The major differences in marital status are between all migrants and migrants living alone in 1991 rather than by gender, as table 11.14 demonstrates. The majority of those migrants living alone in 1991 were single, followed by those

Table 11.14 Distribution of marital status groups for all migrants and migrants living alone, England and Wales and Inner London 1991

Marital status in 1991	*All migrants (%)*		*Migrants living alone in 1991 (%)*		*All migrants (%)*		*Migrants living alone in 1991 (%)*	
	M	*F*	*M*	*F*	*M*	*F*	*M*	*F*
Single	31.6	34.5	63.5	55.2	63.5	64.3	81.4	75.0
Married	61.3	56.0	10.5	9.5	29.8	28.8	5.9	9.4
Divorced	6.7	8.2	24.1	26.7	6.5	6.5	12.4	14.5
Widowed	0.4	1.4	1.9	8.6	0.3	0.5	0.3	1.2

who were divorced. For all migrants, the majority was married, followed by those who were single. The only notable gender difference for any migrant group was amongst those living alone, where a larger proportion of female migrants were widowed – nearly 9 per cent compared with 2 per cent of males living alone.

In-migrants to Inner London living alone are much more likely to be single than the national figure for migrants living alone. There were fewer divorcees among male migrants to Inner London living alone in 1991 compared to all migrants living alone in 1991. Once again, gender differences are less important than the differences between all migrants and migrants living alone.

The varying propensity to migrate between those living alone and the total population is shown in table 11.15 by the proportions of each marital status groups that have migrated during the decade. Single people living alone had much higher migration rates than the total single population, while divorced men living alone had lower migration rates (by ten percentage points) than total divorced male migrants. There was no difference in the proportions of divorced women who migrated. Apart from the single category, women had a lower propensity to migrate than men in each marital status category.

Reasons and attitudes towards migration

The analysis of the characteristics of migrants living alone both nationally and to Inner London shows some gender differences, although more frequently the contrasts are between the migrants living alone and all migrants. Migrants to Inner London emerge as a specific subgroup, while migrants to live alone there are, in many respects, even more distinctive. Unfortunately, detail on motivations and attitudes towards living alone can only be surmised from data such as the LS. Other data sources allow us to explore in a little more depth possible motives for and attitudes towards living alone. The British Household Panel Survey (BHPS) gives some indication of the reasons for migration.

This survey asks people to say whether their move between waves of the BHPS was primarily for employment-related reasons, and to give the first non-employment reason for migration. From table 11.16 we see that a much higher proportion of migrants living alone under pensionable age gave employment as the primary reason for their move – 24 per cent of both males

Table 11.15 Proportion of migrants in each marital status category, England and Wales 1991

	Total migrants (%)		*Living alone migrants (%)*	
Marital status in 1991	*Male*	*Female*	*Male*	*Female*
Single	54.3	58.1	74.9	76.6
Married	67.4	58.5	78.6	70.7
Divorced	80.2	65.8	70.3	65.3
Widowed	48.2	40.7	50.7	40.4

Table 11.16 Reasons for move[1]

	All migrants (%)[2]		*Migrants living alone (%)*[2]	
	M	*F*	*M*	*F*
Gave employment-related reason	15.6	9.9	24.0	23.5
First non-employment reason given:				
Partnership change	19.2	19.6	13.2	8.9
Move to/from family/friend	11.7	8.1	7.9	5.0
Move to/from college	12.4	11.6	26.0	24.5
Job: self/other	5.0	5.5	4.5	3.9
Housing reason (including eviction, larger, smaller or own property)	27.0	29.4	28.9	36.7
Accommodation type	9.2	10.1	7.0	3.3
Environment/area	8.6	9.4	7.0	12.2
Other	7.1	6.3	5.4	5.6
N =	*1880*	*2074*	*242*	*180*

Source: British Household Panel Survey.

Notes
1 Waves 1–4 (1991–94) used. All people who have moved between two consecutive waves have been included. Data include all those who were living alone after migration – not necessarily before.
2 Percentages may not add up to 100 because of rounding.

and females – compared with all migrants. The sample size is too small for Inner London to draw any real inference from the data; none the less, it is interesting to see that 37 per cent of men compared with only 15 per cent of women living alone gave employment as the reason for their move.

Although employment emerges as an important reason for migrating, especially for those living alone, it is not as important as is often assumed. Analysis of the first non-employment reasons given for migrating by all BHPS members and those living alone reveal a number of interesting points. Housing reasons – which include eviction, moving to a larger or smaller property, to buy or to their own property – is the most important category given by everyone, but particularly for females living alone. It seems that it is the purchase or the move to their own accommodation which sets the women living alone apart from both men living alone and all migrants: 22 per cent give this as a reason for their move – double the figure for men and more than three times that of all migrants.

Looking at all men and women and those living alone who give housing as the first non-employment reason for their move separately, there are highly significant differences among them in the type of reasons given. More men than expected gave eviction as a reason, rather fewer than expected men and women gave moving to their own accommodation or buying accommodation as a reason, whereas many more than expected of women living alone gave this as a reason.

The second most important reason listed for all migrants is a partnership change, while for those living alone it is a move to or from college. The third most important reason given by men living alone is a partnership change. For women, it is environmental or area change, and partnership change is fourth. There are significant gender differences in the reasons given for migration both for all migrants (chi-square statistic significant at 0.015) and for those living alone (significant at 0.15).

Tentatively, then, these data suggest that for males and females living alone, although job reasons are important for migration, housing and personal reasons are also important. Relationship breakdown is more important for males living alone, while moving to their own accommodation, buying accommodation and moving to a specific place are more important reasons for migration for women living alone. This can perhaps be summarized by saying that for women the move to live alone is associated to a greater extent with personal choice whereas for men it is the result of a change in job or changing personal circumstances.

Conclusion

Household transitions to live alone are associated with particularly high rates of geographical mobility – slightly more so for males than females. Inner London has the highest proportion of total in-migrants living alone of any region, again with slightly more male than female in-migrants living alone. The age structure of male migrants living alone is similar to that of all migrants, while female migrants living alone have an older age structure. In-migrants living alone in Inner London are much younger than nationally and here there is no gender difference in age structure – women living alone are as young as men. Long-distance migrants are even younger. Large proportions of males and females living alone are in the highest social class with little gender difference, while there is a clear gender difference in the social class structure of all migrants. Migrants to Inner London are even more concentrated in the highest social class with even higher proportions of those living alone in this class, although there is a somewhat greater gender differential. The differential disappears, however, when those living alone have migrated more than 50 miles. There is, though, a much greater contrast between women migrants living alone and all women migrants, compared with men living alone and all male migrants. The move to Inner London is particularly associated with a rise in social class for women living alone and very few women living alone experience a fall in social class. Migrants living alone are much more likely to be privately renting and those to Inner London even more so with only slight gender differences. In-migrants to Inner London living alone are much more likely to be single than divorced with slightly more women than men being divorced.

Migrants living alone do, then, emerge as a distinctive subgroup, although the differences are most apparent among those migrants living alone in Inner London. Here women emerge as a particularly distinctive group, both compared with all migrants, and with women migrants living alone in other regions of the country. We can start to see how young, professional adults are able, and more likely, to migrate to Inner London to obtain better jobs offered by the city, and this is especially true for women. Migration emerges as a key link between professionalization and one-person households with the large amounts of rented property in Inner London providing the necessary flexibility of the housing market to permit high levels of in-migration. People living alone epitomize a professional, independent, mobile lifestyle, facilitated by the nature of the housing and labour markets in cities such as London.

Acknowledgements

The wider project is funded by the ESRC as part of their Population and Household Change Programme (reference L315253011). We should like to thank them, and also the ONS for allowing us to use the Longitudinal Study (LS) and members of the LS Support Programme at the Social Statistics Research Unit (SSRU), City University, for assistance with accessing datA. The views expressed in this publication are not necessarily those of the ONS or SSRU. The LS data presented in this chapter's tables are Crown Copyright.

References

Buck, N., Gershuny, J., Rose, D. and Scott, J. (eds) (1994) *Changing Households. The British Household Panel Survey 1990–1992*, University of Essex: Economic and Social Research Council Research Centre on Micro-social Change.

Cadwallader, M. (1992) *Migration and Residential Mobility: Macro and Micro Approaches*, Madison: University of Wisconsin Press.

Carter, H. and Glick, P. C. (1970) *Marriage and Divorce: a Social and Economic Study*, Cambridge, MA: Harvard University Press.

Dale, A. (1993) 'The OPCS Longitudinal Study', in A. Dale and C. Marsh (eds) *The 1991 Census User's Guide*, London: HMSO, pp. 312–29.

Fielding, A. J. (1989) 'Inter-regional migration and social change: a study of south East England based upon data from the Longitudinal Study', *Transactions of the Institute of British Geographers*, 14: 24–36.

Fielding, A. J. (1993) 'Migration and the metropolis: an empirical and theoretical analysis of inter-regional migration to and from south-east England', *Progress in Planning* 39: 71–166.

Grundy, E. M. D. (1985) 'Divorce, widowhood, remarriage and geographic mobility among women', *Journal of Biosocial Science* 17: 415–35.

Grundy, E. M. D. (1987) 'Household change and migration among the elderly in England and Wales', *Espace, Populations, Sociétés* 1: 109–23.

Grundy, E. M. D. (1992) 'The household dimension in migration', in T. Champion and T. Fielding (eds) *Migration Processes and Patterns. Volume 1. Research Progress and Prospects*, London: Belhaven, pp. 165–74.

Grundy, E. M. D. and Fox A. J. (1985) 'Migration during early married life', *European Journal of Population* 1: 237–63.

Hall, R., Ogden P. E. and Hill, C. (1997) 'The pattern and structure of one-person households in England and Wales and France', *International Journal of Population Geography* 3: 161–81.

Hamnett, C. (1990) 'Migration and residential social change: a longitudinal analysis of migration flows into, out-of and within London, 1971–1981', *Revue de Géographie de Lyon* 65: 155–63.

Hamnett, C. (1994a) 'Social polarization in global cities: theory and evidence', *Urban Studies* 31: 401–24.

Hamnett, C. (1994b) 'Socio-economic change in London: professionalization not polarization', *Built Environment* 20: 192–203.

Harvey, D. (1989) *The Condition of Postmodernity*, Oxford: Blackwell.

Kaufmann, J. -C. (1994) 'Les ménages d'une personne en Europe', *Population* 49: 925–58.

Moore, E. G. and Clark, W. A. V. (1990) 'Housing and households in American cities: structure and change in population mobility, 1974–1982', in D. Myers (ed.) *Housing Demography*, Madison: University of Wisconsin Press, pp. 203–31.

Smith, N. (1996) *Gentrification and the Revanchist City*, London: Pergamon.

Speare A. and McNally, J. (1992) 'The relation of migration and household change among elderly persons', in A. Rogers (ed.) *Elderly Migration and Population Redistribution*, London: Belhaven, pp. 61–76.

Warnes, A. M. (1992a) 'Migration and the life course', in T. Champion and T. Fielding (eds) *Migration Processes and Patterns. Volume 1. Research Progress and Prospects*, London: Belhaven, pp. 175–87.

Warnes, A. M. (1992b) 'Age related variation and temporal change in elderly migration', in A. Rogers (ed.) *Elderly Migration and Population Redistribution: a Comparative Study*, London: Belhaven, pp. 35–55.

Warnes, A. M. (1992c) 'Temporal and spatial patterns of elderly migration', in J. Stillwell, P. Rees and P. Boden (eds) *Migration Processes and Patterns. Volume 2. Population Redistribution in the United Kingdom*, London: Belhaven, pp. 248–70.

Warnes, A. M. and Ford, R. (1995) 'Housing aspirations and migration in later life: developments during the 1980s', *Papers in Regional Science* 74: 361–87.

12 On the journeys of the gentrifiers

Exploring gender, gentrification and migration

Liz Bondi

Introduction

This chapter explores the journeys of men and women who move into urban areas subject to gentrification. I use the metaphor of journey for several reasons. First, it is a metaphor widely used in the narratives people in urbanized western societies construct about their lives, and so I use it to flag up a focus on personal accounts and subjective experiences. Second, in its everyday use, 'journey' suggests movement through time and through space, dimensions of experience I foreground in this chapter. Third, the term 'journey' draws attention to continuing movement, which I wish to evoke in relation to people's current places of residence.

Gentrification necessarily prompts questions about migration: who are the occupants of the new or renovated housing symptomatic of gentrification and where have they come from? Who moves out of neighbourhoods subject to gentrification and where do they go? In the first section of this chapter, I review existing debates and research evidence that address these questions, aspects of which I supplement and deepen later in the chapter. My engagement with these questions is framed by a concern with gender in its widest sense: I am interested not just in whether the people whose journeys I discuss are men or women, but also with what their accounts reveal about the construction of these categories in western urban societies. Therefore, in the second section I review existing debates and research evidence about gender and gentrification, whilst in the third section I outline the case study drawn upon in this chapter, setting the scene for the discussion of interview material presented in the fourth section.

Gentrification and migration: where are we now?

When Ruth Glass (1964) coined the term 'gentrification' in the context of her study of London, she did not address explicitly the question of where the 'invading' middle class incomers to working class neighbourhoods came from. However, through her emphasis on the frustrations of commuting to London, she implied that they had probably moved from the suburbs.

Taking up the observations of a sociologist, and identifying similar processes in many other cities, urban geographers reflected on the implications for theories about the organization and growth of cities (for a full review, see Hamnett 1984). Gentrification appeared to reverse a longstanding demographic decline in inner urban areas, which had been accompanied by steady suburban growth. The resulting challenge to urban theory prompted adaptations to economic models of residential location, which had assumed that the trade-offs between living space and commuting costs would generate residential patterns in which the most affluent would gravitate towards large tracts in suburban locations (Berry 1980, 1985; Evans 1973). This process of reviewing and revising existing models in order to explain why some affluent households chose inner urban locations rather than suburban locations led some commentators to describe gentrification as a 'back-to-the city' movement (Laska and Spain 1980). Combined with a focus on residential locations as expressions of consumer preferences, this implied that gentrification was a manifestation of disillusioned suburbanites choosing to move to inner urban locations.

This interpretation was soon countered, both theoretically and empirically. The focus on consumer preferences emphasized choice and human agency and was therefore questioned and challenged by advocates of theoretical positions that emphasized constraints and social structures. The most obvious example was Neil Smith's hypothesis of a 'rent-gap' (Smith 1979), underpinned by a broader theory of uneven development (Smith 1982, 1984). Commentators such as Smith, Peter Marcuse and others were also deeply critical of depictions of 'urban renaissance' or 'urban revitalization', arguing that gentrification is usually profoundly destructive of pre-existing working class neighbourhoods (Cybriwsky 1978; Deutsche and Ryan 1984; Marcuse 1989; Smith 1986, 1996).

Displacement caused by gentrification is difficult to measure and trace geographically but, directly or indirectly, there can be no doubt that gentrification has increased housing stress and homelessness in many localities (LeGates and Hartmann 1986; Mair 1986; Marcuse 1986, 1989; Palen and London 1984; Smith 1992, 1996). At the same time, empirical evidence from a number of studies indicated that gentrifiers were not, by and large, households moving from suburban residential locations but were people moving from other inner urban addresses (for example, Gale 1980; Hodge 1981; Smith 1979). Several studies have suggested that gentrifiers are, in the main, young professionals who first arrived in inner urban locations as students, struggling artists and so on, often having grown up in suburban neighbourhoods (Gale 1980; Mills 1988; Zukin 1982). Nevertheless, the evidence concerning where gentrifiers come from remains patchy, in the sense that most studies consider the move to a gentrified neighbourhood in terms of the straightforward relocation of a stable household unit, generally consisting of a single young adult or a young couple. This reveals little about the place of gentrification in people's experiences and understandings of their own housing and household trajectories.

A related critique of migration research has pointed to the limitations of behavioural approaches, which assume that discrete migration events can be understood as decisions resulting from the rational evaluation of 'push' and 'pull' factors. Thus, Keith Halfacree and Paul Boyle (1993) argue instead for a perspective that contextualizes migration biographically and that is sensitive to subjective understandings of the moves people make. This kind of approach is illustrated by Forrest and Murie (1987), whose qualitative analysis of the housing histories of a small group of affluent homeowners illustrates rich and complex understandings of the links between the housing and working careers of households headed by well-paid male salary earners.

Forrest and Murie observed that their respondents belonged to a fairly well-defined generation, among whom a particular form of life-course predominates: the majority were born in the 1930s and 1940s and, of seventeen married couples, only one man was in his second marriage and all had at least one child. In contrast, subsequent generations, which are more closely associated with gentrification, display much more diverse life-courses and household forms (for example, Joshi 1990). Consequently, Tony Warnes's (1992) argument that analyses of migration need to acknowledge a wide range of life-course forms and transitions is particularly pertinent in relation to gentrification.

The research I present later in this chapter explores the housing trajectories of gentrifiers by considering how people understand their present housing positions in the context of their routes into gentrified neighbourhoods. In particular, do they see them as the realization of housing ambitions or as staging posts *en route* to somewhere else? And, if the latter, where are they hoping to go? This focus on the place of current residence in a fuller housing career is connected to questions about household membership, household trajectories and life-course. While some residential moves entail the relocation of a household unit without any movement across its own boundary, a large proportion coincide with the creation of new households and the reconstitution of pre-existing ones (see Warnes 1992). These processes are intimately tied up with questions of gender, and it is to this theme that I now turn.

Gentrification and gender: issues of household dynamics

While gentrification is by definition a class process, in that it changes the class composition of the neighbourhoods affected, several commentators have argued that changes in the position of women in the family and in the labour market have been integral to what Damaris Rose describes as the 'production of gentrifiers' (Rose 1984; also see Bondi 1991a; Bridge 1995; Butler and Hamnett 1994; Rose 1989; Smith 1987; Warde 1991). Statistical evidence indicates that inner urban areas in general contain more women than men, many living in poor and disadvantaged households, including lone elderly women and lone mothers with children (Bondi 1991b; Holcomb 1986; Wekerle 1984; Winchester and White 1988). However, there is also some evidence to suggest that women are disproportionately represented amongst gentrifiers in at least some localities (Mills 1988; Rose and Le Bourdais 1986; Rose 1989; Smith 1987; also see Boyle and Halfacree 1995; Duncan 1991).

Moreover, these data, which treat gentrifiers as belonging to one of two mutually exclusive gender categories, reveal only part of the story. The categories 'women' and 'men' are cultural constructions and, as feminist geographers have demonstrated, diverse gender practices and gender representations have been integral to many aspects of urban change (for example, Little, Peake and Richardson 1988; McDowell 1983; Mackenzie and Rose 1983; Nelson 1986). Therefore, the connections between gentrification and constructions of gender also merit attention (see also Bondi 1991a, 1992).

Several studies have identified gay and/or lesbian households as significant agents in the gentrification of particular neighbourhoods (for example, Castells 1983; Knopp 1987, 1990; Lauria and Knopp 1985; Markusen 1980; Rothenberg 1995). Gay men and lesbians challenge dominant constructions of gender in powerful ways, but many other household forms also signal important changes in gender practices. For example, the prominence of dual-career households, professional women living alone or with their children, and other 'non-traditional' household forms in neighbourhoods subject to gentrification, also point towards links with the constitution of gender (Rose 1989; Warde 1991).

Crucial here are processes of household formation and dissolution. Connections between gentrification and household dynamics have received little attention to date but their potential significance is suggested by results from the British Household Panel Survey. This longitudinal survey of approximately 10,000 individuals found that 14 per cent of households to whom these individuals belonged had changed in composition in the two-year period from September 1990 to September 1992 (Buck et al. 1994). Moreover, of the 10 per cent of adults who moved house in this period, 62 per cent reported that these moves were associated with changes in household composition. Below, therefore, I explore decisions about housing moves in relation to the constitution of households, focusing especially on the interpretations offered by the people involved. This approach leads me to consider the interconnections between gender, gentrification and migration in relation to both place and life-course transitions.

Gentrification in Edinburgh: two local area studies

The Edinburgh studies

To explore these decisions I draw on interviews that were conducted in Edinburgh in 1991. During the period between January 1985 and December 1990, the interviewees had bought property in one of two neighbourhoods in the city of Edinburgh.[1] Both neighbourhoods, shown in figure 12.1, had been

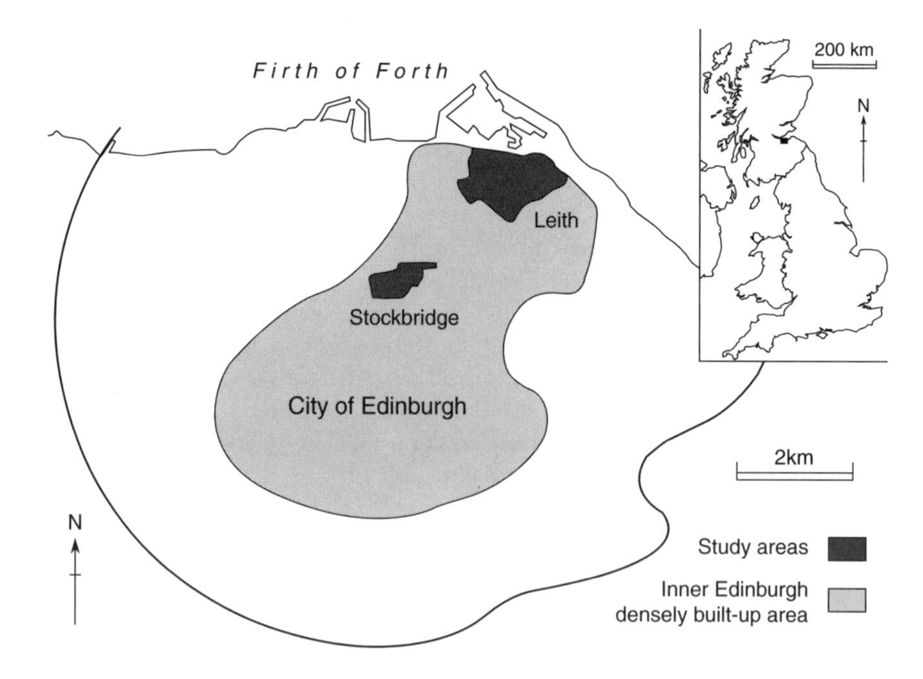

Figure 12.1 Location of the Edinburgh study areas

subject to gentrification, but at different times and taking different forms. In Stockbridge, gentrification had begun in the 1960s, when individual owner-occupiers began to buy and upgrade the largely Victorian housing stock. For the next two decades property continued to come on the market in a condition described by selling agents as 'in need of upgrading'. By the time the interviews were conducted this kind of description had become rather rare and the process of gentrification appeared to be more or less complete. While Stockbridge was being gentrified, the Leith waterfront area had been suffering deep industrial decline, culminating in the closure of the last shipyard in the late 1970s. This decline left a great deal of physical dereliction and was accompanied by considerable out-migration. In 1981, a public-private partnership was launched with the aim of securing local economic regeneration. This provided the context for developers to buy vacant industrial buildings, which they proceeded to convert for residential use, or derelict sites, where they built new residential complexes. This property was sold to owner-occupiers and stimulated the private housing market in the neighbourhood as a whole. By 1991, the neighbourhood was still very much in the process of being gentrified. Moreover, it contains a substantial amount of local authority housing, some of which consists of deeply unpopular system-built deck access flats. Consequently, it looks set to continue to be a very mixed area in terms of its social composition.

Within each of these neighbourhoods, a group of adjacent streets was selected, comprising between 350 and 500 predominantly privately owned housing units.[2] In each case the selection was made using estimates of house prices derived from property advertisements: property in both areas fell within similar price ranges, towards the cheaper end of the market for the whole of Edinburgh. Having selected the streets, all property transactions occurring between January 1985 and December 1990 were identified from the Register of Sasines, which records all changes in property ownership in Scotland. In each area a sample of purchasers was drawn and invited to participate in the project through a single semi-structured interview. No attempt was made to follow up non-respondents.[3] In the end, thirteen interviews were conducted with homeowners in the Stockbridge study area and fourteen with homeowners in the Leith waterfront study area. This small and non-representative sample does not provide a basis for making generalizations about the people moving into neighbourhoods subject to gentrification. However, it provides material through which some of the meanings of particular housing moves may be explored. Before examining some of these accounts, the data collected from the Register of Sasines merits closer scrutiny.

Evidence from the Register of Sasines

The Register of Sasines includes a brief description of the property and its location, the names and addresses of both the seller(s) and the buyer(s), the date on which title changed hands, the price at which it changed hands, and details of mortgages attaching to the property. Entries in the Register appear in date order and are indexed by name and by street (see McCleery 1980; Williams and Twine 1991). The names of purchasers provides some insight into the type of households moving into an area; in particular they provide an indication of the mix of couples and singles, and of men and women. Of course it may be that property purchased in one name is in fact the home of a couple. Equally, forenames cannot always be interpreted with sufficient accuracy to differentiate between women and men. Nevertheless, the list of names indicates broad patterns,[4] shown in table 12.1.

Precise comparisons at national or regional scales are difficult to find, but these data can usefully be compared to information about mortgagees.[5] Using evidence collected by the Council of Mortgage Lenders, Early and Mulholland (1995) estimate that, in 1993, lone women accounted for 17 per cent of all borrowers in the United Kingdom, with lone men accounting for 20 per cent and male and female couples 62 per cent. In both of the Edinburgh study areas, but especially in Stockbridge, the number of single people buying property was much higher than this, and among these, women purchasing property on their own formed a substantial proportion of the total.

It is not possible to differentiate between households any further using the data collected from the Register of Sasines. Thus, there is no evidence of the age of house-purchasers, or of the presence or absence of children, or of the occupations of household members. Nevertheless, this crude categorization

Table 12.1 Property transactions by type of purchasers 1985–90

	Stockbridge		*Leith*	
Heterosexual couples	83	19%	123	30%
Lone men	185	42%	165	40%
Lone women	157	37%	115	27%
Others[1]	12	2%	13	3%
Total	437		416	

Note

1 Including purchasers of the same sex, who might be gay or lesbian couples, but who might also be siblings acquiring property through inheritance.

of household types provides a preliminary indication that these neighbourhoods conform to the image of areas subject to gentrification in the preponderance of single people buying property and to the relative prominence of lone women within this group.

The journeys of gentrifiers

The geographical trajectories of 'typical' gentrifiers

Interview material greatly enriches this preliminary sketch. In both Stockbridge and Leith, two-thirds of the interviewees confirmed the general picture of gentrifiers as young professionals who first experienced inner-city living as students and went on to become owner-occupiers in similar areas. For example, Angela (pseudonyms are used for all interviewees) was brought up in an affluent suburb of Edinburgh and then studied at the University of Edinburgh. As a student she moved into rented accommodation in the centre of the city. After graduating she proceeded to a management traineeship with a prestigious department store, still in Edinburgh. Two years later, at the age of 24, she bought her first home – a small tenement flat in a refurbished block in the Leith study area. When she was interviewed she was 27, working as a buyer for the firm she had first joined as a graduate trainee, and described herself as ambitious:

> To get where I've . . . got today . . . you have to be . . . fairly ruthless . . . I always used to say I wanted my own multi-national by the age of thirty.

While she is well aware of other professional people moving into the area now, she considers that she was something of a pioneer when she moved:

> when I first [moved to the area] I didn't really notice anybody else [from work] who . . . lived down here except for . . . some part-timers or . . . more on the domestic side, the caterers and cleaners whereas

> gradually there's more and more people in the management . . . moving down here.

Others came from similar backgrounds from other cities and, in some cases, other countries: three of the respondents had been brought up in the United States. Geographical trajectories were not always straightforward. For example, David, a 30-year-old American had first come to Britain to study in London:

> and I kept coming up here for holidays and ended up getting married to a Scottish girl and went back to the States for three years and finally came back here where we separated so . . . I'm here and we have a [four year old] son . . .

While the interviewees I have quoted so far first lived in inner urban areas as students, for others their most recent move brought them to live in an inner urban area for the first time. For example, Craig had continued to live with his parents on the outskirts of Edinburgh when he first started higher education. Later he moved into a flat owned by his girlfriend in a middle class Edinburgh suburb, from where he had moved to live on his own in a flat he bought in Stockbridge. Nevertheless, like most of the other young professionals, he had become familiar with the inner urban area as a student.

These accounts serve as a reminder of the rich tapestry of people's lives, which underlies generalizations about gentrification. While the respondents I have quoted so far conform to the description of young professional people already familiar with inner city neighbourhoods, their experiences are far from uniform.

Variations in social trajectories

A few of the interviewees had followed rather different social trajectories from those outlined above and diverged significantly from dominant images of gentrifiers. Mary was very much a marginal homeowner.[6] She did not have a college education but on leaving school trained as a nanny and first left home to work as an *au pair*. Her mother died when she was 20, at which point she returned home. She retrained by taking a typing course and the following year bought a small flat in one of the cheapest parts of inner Edinburgh. This was the first time she had lived in her own home but she chose to buy partly because of the encouragement and practical assistance offered by her father. After struggling to make ends meet, four years later, working as a computer operator, she moved to be closer to her place of work. She found a one-bedroom flat in Stockbridge in need of a considerable amount of renovation. Securing a loan to cover some of the repair costs, she spent the five years after she moved gradually upgrading the flat. Meeting all her bills remains a constant struggle. Although influenced by the location of her workplace and by the availability of cheap housing, Mary's arrival in Stockbridge appears to have had less to do with any *intentional* participation in the process of gentrification and rather more to do with homeownership on a low budget, which restricted housing choice to properties in need of repair. Put another way, Mary's participation in the gentrification process arose through economic necessity rather than as a conscious cultural choice (cf. Rose 1989; Warde 1991).

While Mary came from a homeowning background, several other interviewees were first-generation homeowners. For some, upward mobility came with higher education, but for others this was not the case. For example, Grant had taken voluntary redundancy the previous year from unskilled manual work at a large manufacturing firm in Edinburgh and was now working as a taxi-driver. Brought up in a council flat, until he was married he had always lived in rented accommodation. His wife Irene had left school at 16 years of age and then began to train as a nurse, which brought her from the Scottish Borders to Edinburgh where she stayed in a Nurses' Home. Marriage and homeownership came together, also providing an opportunity for Grant to return to Leith:

> I'm a third generation Leither. From being born round the corner, I moved . . . with my parents . . . I'm quite proud o' bein' . . . [a Leither].

Irene's journey from a working class rural background to the inner city to train for a career in nursing and into homeownership stretches to the limit existing descriptions of routes into gentrification. Grant's certainly lies beyond them. The recent expansion of homeownership in Britain has brought many people from working class backgrounds and in blue-collar employment into owner-occupation for the first time (Forrest, Murie and Williams 1990). Such trajectories are generally considered to be very different from those of gentrifiers. However, Grant's story points to interconnections between gentrification and working class homeownership. Even within the small sample of interviewees in this study, he was not alone. Jim and Judith, interviewed in the Leith study area, were both from working class backgrounds – born and bred in Leith – and had entered homeownership by buying their council house.[7] Two other couples, one in Stockbridge and one in Leith, consisted of a woman in 'pink-collar' work and a man in manual work.[8]

To extend definitions of gentrification to include homeowners like Grant and Irene or Jim and Judith makes little sense given that the term itself refers so clearly and evocatively to a class-based transformation (see Smith 1996).[9] I think it is more appropriate to accept that these households are not themselves gentrifiers but that their house moves have been influenced by gentrification. While many case studies have drawn attention to the displacement of working class people as a more or less direct result of gentrification, these stories point to other possibilities. In Leith, a major displacement of working class people was accomplished long before gentrification began to occur, as a result of the progressive loss of jobs in the area in association with state housing policies, through which many local families moved to council estates in other parts of Edinburgh (cf. Smith 1996: ch. 8). Consequently, gentrification resulted in little if any direct displacement but, in conjunction with a broader shift from public sector housing into owner occupation, set the scene for some working class people to return to Leith.

Household trajectories and constructions of gender

Of the twenty-seven interviews drawn on here, eleven linked their most recent house move to a major change in household membership, demonstrated in table 12.2. Four of these did so in ways that conform to dominant representations of a 'conventional' life course: the new household formed as a result of marriage and entry to homeownership constituted the establishment of the couple's first marital home. Only one of these was a dual-career household; in two the woman was primary child-care provider, and the fourth consisted of Grant and Irene whose story is sketched above. Thus, there is little evidence here with which to elaborate claims that the choice of an inner urban location rather than a suburban one is linked to 'non-traditional' versions of femininity or household forms.

Turning to the six households formed as a result of separation or divorce, none included resident children. In three cases (one divorced woman and two men who had separated from girlfriends with whom they had cohabited), the individuals had no children. A fourth case was David (quoted above) who had a 4-year-old son who stays with him occasionally. The remaining two households formed through separation or divorce consisted of one woman and one man, both of whom were in their fifties and had grown-up children and grandchildren. In both cases, their marriages had broken up after their children had left home. The man had been born and bred in Leith. After working for many years as a bus-driver he had retrained and now worked as a teacher. Throughout his married life he had lived in owner occupation in Leith. When he separated (two years prior to the interview) he bought a flat in one of the new residential complexes in Leith. The woman, by contrast, came from a middle class background. Raised in the South West of England, she had first moved to a middle class suburb of Edinburgh with her Scottish mother after the death of her father. She moved to another affluent Edinburgh suburb when she married. After her marriage broke up she used her share of the proceeds from the sale of the marital home to buy outright a flat in a converted Georgian house in Leith. Her choice was influenced by the fact that one of her daughters already lived in the area. Thus, in both cases, the decision to buy a home in the area had rather more to do with family connections than is usually presumed of gentrifiers. While separation and divorce may be considered to create 'non-traditional' households, these two

Table 12.2 House move and household composition

Setter's query - please supply column header row	*Number*	*Sub-total*	*Total*
No change in composition			16
Single person households		12[1]	
– lone women (first-time buyers)	7		
– lone women (trading-up)	2		
– lone men (first-time buyers)	3		
Couples		4	
– married couple (first-time buyers)	1		
– married couple (trading-up)	1		
– gay couple (first-time buyers)	1		
– house purchased as second home	1		
Changes in household composition			11
Formation of new household on marriage (all first-time buyers)		4	
Dissolution of household on separation/divorce		6	
– women (previously co-owners)	2		
– men (previously co-owners)	3		
– men (previously living in flat owned by girlfriend)	1		
Dissolution of household on parent's death		1	

Note

1 All the first-time buyers had moved into owner occupation from the private rental sector; some had been living alone but several came direct from flats shared with friends.

cases illustrate the significance of family ties associated strongly with the nuclear family form (cf. Finch 1989).

Existing studies of gentrification suggest that 'non-traditional' households, such as lone parents and women living alone, choose inner city residential locations because of the availability of 'supportive services' and the presence of a '"tolerant" ambience' (Rose 1989: 131). To what extent is this view reflected in the personal accounts of gentrifiers in the Edinburgh study areas? I will return to the question of children in due course but, in the absence of lone-parent households with resident children, I focus here on lone women's perceptions of their neighbourhoods.

The class composition of the Leith area is socially mixed and class conflict is evident in several ways, including the use of multiple security devices in the new residential projects and in perceptions of the neighbourhood's reputation. As I demonstrate at greater length elsewhere (Bondi 1998), interviewees in the Leith study area tended to view public space as hostile and as a source of male violence. Thus, there is little evidence of an environment supportive or tolerant of women, whether living in traditional or non-traditional households. Stockbridge was viewed rather differently, and many respondents regarded the area as safer than many other parts of Edinburgh, as 'relaxed' and as slightly 'bohemian'. This may well be linked to the particularly large proportion of lone women among those buying property in the area. Interviewees emphasized the importance of access to amenities such as excellent shops, museums, art galleries and cinemas in their decision to move to the area (Bondi 1998). Consumption of these services is strongly associated with middle class status and does not relate specifically to the needs of women living alone. This suggests that, beyond the numerical prominence of households other than those of a conventional nuclear family variety, there is little evidence to support an association between 'non-traditional' household forms and gentrification (cf. Bridge 1995).

However, the presence of substantial numbers of women owner-occupiers in managerial and professional grades does point to the existence of distinctive versions of femininity among this urban middle class group. This was illustrated by interviewees in several ways. For example, in relation to gender roles, these women expressed strongly egalitarian views, among whom Linda (in a managerial position at an advertising agency) was typical:

> I'm of the opinion that there's no such thing as a woman's role and a man's role.

Career success was associated with a belief that women are at least as capable as men, exemplified by Andrea, who works for a major clearing bank:

> to get to that level takes several years . . . the women are very very good . . . whereas there are a lot of mediocre [men] . . . being promoted.

Angela, quoted earlier, illustrates the ambitiousness of some successful young women. What these viewpoints indicate is a sense of identity in which financial independence through well-paid, satisfying employment is central.

Housing aspirations

So far, I have drawn attention to the diverse geographical, social and household trajectories of those who move into neighbourhoods subject to gentrification, but what of their housing aspirations? Existing studies emphasize gentrifiers' disdain for suburban lifestyles (for example, Mills 1988) and their desire for all that is offered by city living (for example, Caulfield 1989). Certainly, some of those interviewed for the study reported here considered themselves to be intrinsically 'city' people. For example, when asked if he could see himself ever living in the suburbs, Derek, a university lecturer in his early thirties, replied:

> no, not the suburbs of Edinburgh; I would stay in the town centre . . . I just think it's boring [in the suburbs].

Andrea, who had journeyed from a working class background on a council estate in Glasgow via higher education to a position as a computer systems analyst and developer in the financial sector, and owner occupation in Stockbridge, gave a similar reply:

> I wouldn't want to leave the town. I have friends who've bought houses in places like . . . [a suburb of Edinburgh], and I think that's bloody madness.

Others would contemplate moving to ex-urban locations well beyond the city as the only alternative to living in a central location. These respondents associated suburban living with traditional family units and with a risk of isolation for women. For example Andrea said:

> I think it [suburban living] would be a very lonely choice. I mean, obviously I don't 'know', but these days you know that they don't stand around talking over the garden fence all day.

None the less, this kind of response was by no means universal. For example, having lived in the area for three years, Angela (quoted earlier) is ready to move on. Interviewed in October, she planned to put her flat on the market early in the following spring with a view to moving to a larger house with a garden. Getting a dog looms large in her account:

> Somewhere with a garden because I want a dog . . . probably further away from town . . . because that'll be the only place to get a garden, you have to go more to the outskirts.

However she rebels against the association between family lifestyles and suburban living:

> I think the chances of me pushing a pram are fairly minimal it has to be said . . . I find children more of a hindrance than anything . . . [I just want] a garden for the dog . . .

Several other respondents associated a move to the suburbs with marriage or with childrearing, or with children above a particular age. Some discussed this as a distant possibility, with no clear connection to their own lives. For example, Ruth, a design consultant in her mid-twenties living in Leith said that:

> the word 'suburb' fills me with dread,

but acknowledged that:

> there might be a time when I might crave that sort of thing.

For others, the issue was one to which they gave serious thought. For example,

Susan and Jack, both legal professionals in their late twenties and living in a flat in Stockbridge, consider that they would be likely to move next after they start a family. Susan was more specific, suggesting that their current flat would be fine until a child is about 3 years old, after which she would prefer to be in a house with a garden. Finally, Mary (cited above) made it clear that she aspires to a conventional suburban family lifestyle, which she would certainly prefer to her current position as a single woman living in the inner city.

These aspirations suggest that for many young professional people, gentrification is less of a lifetime alternative to suburban lifestyles and more of a staging post on a journey likely to proceed towards parenthood and suburban or ex-urban living. Equally, these aspirations suggest that 'suburbs' may be imbued with multiple meanings (cf. Dyck 1990). What, therefore, of respondents in other kinds of household?

Three households were at later stages in their life courses: they were 'empty-nesters' who had raised families to adulthood. None expressed any thoughts of moving elsewhere. Only three of the households interviewed included dependent children. These were Judith and Jim (referred to above), who were living with two teenage sons in the house in Leith they had bought from the council; Tina and Joe, a couple in their early thirties with a baby daughter, who were living in a flat (their first marital home) in one of the new residential complexes in Leith; and Ian and Noreen, also in their early thirties, who were living with their 3-year-old son in a flat in Stockbridge, also their first marital home. These were all 'conventional' families in the sense that they consisted of nuclear families in which the husbands were in full-time employment and the wives were working part-time (Judith and Noreen) or on maternity leave (Tina). As Jim explains, he and Judith have in mind to move house quite soon:

> when the boys go there's every chance that we'll move into a smaller accommodation. Judith always wanted a house with a garden . . . if we could get a house with a garden in Leith . . . we'd stay in Leith.

Like Judith and Jim, Noreen and Ian would like a house with a garden, preferably still in the Stockbridge neighbourhood, where they now live. In contrast, motherhood has prompted Tina to consider suburban living:

> Two years ago I would have told you I'm very happy to live here and I wouldn't want to move anywhere because it suited our lifestyle and now I've got a baby and life changes . . . I'd rather be with people my own age with families [for example on a suburban estate] . . . I'm very isolated here now that I've got a baby.

In summary, these aspirations are varied but include a strong strand linking current housing choice with a particular phase of life courses associated predominantly with young adulthood. It is also noteworthy that all those from middle class backgrounds and in middle class occupations viewed suburban or ex-urban environments as more appropriate than inner city environments for school-aged children.

Conclusion

The interview material presented in this chapter provides a basis for refining our understandings of the interconnections between gender, gentrification and migration. Being based on a small and non-representative set of interviews, it does not provide insight about general patterns. However, it does illuminate processes, particularly at the level of individual households and it points to some of the limits of generalizations made about gentrification. I draw out four key points.

First, I have illustrated that at least some areas subject to gentrification attract a range of house purchasers, not all of whom are gentrifiers in the sense of contributing to changes in the class composition of an area. Even among those that can unambiguously be described as gentrifiers, there is a good deal of diversity. Alongside those in their twenties and thirties are some who are considerably older. While many households contain no resident, dependent children, some do, including some that conform to the conventional nuclear family form.

Second, while some inner urban areas do appear to hold particular attractions for women, it is not necessarily the case that they are viewed as more tolerant or more supportive of non-traditional household forms. Data from the Register of Sasines suggest a clear link between gender and gentrification in that substantial numbers of lone women and lone men choose to live in gentrified areas. This points to an association between gentrification and alternatives to nuclear family living as well as between women and gentrification. In terms of the cultural construction of gender, it is important to conceptualize gentrification as a class process. The accounts presented here illustrate the presence of financially independent, middle class versions of femininity, within which equality of opportunity is taken for granted and gender difference is played down.

Third, the interpretations offered by respondents in the research reported here suggest complex connections between gentrification as a class-and-gender process, and both place and life course. On the one hand, some neighbourhoods subject to gentrification provide environments the ambience of which is consonant with particular middle class cultural constructions of gender, but other neighbourhoods in which gentrification is also marked do not. Similarly, while many of those interviewed linked their move to a gentrified neighbourhood with a period of young adulthood prior to childrearing or at least to the pre-school phase of childrearing, others illustrated the potential significance of other life-course transitions.

While the latter point is a negative one in terms of explanations of gentrification, it illustrates the importance of qualitative and biographical analyses of gender and gentrification (cf. Halfacree and Boyle 1993). This brings me to my fourth and final point, which is to reaffirm that the metaphor of journey with which I opened this chapter provides a way of framing the complex tapestries of the movements people make through time, space and household membership.

Notes

1 The project included three neighbourhoods, two subject to gentrification and one a suburban neighbourhood. This chapter deals only with the former.

2 The boundaries were constructed to coincide with those of enumeration districts used in the 1981 Population Census. Each study area consisted of four enumeration districts. The 1981 Population Census recorded 471 private households in the Stockbridge study area and 352 in the Leith study area. The Stockbridge area included some mews flats as well as tenement flats. The Leith area included some of the new residential complexes together with some Victorian tenement flats and some subdivided Georgian townhouses.

3 Initially, 30 households were contacted in each area. This yielded insufficient interviews so further sweeps were implemented. Altogether 62 households were contacted in Stockbridge and 84 in Leith. This yielded a total of 14 interviews in Stockbridge and 16 in Leith. One of the Stockbridge households and two of the Leith households turned out to be renters.

4 A partial check was provided through the interviews. In all cases we had categorized the gender of the purchaser(s) correctly. In one case a property purchased in the name of a man only turned out to have been bought as the home of a couple. Other interviews revealed changes in household composition that occurred after the house purchase.

5 Some of the owner-occupiers identified in the Register of Sasines had bought their homes outright. However, the majority of owner-occupiers who own their homes outright have paid off their mortgages *in situ* and/or have inherited from their deceased spouses. This group includes many lone adults, with substantial numbers of lone women. Indeed inheritance is the main route by which women become lone homeowners (Gilroy 1994). This group of people are not house-purchasers. Mortgagees, therefore, provide a more reliable comparator group.

6 Although tape-recording equipment was set up for this interview, the tape did not record. Consequently it is not possible to quote Mary's own words. The account presented here is based on detailed notes written soon after the interview.

7 As explained in the previous section, the streets selected for this study consisted predominantly of private housing. Inevitably, some of the occupants turned out to be renting privately. In addition, one of the streets in the Leith study area included a small low-rise block of flats built by the local authority in the 1970s. The data extracted from the Register of Sasines included any that passed into owner-occupation or changed hands within the private sector between 1985 and 1990.

8 The presence of couples such as these in owner-occupation in the two study areas illustrates Pratt and Hanson's (1988) argument about how gender cuts across class and undercuts the logic of urban models that predict socially homogeneous neighbourhoods.

9 Gentrification has been described as a 'chaotic conception' (for example, Rose 1984; Beauregard 1986). Examples of this kind illustrate one aspect of this 'chaos'.

Acknowledgement

The research reported in this chapter would not have been possible without a great deal of assistance for which I am deeply grateful. I wish especially to thank all the people who so generously volunteered to take part in the interviews drawn on here; Nuala Gormley, who conducted all the interviews and contributed much more besides; and the Economic and Social Research Council, which provided a research grant (reference R000232196).

References

Beauregard, R. (1986) 'The chaos and complexity of gentrification', in N. Smith and P. Williams (eds) *Gentrification of the City*, Boston: Allen and Unwin, pp. 35–55.

Berry, B.J.L. (1980) 'Inner city futures: an American dilemma revisited', *Transactions of the Institute of British Geographers* 5: 1–28.

Berry, B.J.L. (1985) 'Islands of renewal in seas of decay', in P.E. Petersen (ed.) *The New Urban Reality*, Washington DC: Brookings Institute, pp. 69–96.

Bondi, L. (1991a) 'Gender divisions and gentrification: a critique', *Transactions of the Institute of British Geographers* 16: 190–98.

Bondi, L. (1991b) 'Women, gender relations and the inner city', in M. Keith and A. Rogers (eds) *Hollow Promises: Rhetoric and Reality in the Inner City*, London: Mansell, pp. 110–26.

Bondi, L. (1992) 'Gender symbols and urban landscapes', *Progress in Human Geography* 16: 157–70.

Bondi, L. (1998) 'Gender, class and urban space', *Urban Geography* 19: 160–85.

Boyle, P.J. and Halfacree, K.H. (1995) 'Service class migration in England and Wales, 1980–1981: identifying gender-specific mobility patterns', *Regional Studies* 29: 43–57.

Bridge, G. (1995) 'The space for class? On class analysis in the study of gentrification', *Transactions of the Institute of British Geographers* 20: 236–47.

Buck, N., Gershuny, J., Rose, D. and Scott, J. (1994) *Changing Households: the British Household Panel Survey*, Colchester: University of Essex, ESRC Research Centre on Micro-Social Change.

Butler, T. and Hamnett, C. (1994) 'Gentrification, class, and gender: some comments on Warde's "Gentrification as consumption"', *Environment and Planning D: Society and Space* 12: 477–94.

Castells, M. (1983) *The City and the Grassroots*, Berkeley: University of California Press.

Caulfield, J. (1989) 'Gentrification and desire', *Canadian Review of Sociology and Anthropology* 26: 617–32.

Cybriwsky, R.A. (1978) 'Social aspects of neighborhood change', *Annals of the Association of American Geographers* 68: 17–33.

Deutsche, R. and Ryan, C.G. (1984) 'The fine art of gentrification', *October* 31: 91–111.

Duncan, S. (1991) 'The geography of gender divisions of labour in Britain', *Transactions of the Institute of British Geographers* 16: 420–39.

Dyck, I. (1990) 'Space, time and renegotiating motherhood: an exploration of the domestic workplace', *Environment and Planning D: Society and Space* 8: 459–83.

Early, F. and Mulholland, M. (1995) 'Women and mortgages', *Housing Finance* 25: 21–7.

Evans, D.J. (1973) 'A comparative study of urban spatial structures', *Transactions of the Institute of British Geographers*, Special Publication 5: 87–102.

Finch, J. (1989) *Family Obligations and Social Change*, London: Polity.

Forrest, R. and Murie, A. (1987) 'The affluent homeowner: labour-market position and the shaping of housing histories', in N. Thrift and P. Williams (eds) *Class and Space. The Making of Urban Society*, London and New York: Routledge and Kegan Paul, pp. 330–59.

Forrest, R., Murie, A. and Williams, P. (1990) *Home Ownership: Differentiation and Fragmentation*, London: Unwin Hyman.

Gale, D.E. (1980) 'Neighborhood resettlement: Washington DC', in S. Laska and D. Spain (eds) *Back to the City: Issues in Neighborhood Renovation*, New York: Pergamon Press, pp. 95–115.

Gilroy, R. (1994) 'Women and owner occupation in Britain: first the prince, then the palace?', in R. Gilroy and R. Woods (eds) *Housing Women*, London and New York: Routledge, pp. 31–57.

Glass, R. (1964) *London: Aspects of Change*, London: Centre for Urban Studies and MacGibbon and Kee.

Halfacree, K.H. and Boyle, P.J. (1993) 'The challenge facing migration research: the case for a biographical approach', *Progress in Human Geography* 17: 333–48.

Hamnett, C. (1984) 'Gentrification and residential location theory: a review and assessment', in D.T. Herbert and R.J. Johnston (eds) *Geography and the Urban Environment*, London: Wiley, pp. 283–319.

Hodge, D.C. (1981) 'Residential revitalization and displacement in a growth region', *Geographical Review* 71: 188–200.

Holcomb, B. (1986) 'Geography and urban women', *Urban Geography* 7: 448–56.

Joshi, H. (ed.) (1990) *The Changing Population of Britain*, Oxford: Blackwell.

Knopp, L. (1987) 'Social theory, social movements and public policy: recent accomplishments of gay and lesbian movements in Minneapolis', *International Journal of Urban and Regional Research* 11: 243–61.

Knopp, L. (1990) 'Exploiting the rent gap: the theoretical significance of using illegal appraisal schemes to encourage gentrification in New Orleans', *Urban Geography* 11: 48–64.

Laska, S. and Spain, D. (eds) (1980) *Back to the City: Issues in Neighborhood Renovation*, New York: Pergamon Press.

Lauria, M. and Knopp, L. (1985) 'Toward an analysis of the role of gay communities in the urban renaissance', *Urban Geography* 6: 152–69.

LeGates, R. and Hartmann, C. (1986) 'The anatomy of displacement in the United States', in N. Smith and P. Williams (eds) *Gentrification of the City*, Boston: Allen and Unwin, pp. 178–200.

Little, J., Peake, L. and Richardson, P. (eds) (1988) *Women in Cities*, Basingstoke: Macmillan.

McCleery, A. (1980) 'The Register of Sasines as a source of migration data', *British Urban and Regional Information Systems Association Newsletter* 46: 16–17.

McDowell, L. (1983) 'Towards an understanding of the gender division of urban space', *Environment and Planning D: Society and Space* 1: 59–72.

Mackenzie, S. and Rose, D. (1983) 'Industrial change, the domestic economy and home life', in J. Anderson, S. Duncan and R. Hudson (eds) *Redundant Spaces in Cities and Regions*, London: Academic Press, pp. 155–99.

Mair, A. (1986) 'The homeless and the post-industrial city', *Political Geography Quarterly* 5: 351–68.

Marcuse, P. (1986) 'Abandonment, gentrification and displacement: the linkages in New York City', in N. Smith and P. Williams (eds) *Gentrification of the City*, Boston: Allen and Unwin, pp. 153–77.

Marcuse, P. (1989) 'Gentrification, homelessness and the work process: housing markets and labour markets in the quartered city', *Housing Studies* 4: 211–20.

Markusen, A. (1980) 'City spatial structure, women's household work, and national urban policy', *Signs* 5: 23–44.

Mills, C. (1988) '"Life on the upslope": the postmodern landscape of gentrification', *Environment and Planning D: Society and Space* 6: 169–89.

Nelson, K. (1986) 'Labor demand, labor supply and the suburbanization of low wage office work', in A. Scott and M. Storper (eds) *Production, Work and Territory*, London: Allen and Unwin, pp. 149–71.

Palen, J.J. and London, B. (eds) (1984) *Gentrification, Displacement and Neighborhood Revitalization*, Albany: State University of New York Press.

Pratt, G. and Hanson, S. (1988) 'Gender, class and space', *Environment and Planning D: Society and Space* 6: 15–35.

Rose, D. (1984) 'Rethinking gentrification: beyond the uneven development of Marxist urban theory', *Environment and Planning D: Society and Space* 2: 47–74.

Rose, D. (1989) 'A feminist perspective of employment restructuring and gentrification: the case of Montréal', in J. Wolch and M. Dear (eds) *The Power of Geography*, Boston: Unwin Hyman, pp. 118–38.

Rose, D. and LeBourdais, C. (1986) 'The changing conditions of female single parenthood in Montréal's inner city and suburban neighborhoods', *Urban Resources* 3: 45–52.

Rothenberg, T. (1995) '"And she told two friends": lesbians creating urban social space', in D. Bell and G. Valentine (eds) *Mapping Desire*, London and New York: Routledge, pp. 165–81.

Smith, N. (1979) 'Toward a theory of gentrification: a back to the city movement by capital not people', *Journal of the American Planners Association* 45: 538–48.

Smith, N. (1982) 'Gentrification and uneven development', *Economic Geography* 58: 139–55.

Smith, N. (1984) *Uneven Development: Nature, Capital and the Production of Space*, Oxford: Blackwell.

Smith, N. (1986) 'Gentrification, the frontier, and the restructuring of urban space', in N. Smith and P. Williams (eds) *Gentrification of the City*, Boston: Allen and Unwin, pp. 15–34.

Smith, N. (1987) 'Of yuppies and housing: gentrification, social restructuring, and the urban dream', *Environment and Planning D: Society and Space* 5: 151–72.

Smith, N. (1992) 'New city, new frontier: the Lower East Side as Wild West', in M. Sorkin (ed.) *Variations on a Theme Park: the New American City and the End of Public Space*, New York: Hill and Wang, pp. 61–93.

Smith, N. (1996) *The New Urban Frontier*, London and New York: Routledge.

Warde, A. (1991) 'Gentrification as consumption: issues of class and gender', *Environment and Planning D: Society and Space* 9: 223–32.

Warnes, T. (1992) 'Migration and the life course', in T. Champion and T. Fielding (eds) *Migration Processes and Patterns. Volume 1*, London and New York: Belhaven, pp. 175–87.

Wekerle, G. (1984) 'A woman's place is in the city', *Antipode* 16: 11–19.

Williams, N. and Twine, F. (1991) *A Research Guide to the Register of Sasines and the Land Register of Scotland*, Edinburgh: Scottish Homes.

Winchester, H. and White, P. (1988) 'The location of marginalised groups in the inner city', *Environment and Planning D: Society and Space* 6: 37–54.

Zukin, S. (1982) *Loft Living: Culture and Capital in Urban Change*, London: Century Hutchinson.

13 Gender issues in Irish rural out-migration

Catriona Ní Laoire

Introduction

Recent trends in human geography towards recognition of 'neglected geographies' have illuminated the processes by which certain social groups become defined as 'the other' in popular as well as academic discourses. Drawing on literature in the fields of Irish migration, Irish feminist studies and rural studies, it could be argued that the young person, the migrant and the Irish have been variously constructed as 'others' in dominant discourses, and part of this process has involved the silencing and marginalization of these groups. Moreover, this process occurs differentially and is highly gendered, with the subordination of Irishwomen in Britain being related to the marginalization of women in Irish society generally.

In this chapter, I explore the ways in which this gendered subordination occurs in rural out-migration specifically, and the ways in which gender relations intersect with other power relations in the process of life-path formation. In the process, I hope to highlight the importance to migration studies of adopting a biographical approach, which can place migration in its political context and, by integrating quantitative and qualitative methods, can illuminate the choices and constraints which are part of any migration decision-making process.

Research framework

This chapter is based on a research project which focused on life-path formation among Irish rural youth, with the aims of untangling the different power relations which are involved in the decision-making process and exploring the experience of Irish migration among this internally diverse social group (Ní Laoire 1997). The research utilized a biographical approach to explore how individuals decide whether to migrate or to stay, by contextualizing their decision in the long-term time-space context of their lives: 'a specific migration exists as a part of our past, our present and our future; as part of our biography' (Halfacree and Boyle 1993: 337). Therefore, understanding why people decide *not* to move is just as important to migration study as understanding why they do. This type of approach is concerned with the ways in which people negotiate their life paths between the choices and constraints they meet as they move from childhood to adulthood. Giddens's (1984) conceptualization of power as the means of getting things done, of enablement as well as constraint, is useful in understanding how particular groups such as young people and women are marginalized during life-path formation. Of particular relevance for this study is Pred's (1984) concept of authoritative resources, such as those which influence an individual's life chances, through a range of capabilities and aptitudes. The uneven distribution of material, cultural and human resources among even a small cohort of young people is evident in this research, and is strongly related to levels of mobility. The ability to control one's own life path can be seen as the ultimate expression of the power relations in which one is involved. This raises questions regarding the role of power relations in migration. If levels of mobility are closely related to the uneven distribution of resources, are migrants therefore passive victims of dominant processes beyond their control? Or, can migration be seen as a means by which marginalized groups resist domination? The research therefore was concerned specifically with the ways in which Irish rural youth cope with the power relations in which they are involved.

The methodology of the research involved three main strands of investigation within a locality-based study. First, ethnographic methods were used to explore the identities of young people living in the study area in north Cork in the 1990s. Second, a life-history survey attempted to trace, retrospectively, the biographies of the 25–29-year age group from the study area between the late 1970s and early 1990s. Third, in-depth interviews were used to explore the life stories of some of those individuals in more detail. (For reasons of confidentiality, the names of all interviewees have been changed in the text.)

The place

The study area, shown in figure 13.1, is focused on the villages of Boherbue, Kiskeam and Ballydesmond, in the Duhallow area of north Cork, with a combined population of 3,000 people. Young people grew up here in the 1970s to 1990s in the context of rural restructuring and the contradictions of capitalist processes in the areA. Indeed, it is an area affected by agricultural decline, a lack of alternative job opportunities, an over-reliance on the manual sector of the economy, population decline and high youth out-migration. It is also characterized by high levels of education and relatively low unemployment. Young people in the area are usually faced with a choice between local employment in agriculture, nearby factories or shops, the construction industry, or migration. Those who move tend to be those who gain the highest levels of education, who are educated out of local and rural labour markets, while those with low levels of education tend to be restricted to the local area. In a general sense, then, migration is a necessary process given the contradictions and tensions of the local experience of capitalist processes, contradictions between skill levels and employment opportunities, between aspirations and realities.

Gender relations in rural society

Local migration is also of course highly gendered. Other researchers (for example, Stebbing 1984; Little 1987) have highlighted the subordinate role of women in rural society in general, pointing to their restriction to the domestic sphere and the dominance of particular constructions of femininity and masculinity in rural society. The subordinate role of women in Irish society has also been well documented (Beale 1986; Nash 1993; Walter 1995) and, in this context, the marginalization of women in rural Ireland is not surprising. Beale (1986) documents the historically subordinate role of

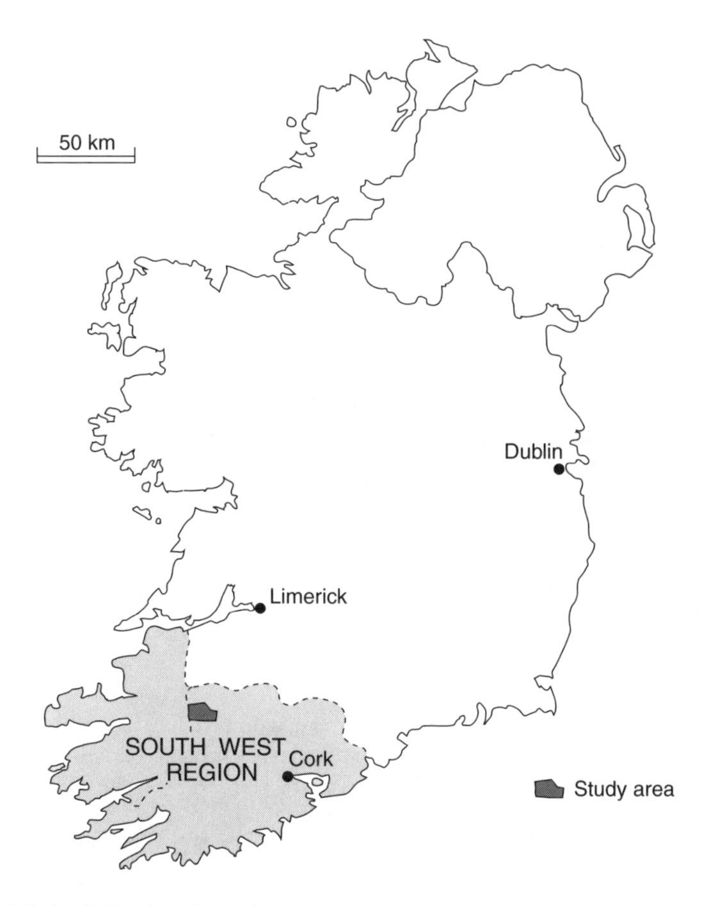

Figure 13.1 Ireland, showing the study area

women in Irish rural society since the last century, which was bound up with strict divisions of labour and male primogeniture. This changed with the modernization of rural Ireland from the 1960s, particularly the onset of rural industrialization and the commercialization of agriculture, but the subordinate position of women did not change, as women were increasingly restricted to the domestic sphere. This marginalization of women is related to the dominant role of the Catholic Church in the Irish state, and to the central position of 'the family' in Irish society, enshrined in the 1937 Constitution (Walter 1995), whereby the woman's role as mother and reproducer is affirmed. Many studies of Irishwomen have found that a prime motivating factor behind the decision to migrate has been the intolerant climate and subordinate position of women in Irish society (for example, Kelly and Nic Giolla Choille 1990).

Writing about young women in rural Norway, Dahlstrom (1996) proposes the idea of a 'male periphery' to signify the way in which a rural peripheral area is dominated by male economic and leisure activities. Male activities, she argues, such as fishing and cruising in cars are visible and highly valued. My research in north Cork, particularly my discussions with 15–16-year-old boys and girls, has revealed a picture of a society in which marginalization of women occurs from a young age. Massey (1994) writes of her recollection of the huge tracts of land around where she grew up which were divided up into football and rugby pitches, and her realization at the time that all this land had been entirely given over to boys. The reality for girls growing up in the study area in the 1990s is not very different. The male-dominance and masculinity of the sport culture is reflected in male control of space. The theme of sport is very much to the forefront for young people in the area and is one that recurs frequently. Young people and adults alike generally accept that sport is the main or perhaps the only recreational outlet available to the youth of the area. The sport culture is therefore a dominant one and those who do not belong to it are to a large extent marginalized. The group of people most obviously excluded is women. The sport culture is a particularly male-dominated one, and the young women of the area consciously feel this:

> The boys have handball, soccer, volleyball and football during the summer . . . The boys get handed [everything] and the girls have to do it themselves.
>
> (15-year-old female student)

Some try to deal with it through having a women's football team but receive very little local support. This is a particularly emotive issue for many young women. The outcome of this is that young women lack the social space that many young men can enjoy through sport.

The issue of sport is especially important because this sport provides a powerful way for rural youth to regain some control in their lives, through establishing a focus of identity. It contributes perhaps more than anything else to a sense of place, and a sense of value and worth in that place. This emphasizes the lack of such a powerful resource for those who are not directly involved. Such a lack can often get translated into a general sense of alienation from local society and from place by many young people, most particularly the young women. This can influence future decisions to leave or to stay through its influence on levels of attachment and belonging to place. It contributes to the perception among women that there is very little for them in that place.

In addition, many young people who do leave the area, both men and women, also speak of the overbearing influence of the Church and the general conservatism and restrictiveness of local society. For example, James (29-year-old migrant, living in London) links the Church with the pub and football as the three cornerstones of life in Kiskeam. The gender connotations of this are obvious.

These dimensions of social marginalization are related, in turn, to the economic marginalization of young women. For boys who leave school early, or want part-time jobs, the main options are farming, construction work or employment in the local joinery. For the girls, local employment opportunities are even more restricted. Some can obtain jobs in the joinery, while summer or part-time jobs tend to be restricted to shop-work or baby-sitting, unless travel outside the area is a possibility. Small Area Census Data show that the proportion of females who do enter the local labour force is increasing, however, rising from 20 per cent in 1981 to 27 per cent in 1991.

While the economic and social structures of rural Ireland are changing all the time, some of the realities of growing up there have changed very little. In the study area in the 1990s, young women are still more likely to leave the area than are young men, continuing the long tradition of female-dominated out-migration from rural Ireland. This is intimately bound up with the subordinate role of women in society, but is also closely related to their economic marginalization. Analysis of the results of the life-history survey, aimed at all 25–29-year-olds from the study area, shows that among the group, females are much more likely than males to have educational qualifications and to migrate from the area. Figures 13.2 and 13.3 show that 48 per cent of males as against 67 per cent of females have third-level qualifications and that 58 per cent of males as against 28 per cent of females are currently living in the study area. The two findings are not unrelated: generally those with the highest qualifications, male or female, tend to move furthest away, while those with none tend to remain in the area. Table 13.1 displays the odds ratios calculated for educational qualifications – the proportion of the sample with a particular educational qualification – in each present location. A ratio above one signifies over-representation. Those with post-primary qualifications only or certificate qualifications are over-represented in the study area and other parts of Cork and Kerry; those with diplomas or nursing qualifications are over-represented in Dublin and Cork; and those with degrees are over-represented in Dublin and other parts of

Ireland. What clearly emerges is a pattern of increasing spatial mobility in accordance with increasing educational qualifications. The proportion of past-pupils with only post-primary qualifications who are now living outside the Cork-Kerry region is very small. Only a small number with Leaving Certificates are now in England. Dahlstrom's (1996) study in northern Norway also found that more young rural women than young rural men are involved in higher education. Other studies (for example, Clancy 1988) have shown that, in Ireland as a whole, farmers' daughters have higher participation rates in higher education than farmers' sons. Given that the majority of the sample in this study are the sons and daughters of farmers, the findings can indicate why there may be a predominance of rural women in higher education.

In the study area, young men who leave school early, or who choose not to enter third-level education have a number of options: they can go into agriculture, manual work or try to obtain an apprenticeship. Young women have fewer options, so they tend to continue in education, which involves moving away and usually moving up the occupational ladder to the non-manual or professional sectors, as shown in figure 13.4. While 32 per cent of both males and females enter the professional sector, the majority of remaining females (44 per cent) are in the non-manual or 'other non-manual' sector, while the majority of remaining males (46 per cent) are in farming or manual occupations. Young women, therefore, tend to be more mobile, both spatially and

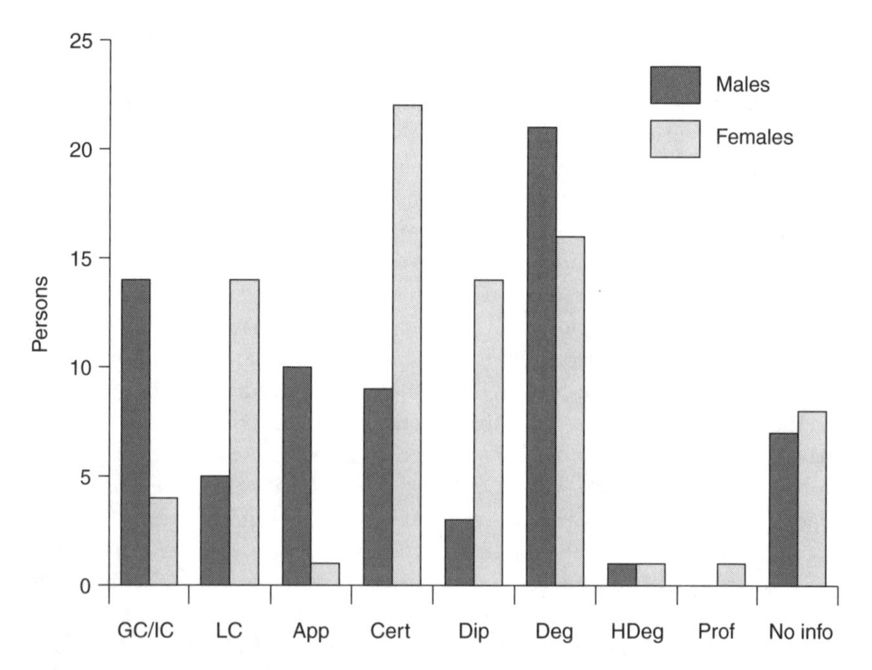

Figure 13.2 Education levels achieved

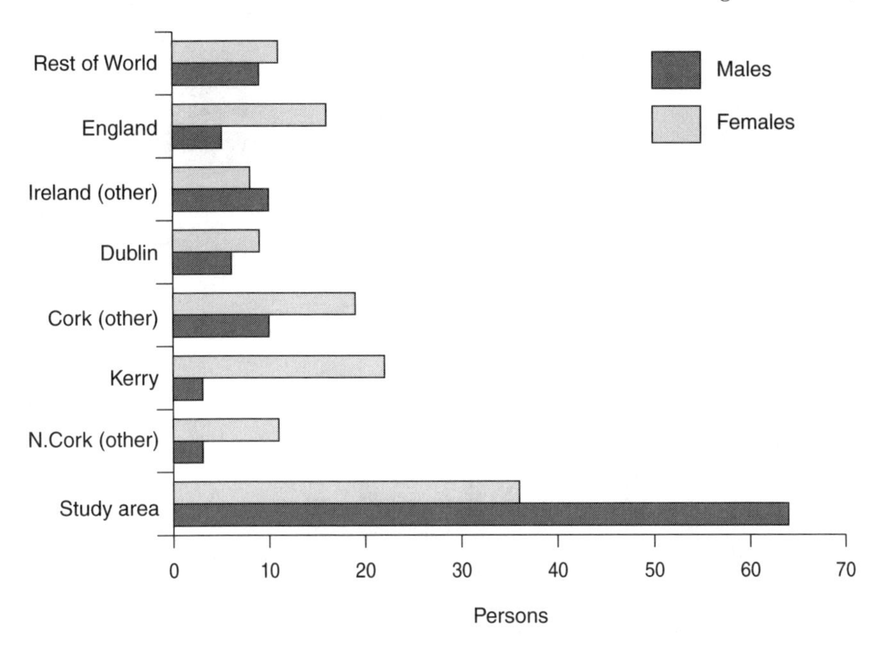

Figure 13.3 Present locations

socially, than young men. Closer examination, however, reveals that young women are more successful educationally than their male counterparts, but only up to a point. Those males who do continue with their education are more likely than females to get degrees and to enter the professional sector. There is also a significant tendency towards a narrow range of courses of study. Women are still very much restricted to the secretarial, nursing and teaching sectors, with nurses and secretaries being among the most mobile groups of all. There is a significant polarization in terms of spatial mobility between the highly

Table 13.1 Odds ratios for educational qualifications in each present location

Location	*Group Certificate/ Intermediate Certificate*	*Leaving Certificate*	*Certificate*	*Diploma*	*Degree*	*Higher Degree*	*Unknown*
Study area	1.50	1.10	1.21	0.44	0.65	1.00	0.25
North Cork	0	2.30	1.21	0	0.40	1.00	1.50
Kerry	1.77	1.20	0.42	0.88	0.95	0	1.29
Cork (other)	0.33	1.10	0.95	2.77	1.05	0	0.58
Dublin	0	0	0.36	2.22	2.00	0	0.87
Ireland (other)	0	0	1.05	0.55	2.00	0	1.12
England	0	0.50	1.00	1.00	0.95	0	1.50
Rest of world	0	0.10	0	0.55	0.25	0	2.70

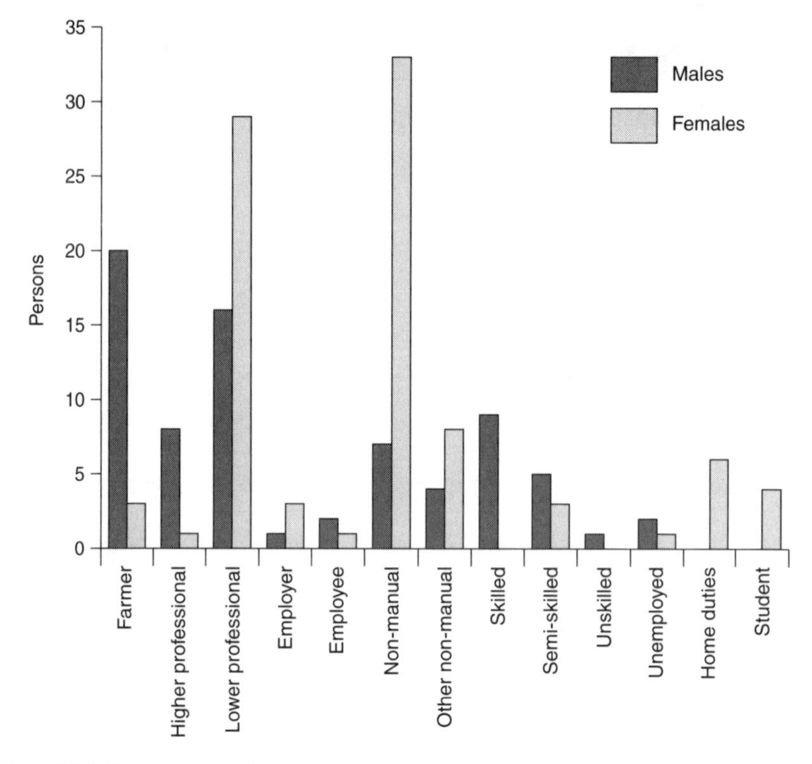

Figure 13.4 Present occupations

educated and those with few educational qualifications, and a predominance of females among the highly educated and the spatially mobile.

Experiences of moving and staying

The in-depth life-history interviews conducted with a small number of individuals from the questionnaire survey have enabled an analysis of the decision-making processes that underlie biography formation. In other words, it is possible to tell some of the stories that lie behind the patterns described in the previous section. The objective is to achieve a greater understanding of the strategies adopted by young people to cope with the circumstances of growing up in rural Ireland. The main aim of each interview was to get below the surface layer of facts and reasons, to the underlying layers of feelings, values and processes. This provides an insight into experiences of migration and staying.

The decision to migrate, although often taken for granted, always carries with it a sense of being a challenge, whether positive or negative. Even a move from the study area to Cork city can be a major upheaval in a young person's life. Emigration, in particular, carries with it the anticipation of something new. Initially, departure can be traumatic, although it is arrival at a new place that generates a more dramatic response among interviewees. Many of them describe in detail the experience of arrival in London, or Dublin, or an American city. The newness of the whole experience, as well as the place itself, is a shock to an individual whose life experiences have been quite different. They speak of city shock and culture shock, of unexpected encounters and experiences. Duffy (1995) writes of the trauma of migration to a large British city from the west of Ireland in the early part of this century. What is evident here is that it is a move that is still traumatic in the 1990s. Homesickness, loneliness and loss are very real emotions among migrants. Helen's story of living in England exemplifies the pain of displacement and of not being in harmony with place:

> I just really longed to be at home, in, I don't know what, I missed everything about it, I missed my family and I missed friends, and it's, as you know, it's a totally different way of life . . .
>
> (Helen, 26-year-old migrant)

However, the experience of staying also carries with it particular stresses and pressures. For many young men and women, the family farm may be a constraint, tying them to the home and the local area, on a daily basis as well as in the long term. Usually this means that they leave the parental home at a later stage than average, if at all. High female out-migration means there is a surplus of young single males in the area and a lack of young married women. The unbalanced sex ratio in the area, especially among the younger age groups (1.21[1] in favour of men for those aged 20–29 years), means that there is a lack of potential marriage partners for young men in the area, resulting in late marriage ages, a high proportion of single males (82 per cent of males in the 20–29-year age group are single, falling to 63 per cent for the 20–34-year group), and the general sense of demoralization which is associated with high out-migration. Together with the tendency towards smaller families, this is contributing to a reduction in birth rates (the 0–4-year age group fell from 319 to 248 between 1981 and 1991). Those who do marry tend to move within the Cork-Kerry-Limerick region to the larger towns, and are less likely to remain in the study area. Those who stay, predominantly single males, but also single females (44 per cent of single females and 65 per cent of single males from the study sample are currently living in the area), are likely to remain in the parental home until migration or marriage. Hannan and Ó Riain (1993) argue that prolonged parental dependence is very distressing if the young person is trapped in the parental home, as in the case of unemployment or return migration. They do not address the case of young farmers, but there is evidence to suggest that they could also be included in this group.[2]

Young men are enabled and encouraged to stay in the area by the economic opportunities and duties of family farming and a male-dominated local labour market. This is bound up with a sense of belonging to a masculine 'pub-and-football' culture. However, this is not to suggest that they are in a privileged position, as their often poor educational qualifications and their restriction to low-paid employment, often temporary or casual, and their spatial confinement combine to suggest a scenario whereby the economic and social marginalization of young men in rural areas is a distinct reality.

Women who migrate, men who wait

What appears to be happening, therefore, is the persistence of a dichotomy between the necessary spatial mobility of young women and the 'spatial entrapment' (England 1993) of many young men. England's research shows that the spatial entrapment thesis is not actually applicable to the everyday lives of female clerical workers living in suburban locations in the United States, as previously assumed. Perhaps it would be more appropriate to apply the thesis to longer-term biography formation among particular groups, such as rural youth with low educational qualifications, or with family ties. I argue, therefore, that the conventional understanding of rural out-migration as 'men who migrate, women who wait' is subverted in the Irish rural context. In reality, it is the women who resist the constraints of rural life, who obtain good educational qualifications and who comprise the larger (though silent or invisible) part of the Irish rural diaspora in Irish cities and worldwide. Although emigration from the state as a whole has been higher for males than for females in the 1980s and 1990s, it has traditionally been dominated by women (Walter 1991). This recent male-dominated emigration also conceals a sharp urban–rural divide, with females dominating the rural exodus, among internal migrants and emigrants. This is reflected in national sex ratios for the 20–29-year age group, at 100.5 for the state as a whole, and 115.5 for the state's aggregate rural areas. It is the men who tend to accept the parameters of life in rural Ireland and stay behind. Young women are more likely to see migration as an empowering escape mechanism and as a means of redefining their identities.

Such a conclusion may seem contradictory, given the conventional association of femininity with home. Massey (1994) argues that patriarchy and capitalism work together to restrict women's mobility and to confine them to the home, constructing home as a woman's place, and men as independent movers:

> Home is where the heart is (if you happen to have the spatial mobility to have left) and where the woman (mother, lover-to-whom-you-will-one-day-return) is also.
>
> (Massey 1994: 180)

This point finds interesting parallels in Nash's (1993) argument that the notion of the rural in Ireland is associated with femininity, nurturing and home. It is also reflected in the following definition of 'home', given by Michael, one of the 25–29-year-olds who had migrated from the study area:

> a nurturing home . . . a cradle to return to if need be . . . some sort of ultimate refuge . . . as distinct from an everyday refuge.

The marginalization of women and girls in Irish rural society can be related to dominant traditional notions of Irishness, constructed as Gaelic, Catholic, masculine and rural (Nash 1993). This masculinization of Irish rural identity is, ironically, partly an outcome of the disappearance of young women from traditional imagery due to high female out-migration as well as the problematic nature of the eroticized representation of young women. They have been replaced in imagery, and largely also in reality, by men and the figure of the woman as mother, as represented in the 'Mother Ireland' image (Nash 1993). In summary, from the earlier part of this century, young women have been leaving rural Ireland, both literally and metaphorically, making their way to the cities of Ireland, Britain and the United States, while their male counterparts stay behind.

Finally, although woman and femininity have been central to the symbolism of Irishness and rurality, women have at the same time been largely absent from power structures outside the home. What is clear is that women in rural Ireland are resisting patriarchal attempts to confine them spatially to the home and to pin down their identities to a particular vision of rurality and of Irishness.

Discourses of home and mobility

In-depth analysis of the discourses that surround Irish migration reveals that they are closely associated with this conventional gendered discourse of Irishness and rurality. Many analyses of Irish migration (Greenslade 1995; Duffy 1995) identify two distinct discourses which help to shape the ways in which we think about Irish migration. The first of these is the traditionalist 'migration-as-exile' discourse. This is often associated with stable and exclusive notions of place and identity, and represents the painful tearing apart of place and self. It has been a common theme in Irish emigration, symbolized in the 'American Wake' of the last century, and reflected in sentimental emigrant songs and ballads. The values that are important in the discourse include family, community and belonging. The influence of traditional ideologies, such as what Miller (1990) terms Catholic nationalism, and community is evident. There is of course an inherent contradiction in this ethos that promotes love of country but cannot promise a place in that country for everyone. Many of the interviewees are aware of this contradiction as they experience its reality:

> the big difference [in my life] was leaving Ireland . . . it was heartbreaking really . . . Not knowing when you're going to go back or what's

> going to happen . . . there's no appreciation for the people coming out of college, y'know . . . they're losing a lot of young people . . . the government are paying for them to go to college, grants and stuff, and they're throwing the money down the drain by leaving them go again.
>
> (James, 29-year-old emigrant)

The emotional heartbreak of James at leaving Ireland co-exists with his angry criticism of a political and economic system that requires the mass emigration of its youth. This apparently innately conservative discourse of migration-as-exile, therefore, can be translated into a radical criticism of the structures that support it. This becomes possible or inevitable by the reality and pain of living out this contradiction.

The experience of migration as exile and hardship, although very real, has also become part of the mythology surrounding emigration. Thomas-Hope (1992), writing about Caribbean migration, suggests that hardship in migration gives rise to a sense of heroism in the imagery surrounding it, and that the successes are valued even more because of the difficulties undergone. This sense of heroism is particularly evident in the language of this male interviewee, significantly a return migrant:

> Good learning experience for any fellow. Make him grow up a bit anyway! I'd recommend it for any fellow, to go away . . .' twould make you grow up when you're away from home and doing your own, you've to do your own thing and paddle your own canoe then.
>
> (Finbarr, 29-year-old return migrant)

In opposition to this exile discourse, which is largely unofficial but very real for many migrants, is the migration-as-opportunity discourse. This is a dominant discourse in government policy and media constructions, particularly with regard to Irish migration in the 1980s and 1990s. In this discourse, migration is constructed as an escape from constraints and represents freedom and individual liberation. Attention is drawn to the successes of Irish emigrants abroad, to the flow of skilled and educated young people availing themselves of opportunities all over the world. The role of this discourse in justifying a pragmatic economic function that maintains equilibrium in an otherwise unstable system is highlighted. Inherent in this discourse is a fundamental contradiction. A transformational and liberationist discourse of migration, which is associated with the rejection of social norms and constraints, serves ultimately to support existing structures. The language of modernization is used to represent migration as progressive, rational and beneficial; as travel rather than migration.

The dominance of the discourse of migration as travel – as progress and modernization – is particularly evident in the language of two of the interviewed migrants: Richard and Michael. Although there are important differences between them, both construct their migrations as a form of escape from the claustrophobia and conservativeness of rural Ireland. Terms such as backwardness, hypocrisy and narrow-mindedness recur in their descriptions of the study area of rural Ireland, which are opposed to positively-charged terms such as change, new, freedom, vibrancy and difference, referring to Dublin or London. Despite the strong criticism of Irish rural society which lies behind this discourse, therefore, it is communicated as part of a neo-liberal modernization ethos which serves to construct migration as a valuable element of the modernization process and thus to maintain existing structures.

These two dominant migration discourses can be seen as polarizations of the reality of migration, and tend to be associated mainly with the language of the male interviewees. Both discourses, although poles apart, reflect masculine constructions of home and mobility. Both are characterized by the conceptualization of home as feminine, rural and 'nurturing', whether in the traditionalist sense of an idealized utopia (the rural idyll), or in the modernist sense of a society that is slightly backward and primitive. Home is constructed as a place either to escape from, or to return to, associated with the 'migrant-as-hero' image, which characterizes both discourses. The migrant is seen therefore as someone who exerts 'his' independence from home by leaving it, or as someone who is a 'hero' because of the hardships undergone through the experience of migration.

McDowell (1996) criticizes the heroic masculine idea of mobility that is a common characteristic of metaphors of travel. She, too, challenges the assumption that it is women who remain at home and men who are the travellers. She points out, on the one hand, that home can also be a site of resistance. This is evident in the study areA. Analysis of the life strategies of young people from the area who choose to stay there or to return to it, shows that it is often as difficult to stay as it is to go. Some individuals, particularly young women, consciously resist the well-worn path of education-and-migration, and instead use their material and human resources to enable themselves to stay in the area successfully and 'make a go of it'. McDowell also argues that mobility as resistance is not a male preserve, and that the associations of mobility with masculinity have served to conceal the experiences of women travellers. Certainly the emergence of a growing body of literature by feminist/post-colonialist critics, such as Trinh (1989), hooks (1990) and Boyce Davies (1994), asserts the role of migration as a form of resistance to patriarchal and imperialist hegemonies. The migrant disrupts hegemonic discourses of identity as place-bound, by embodying mobility, multi-locality and boundary-crossing. Boyce Davies's (1994: 4) writings on black women use 'cross-cultural, transnational, translocal, diasporic perspectives' to redefine identities.

The migration experiences of women tend to get lost in the chasm within the masculinist dualism of 'exile or opportunity'. Talking to young Irish women migrants from the study area, it appears that their interpretations of their own migrations display more clearly the tensions, struggles and contradictions of migration. The women tend to interpret their own migrations in less polarized ways than the men, drawing on both discourses. They emphasize the excitement and opportunity of travelling to other continents above the escape dimension, which is significantly underplayed. Travel is constructed as 'good', in the sense of broadening one's mind, gaining maturity and independence, and the opportunities it offers. Nevertheless, it is usually balanced by a strong attachment to home, the need for family, community and a sense of belonging. The terms that are positively charged include family, community and people. Perhaps the sense of guilt associated with leaving behind home and family is stronger among women than men due to socialization into gendered roles, thus modifying the 'desire to escape' motif where it exists. Among the women, therefore, the tensions between opposing ideas such as home and reach, exile and opportunity are clearly evident in their constructions of migration.

Conclusions

Irish rural out-migration is a highly gendered process, associated with the male dominance of rural labour markets and rural society. One of the outcomes of this is a dichotomy between the high spatial and social mobility of young women, and the 'spatial entrapment' of many young men in rural areas. Both scenarios can involve situations of high stress and trauma. This subverts the conventional associations of migration and masculinity, and of home and femininity. The realities of the lives of young people from rural Ireland challenge these stereotypes, although these realities are generally concealed by traditional migration analyses. The migration experiences of women, and the staying experiences of men, do tend to be surrounded by a 'strange silence', which needs to be addressed. Locating migration between the polarizations of exile and opportunity, in the space of contradictions and struggles and contested identities, can open up a space whereby these silences can begin to be broken, and our understandings of the gender relations which comprise rural societies and migrations can be deepened.

Notes

1 The statistics cited in this section and the next were obtained from the Small Area Census Data for 1991.

2 Indeed, there is a general lack of research on the situations of young farmers in rural Ireland in the 1990s, which is surprising given that they are an important group at the centre of a changing society.

References

Beale, J. (1986) *Women in Ireland: Voices of Change*, Basingstoke: Macmillan Education.

Boyce Davies, C. (1994) *Black Women, Writing and Identity: Migrations of the Subject*, London: Routledge.

Chambers, I. and Curti, L. (eds) (1996) *The Post-colonial Question: Common Skies, Divided Horizons*, London: Routledge.

Clancy, P. (1988) *Who Goes to College? Second National Survey*, Dublin: Higher Education Authority.

Dahlstrom, M. (1996) 'Young women in a male periphery – experiences from the Scandinavian north', *Journal of Rural Studies* 12: 259–71.

Duffy, P. (1995) 'Literary reflections on Irish migration in the nineteenth and twentieth centuries', in R. King, J. Connell and P. White (eds) *Writing Across Worlds: Literature and Migration*, London: Routledge, pp. 20–38.

England, K. (1993) 'Suburban pink collar ghettos: the spatial entrapment of women?', *Annals of the Association of American Geographers* 83: 225–42.

Giddens, A. (1984) *The Constitution of Society*, Cambridge: Polity Press.

Greenslade, L. (1995) '(In)dependence, development and the colonial legacy in contemporary Irish identity', in P. Shirlow (ed.) *Development Ireland: Contemporary Issues*, London: Pluto Press, pp. 94–109.

Halfacree, K. and Boyle. P. (1993) 'The challenge facing migration research: the case for a biographical approach', *Progress in Human Geography* 17: 333–48.

Hannan, D. and Ó Riain, S. (1993) 'Pathways to adulthood in Ireland: causes and consequences of success and failure in transitions among Irish youth', *ESRI General Research Series*, Paper 161.

hooks, b. (1990) 'Talking back', in R. Ferguson, M. Gever, M. T. Trinh and C. West (eds) *Out There: Marginalisation and Contemporary Culture*, New York: New Museum of Contemporary Culture, pp. 337–40.

Kelly, K. and Nic Giolla Choille, T. (1990) *Emigration Matters for Women*, Dublin: Attic Press.

Little, J. (1987) 'Gender relations in rural areas: the importance of women's domestic role', *Journal of Rural Studies* 3: 335–42.

McDowell, L. (1996) 'Off the road: alternative views of rebellion, resistance and "the Beats"', *Transactions of the Institute of British Geographers* 21: 412–19.

Massey, D. (1994) *Space, Place and Gender*, Cambridge: Polity Press.

Miller, K. (1990) 'Emigration, capitalism and ideology in post-Famine Ireland', in R. Kearney (ed.) *Migrations: the Irish at Home and Abroad*, Dublin: Wolfhound Press, pp. 91–108.

Nash, C. (1993) 'Remapping and renaming: new cartographies of identity, gender and landscape in Ireland', *Feminist Review* 44: 39–57.

Ní Laoire, C. (1997) 'Migration, power and identity: life-path formation among Irish rural youth', PhD thesis, Department of Geography, University of Liverpool.

Pred, A. (1984) 'Place as historically-contingent process: structuration and the time-geography of becoming places', *Annals of the Association of American Geographers* 74: 279–97.

Stebbing, S. (1984) 'Women's roles and rural society', in T. Bradley and P. Lowe (eds) *Locality and Rurality*, Norwich: Geo Books, pp. 199–208.

Thomas-Hope, E. (1992) *Explanation in Caribbean Migration*, London: Macmillan Press.

Trinh, Minh-ha (1989) *Woman, Native, Other: Writing Postcoloniality and Feminism*, Bloomington: Indiana University Press.

Walter, B. (1991) 'Gender and recent Irish migration to Britain', in R. King (ed.) *Contemporary Irish Migration*, Dublin: Geographical Society of Ireland, pp. 11–20.

Walter, B. (1995) 'Irishness, gender and place', *Environment and Planning D: Society and Space* 13: 35–50.

14 Gender relations and identities in the colonization of 'Middle England'

Martin Phillips

Introduction: gender relations, migration and rural colonization

While gender relations are becoming an increasing focus of attention in the social sciences, there has been relatively little attention paid to their role in migration, particularly in the so-called 'developed world' (cf. Chant 1992). As Boyle and Halfacree (1995: 44) comment:

> Whilst there is often a recognition of gender differences in migration data sets, such as in propensity to migrate . . . we have yet to highlight these differences in migration research.

This uneven development in concern over gender relations is very much mirrored in the study of rural migration, counterurbanization and rural colonization that form the principal focus of this chapter. For example, in rural studies generally there has been a growing interest and concern with gender issues, witnessed by the work of people such as Brandth (1994, 1995), Hughes (1997), Little (1986, 1987, 1991, 1997; Little and Austin 1996; Little, Ross and Collins 1991) and Whatmore (1991). There has, however, been little substantive research on the role of gender in rural migration and counterurbanization. Gender relations are a notable omission from Champion's (1989) discussion of alternative explanations of counterurbanization, for example.

This neglect has also appeared in studies which have adopted a more 'critical' focus (Phillips 1994), examining the power relations which are bound up in the ability to move into the countryside and in the practices adopted by those living in the countryside. One illustrative example is the notion of 'service class led rural restructuring'. Here, it has been argued that a powerful social group – the service class – gains preferential access to the countryside and actively moulds the physical and social fabric of areas in which its members live (see Cloke and Thrift 1987, 1990; Thrift 1987). Whilst notions of class colonization and restructuring are valuable, it is possible to argue that much of this work has been rather gender-blind, not least in that it tends to speak in genderless terms: it talks about rising numbers of service class 'people' and the migration of service class 'members' into the countryside. Agg and Phillips (1998) have suggested that there are at least three lines of objection which can be raised about the neglect of gender in the study of rural colonization and restructuring:

- Empirical work has highlighted clear gender differences within the service class in terms of employment and consumption power.
- A series of theoretical studies have suggested that gender relations are themselves key components of class formation.
- A series of migration studies have emphasized that many people move not as individuals but as households, and that household structures are changing, in part due to changes in gender identities and relations.

The present work explores these issues in the context of research conducted in five villages in Leicestershire and Warwickshire. It also considers the role that gendered identities and cultural constructions of the countryside may play in migration processes and how migration may be a constituent of the countryside as a 'hegemonically masculinist ordered space'.

Gendered class divisions in rural society

There is in rural studies at present something of a prevailing orthodoxy that most of the people living in villages are middle class, and indeed maybe even predominantly service class. Some studies (for example, Cloke, Phillips and Thrift 1995, 1998; Hoggart 1997; Phillips 1993, 1998a, 1998b; Philo 1992) have in a variety of ways come to question this. In a survey of studies of class analysis of rural space, Murdoch (1995: 1226) argued that 'a description of the rural as "middle class territory" may be overdrawn to capture the range of social practices and axes of collective action', while Murdoch and Marsden (1994: 17) have called for those engaged in rural class analysis to recognize the way 'the processes of class formation may be part and parcel of activities which ostensibly have no class complexion', and that class formation is economically, politically and culturally constituted. More generally, Cloke, Phillips and Thrift (1995) have noted change in the theoretical representation of class, such that it has become 'the convention' to see social divisions and relations such as gender, race, sexuality and ethnicity as not simply 'fractionalizing' class but crucially involved in the very formation of class.

This convention was the perspective adopted in a study of social change in Leicestershire and Warwickshire which sought to explore the interconnections between, amongst other things, gender relations and class formation. The research involved a questionnaire survey, interviews and group discussions in five villages. A major focus of the questionnaire survey was the class composition of these villages. To investigate this, questions relating to a

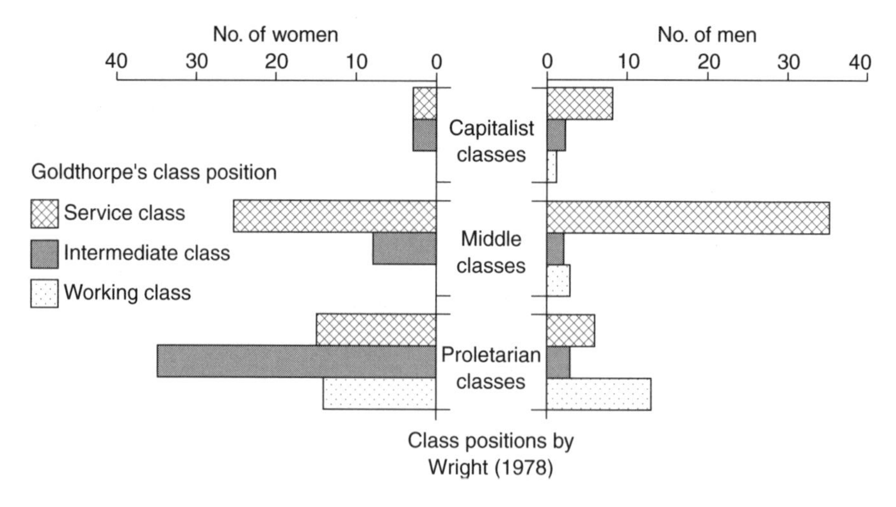

Figure 14.1 Class classification and gender – Leicestershire and Warwickshire villages

range of class classifications were employed, including the 'relations of production'-centred classifications of Erik Ohlin Wright (1978, 1979) and the 'work and market life-chance' perspective of John Goldthorpe (Goldthorpe, Llewellyn and Payne 1980). Analysis of the responses revealed marked gender differences in the way these classifications cross-cut each other, shown in figure 14.1. In particular, while Wright's middle classes showed a strong female and 'service class' presence, his proletarianized classes were predominately female and also include a strong 'service class' presence, particularly when one considers that many people in Goldthorpe's 'intermediate class' are retail and personal service workers.

These patternings may point to important differences within what is commonly taken to be the service class (see Phillips 1998b). Esping-Andersen (1993) has criticized theorists of the service class, such as Gouldner (1979) and Goldthorpe (1982), for focusing on features which are taken to unite – both conceptually and to some degree practically, through providing some basis for collective action – the elements of the service class. Esping-Andersen argues that while this emphasis on relationships such as autonomy, human capital assets and trust relations is 'clearly important', there are also other attributes which serve to differentiate members of the so-called service class. He suggests that there are significant differences within what has been called the service class between professional workers and managers (see also Savage et al. 1992). Moreover, differentiation may also be significant amongst 'the less-exalted occupational groups' (Esping-Andersen 1993: 26). In particular, Esping-Andersen suggests that an important 'new class' in contemporary 'post-industrial'/'post-Fordist' economies is a 'service proletariat' consisting of uncredentialled workers in consumer services. While these workers share with the industrial proletariat low wages and a lack of formal skills, their social situation is quite distinct:

> an unskilled factory worker and a fast food counter worker occupy two distinct worlds of work; the former operates machines in subordination to a managerial hierarchy with a relatively clear productivity-reward nexus; the latter services person in a setting with blurred hierarchies, usually a fair degree of autonomy and discretion, and only a vague link between productivity and work.
>
> (Esping-Andersen 1993: 14)

Esping-Andersen argues that contemporary western societies are structured through two distinct divisions of labour and class positions: a 'Fordist' occupational structure involving a bureaucratic/managerial hierarchy and a 'post-Fordist' structure based on a looser command structure. Furthermore, he suggests that these divisions of labour are differentially gendered. The Fordist division is built around a patriarchal division of labour, in which

> families were both able to and encouraged to split their productive efforts between the male's full-time industrial wage employment and the female's full-time dedication to household reproduction.
>
> (Esping-Andersen 1993: 17)

By contrast, the 'revolutionary essence' of the post-Fordist division of labour is the way it has dissolved the patriarchal gender logic with female and male 'life-cycle profiles' increasingly converging and many of the new 'job slots' being feminized and filled by women. While many of these jobs are professional and managerial in character, many others are in the 'service proletariat', a point which has also been made by Crompton (Crompton 1986; Crompton and Sanderson 1990), who has suggested that there is a distinctly feminized class or class fraction which acts as 'servicers of the service class'.

On the basis of such arguments, it is possible to suggest that the people who appeared in Goldthorpe's service and intermediate classes and in Wright's proletarianized classes may be placed in a 'service proletariat class' (see also Cloke, Phillips and Thrift forthcoming; Phillips 1998b). Such a class constitutes a sizeable element of the population of the study villages, shown in figure 14.2, being almost as large as the service class of professionals and managers which has been the subject of such recent attention.

The two service classes have important lines of difference, including discernibly different workplace situations and gender compositions. For example, while the 'service class' is relatively balanced in terms of the number of men and women, the composition of the 'service proletariat' in the five villages was heavily skewed towards women. Furthermore, these gender differences appear to extend beyond the people present in the sample when respondents were asked to indicate the gender of their work colleagues. As figure 14.3 indicates, it appears that a large proportion of the people classified as working as service proletarians have predominantly women working alongside them. By contrast, under one-third of those classified as working in the

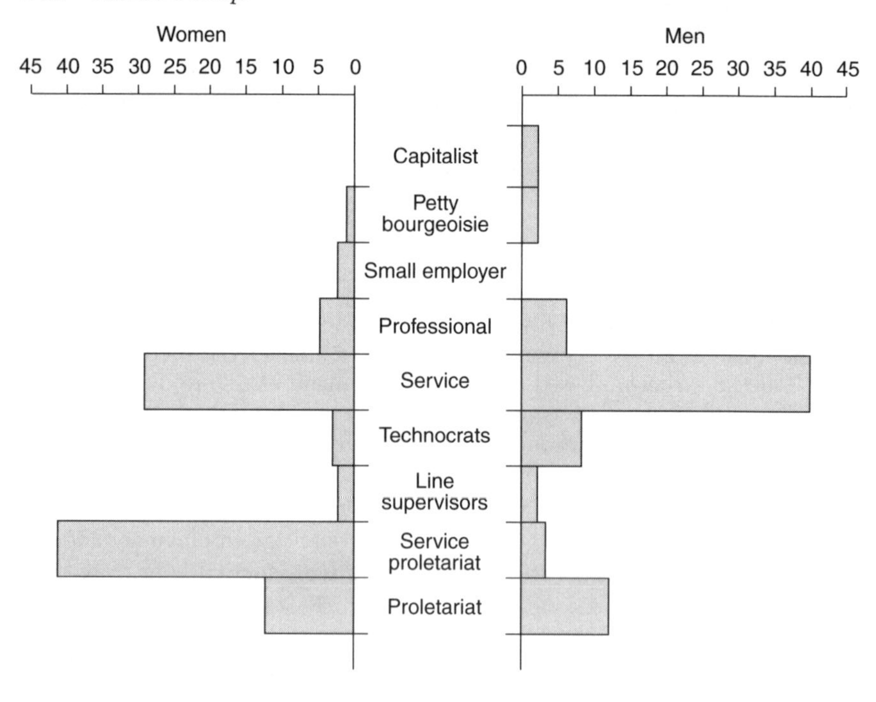

Figure 14.2 Gender and class in rural Leicestershire and Warwickshire

service class have a largely female peer group in their place of work. Rather unsurprisingly, the majority of people in both classes had men as their superiors, although the male dominance was more marked in the service class, reflecting perhaps the degree to which women in the service proletariat have other women acting as line-supervisors or managers above them.

There were clear differences in attitudes to remunerative work between the two groups. For example, only 30 per cent of the service proletariat felt that their occupation had a career structure, while some 67 per cent of the service class felt this way. Furthermore, as table 14.1 demonstrates, a relatively high proportion of female members of the service proletariat viewed remunerative work as being of little importance to them or only of financial relevance. This suggests that they may have only marginal connection with the official market economy, a point that was supported when attention was paid to the current employment situation of the people interviewed.[1] As figure 14.4 shows, there were a large number of women in the service proletariat who were either working part-time or who had at the time of the interview left 'remunerative employment' and were working at minding the house and family. The low level of engagement by women in the service proletariat with the 'official' money economy was further highlighted when a comparison was made between the proportion of household incomes earned by women in the service proletariat as against that earned by men in the service class. While

Table 14.1 Attitudes to working in the official economy

Attitude to work	*Illustrative comments*	*Proportion making statements (%)*					
		Service class		*Service proletariat*		*General*	
		Men	*Women*	*Men*	*Women*	*Men*	*Women*
Enjoyable	Job satisfaction – enjoyed it, otherwise wouldn't have stayed for 22 years; always thought my work was important and enjoyed it; very important, I enjoyed it; enjoyable plus extra money for the children; oh I loved it; I liked to get it right; very important, gives lot of satisfaction; very important career, very interesting.	2.8	2.8	0	11.4	3.5	8.1
Life-centring	It's my life as well as my job; very committed to it, want to do it for the rest of my life; it keeps me sane; it's a big part of life; very important, gives you a sense of value.	4.2	1.4	0	2.3	4.1	2.3
Socially important	Meeting people; gets me out of the house; get out of house with girls; enjoyed, had good friends.	0	0	0	6.8	0	2.3
Personal independence	Personal space quite important; work is very important, interdependence as well for the house.	0	0	0	2.3	0	1.2
Important for other reason	Important as family business – but didn't like job; keeps us here.	1.4	0	0	0	0.6	0.6
Important (reason unspecified)		36.6	23.9	4.5	31.8	23.3	21.1
Financial pragmatism	Most important is the money; financial not vocation; work pays the mortgage; work is a means of existing; money is very important, work hard before have children; work to get paid; work to provide a lifestyle; no job satisfaction, did it entirely for the money.	9.9	4.2	0	11.4	7.6	6.4
Not of central importance	Enjoy it, but not the be all and end all; quite important; fairly important; it's something to do; bored without it	1.4	5.6	0	11.4	2.3	6.4
Unimportant	Not very important; not particularly important; secondary now we've got children: less important – kids etc.	2.8	0	2.3	15.9	2.3	6.4

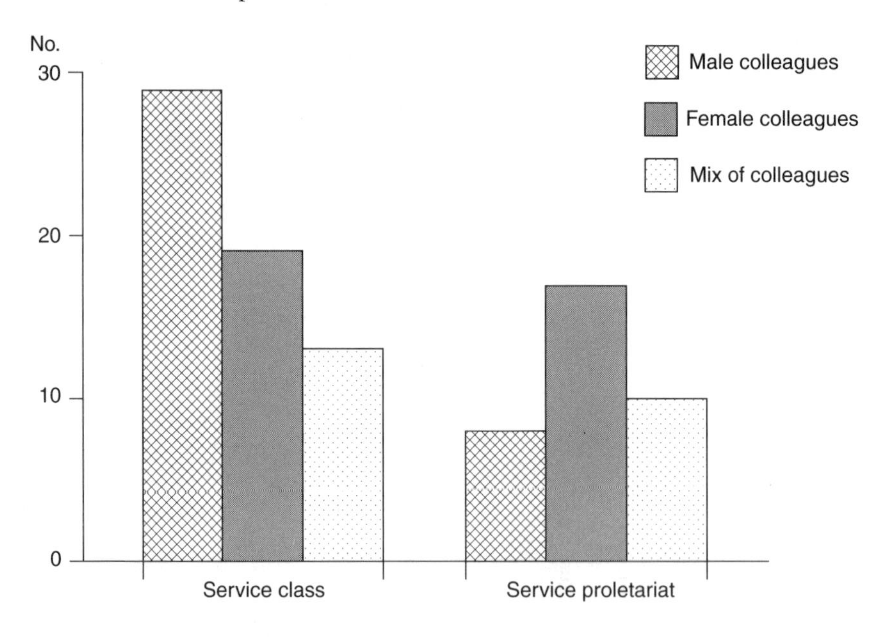

Figure 14.3 The gender of work colleagues

almost 54 per cent of the men in service class occupations earned over three-quarters of total household income, almost 69 per cent of the women service proletariat earned under a quarter of the household income.

One consequence of these findings is to suggest that the characterization of the rural population as being an affluent 'middle class territory' needs to be treated with some caution. On the one hand, it seems to be the case from this research that there are a large number of men and women living in rural Middle England and working in middle/service class jobs, many of whom seem to be able to earn enough money to support a household in a high degree of comfort. However, on the other hand, there also appears to be, at least within the villages studied, and possibly beyond (see Phillips 1998b), a large presence of women working, often part-time or on a temporary basis, in a feminized service proletariat with generally poor career prospects and wages insufficient to support a household. One might also add that Wright's classification, in particular, suggests the presence of a range of classes and class fractions which fall outside the service classes altogether, including a bourgeoisie, a petite bourgeoisie and a traditional proletariat. However, for the purposes of this chapter, attention will continue to focus of the issue of gender so starkly highlighted by distinguishing between a service class and a service proletariat.

The gender order of rural Middle England

So far the issue of gender has been discussed solely in relation to class. It is,

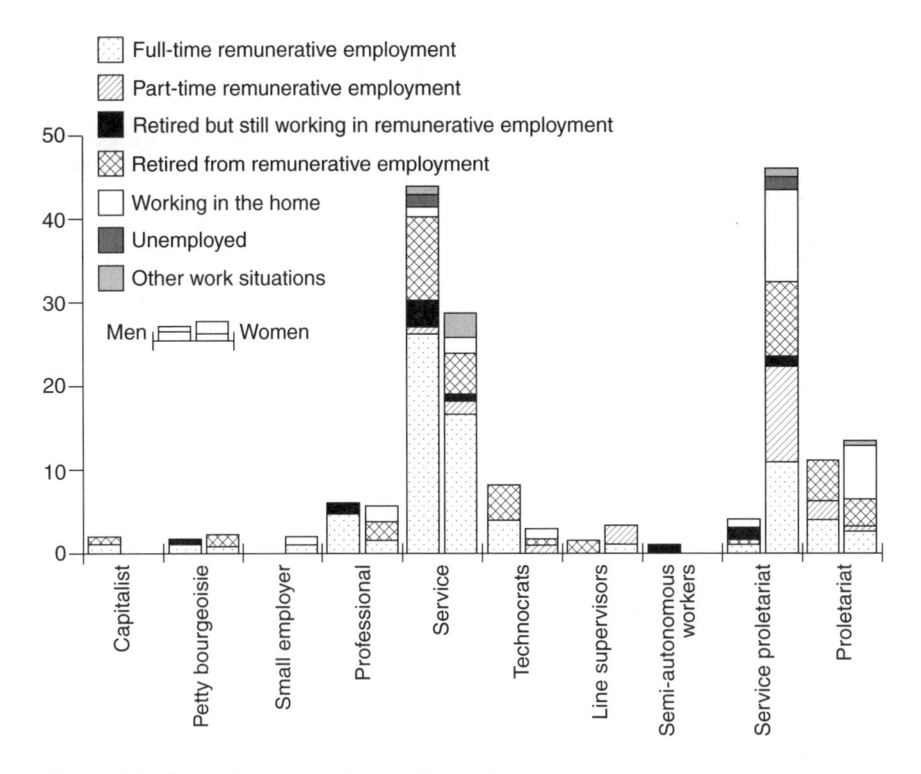

Figure 14.4 The work situation of classes by gender

however, important to recognize that gendering of social life extends far beyond what might reasonably be considered to constitute class relations. Connell (1987), for example, has argued that gender relations can be seen to involve relations of power, divisions of labour and sexually charged relations between people, together with a sexual politics in which people try to order the other three elements of what he terms the 'gender order'. Connell argues that the generally prevailing gender order is a patriarchal or 'hegemonically masculinist' one in which there is:

- a masculinized power regime in social and economic institutions;
- a gendered separation of reproductive labour from the money economy and the political world;
- institutionalized heterosexuality of 'a highly masculine form';
- a sexual politics in which men dominate women.

Connell suggests, however, that the various elements of this gender order have their own 'crisis tendencies' and may be subject to change, albeit at present only within a particular social and geographical milieu, namely within 'the younger intelligentsia of large Western cities' (1987: 163).

Connell effectively argues for the recognition of various 'gender orders' constructed through an ensemble of elements which come together through complex and often difficult processes of 'composition'.[2] His suggestion that certain gender orders may be located in particular localities, although made almost as a throwaway remark, is an interesting one. It raises, for instance, the possibility that there may be something of a 'rural gender order' (Agg and Phillips 1998) and, in the course of studying the five villages in Middle England, there were certainly some features which seem to fit in well with Connell's notion of a 'patriarchal' or 'hegemonically masculine gender order'.

One of the features of this gender order has already been identified as being present in the Middle England villages studied, namely the patriarchal gendering of divisions of labour, work situations and remuneration in the official economy. However, in addition to the gendering of class positions in the official economy, there was also evidence of stark, even stereotypical, gender differences in the performance of reproductive labour. As figures 14.5 and 14.6 show, women in our sample appeared to be doing most cooking and ironing, while men were more likely to change the plug and deal with the bank. It was

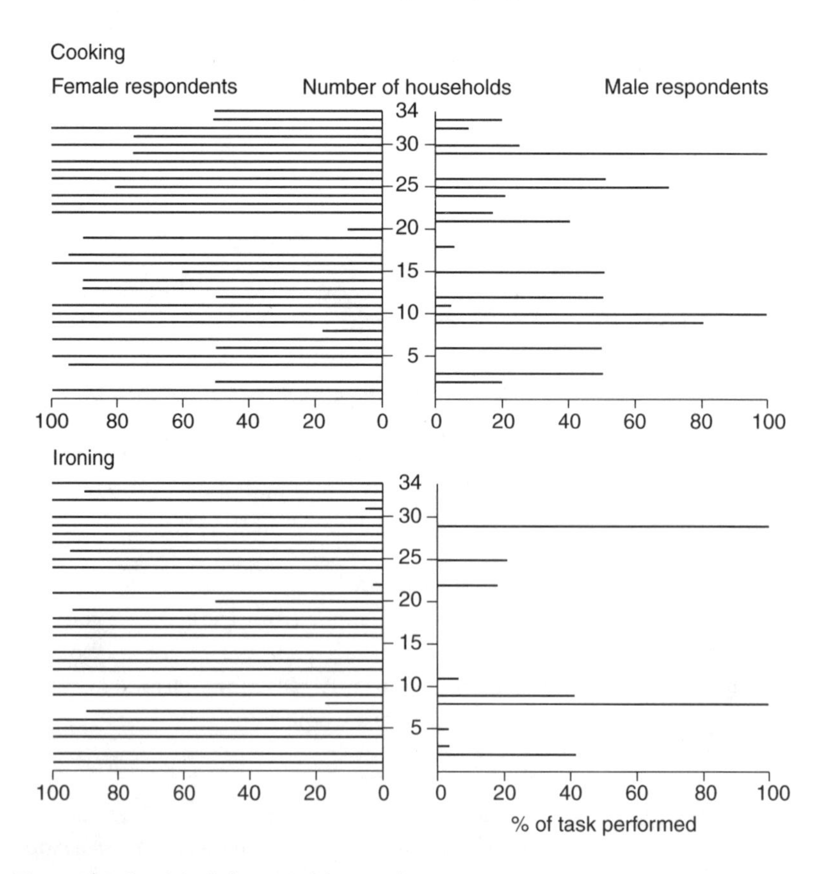

Figure 14.5 Feminized domestic labour tasks

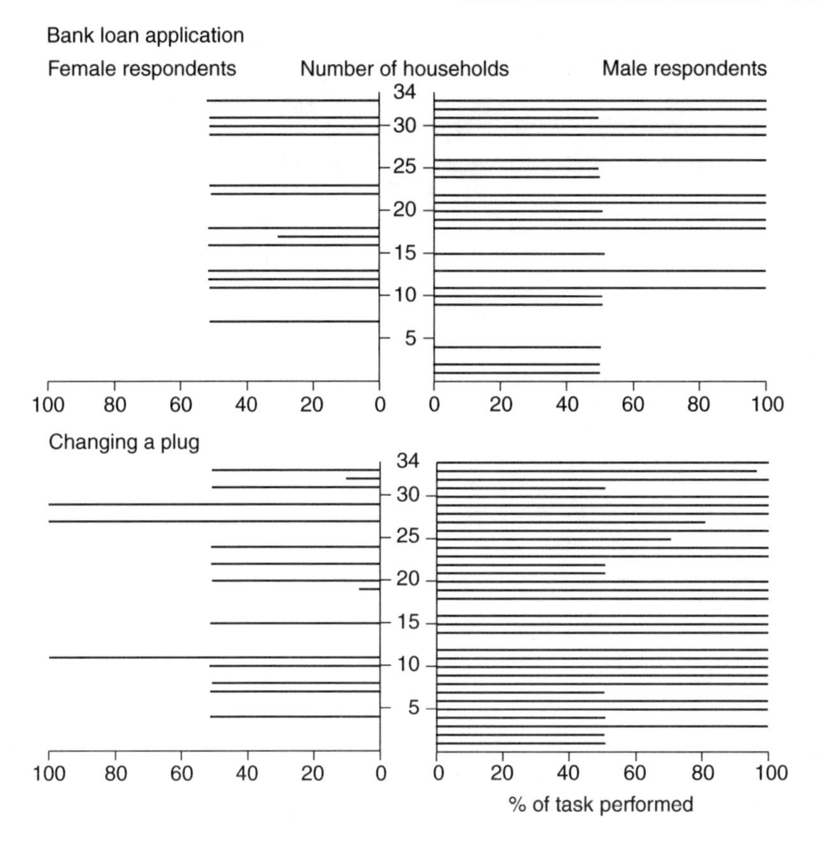

Figure 14.6 Masculinized domestic labour tasks

also clear that across the range of reproductive tasks, women, on the whole, contributed far more than men. This was often the case even when both partners were working outside the home.

There were also clear gender differences in the participation in leisure activities. Interviewees were asked to indicate which leisure activities they pursued. These were then classified as to whether the pursuit drew upon the performance of 'hegemonic masculinity' or its converse, 'emphasized femininity' (Agg and Phillips 1998). Activities were classified as hegemonically masculine if they appeared to promote identities of masculine body power, of man-the-hunter or man-the-protector. Activities were classified as emphasized feminine if they seemed to draw upon notions of women as housekeepers and homemakers, or as body trimmers. There were also a number of activities that were felt to have little or ambiguous gender identities, such as doing crosswords, bell-ringing, walking and horse-riding. Figure 14.7 shows that there appears to be a clear division between the leisure pursuits of men and women, with men favouring masculine or gender-neutral pursuits over feminine ones and women favouring feminine pursuits.

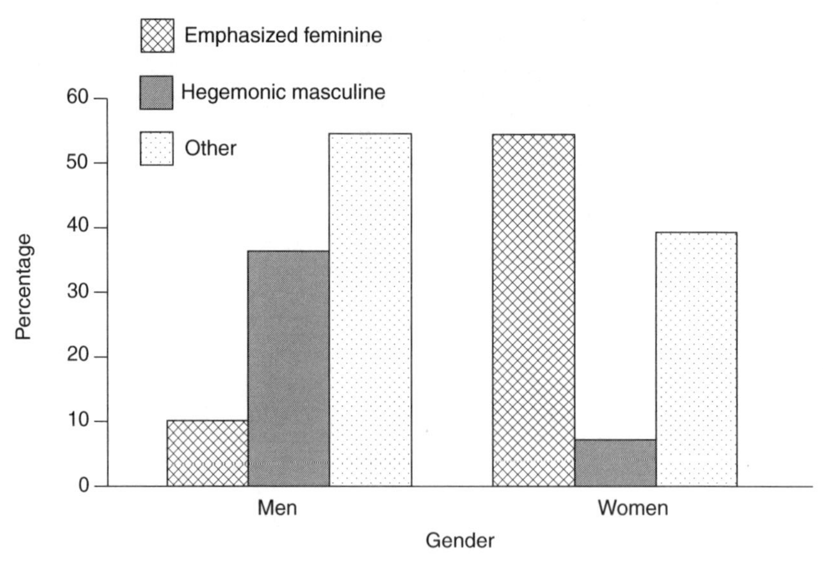

Figure 14.7 Gender identities in leisure activities

Discussions with the residents of these villages revealed quite clear expressions of patriarchal attitudes towards men and women, and clear evidence of an 'othering' – making different and devaluing – of people who did not conform to the behaviour prescribed by the gender identities of 'hegemonic masculinity' and 'emphasized femininity'. For example, one person in a discussion of men performing domestic work and bringing up children remarked:

> I don't know of any eccentrics like that in this village, that are into role reversals, women that go out and do outrageous things like laying tarmac . . .

There were also clear instances of the othering of women who did not (or perhaps could not) perform the tasks associated with food preparation or child care:

Clare: A lot of young women these days don't know how to put a meal on the table. They buy ready-prepared food and stick it on a plate and hand it out, in front of the TV.
Olive: As soon as they've got the children off their hands they go to work, most of them, don't they . . .
Zoe: I know, I'll mention no names, I've had to show {her} how to make marmalade.

None the less, not all men and women interviewed accepted the identities of the patriarchal gender order. Several respondents were evidently actively resisting and contesting it, and many more were certainly reflexive about gender issues and changes in gender relations. Here, for example, is an extract from one group discussion:

Grace: I know a couple expecting a baby and they're talking about the man stopping at home to look after the baby.
Paul: Is that because she earns more than him?
Grace: I think so, she has a better job than he's got, so it makes sense. I've never seen any reason why everything shouldn't be shared. I know it depends on the individuals and I must admit to a certain extent I don't like a man around in the kitchen when I'm cooking and what not. I don't mind him cooking if I can keep out of the way [*laughter*], and I don't mind doing the cooking if he keeps out of my way. But, I don't see why everybody shouldn't dip in and do everything. I mean, I've always done the man's job [*laughter*].
Bert: And I've always done the women's jobs [*laughter*] and most of the man's jobs. We get on quite well don't we [*laughter*] . . .
Maggie: Yes, we've both always bowed to each other's careers and cooperated, you know, supported.

Even the respondent (above) who characterized men staying in the home to look after the family as 'eccentrics' was prompted to comment: 'There's certainly been a massive change in the women's role in the workplace over the last twenty-five years.'

Having said all this, the interviews and group discussions suggested that the dominant discourses of gender in the villages were those of hegemonic masculinity and emphasized femininity and that, taken in conjunction with participation in the official economy and the performance of domestic labour, it is reasonable to suggest that there was in these villages a patriarchal or hegemonically masculine gender order.

Gender, migration and rurality

In this final section, attention focuses on the connections between the dominant gender order, migration and rural space. In particular, three issues will be discussed: the way the rural patriarchal gender order might be created, at least in part, through particular patterns of migration; the influence of patriarchal gender orders on the process of migration; and the ways in which the rural patriarchal gender order may be an outcome of rural space.

Migration and the formation of gender orders

When considering the role that migration may play in the formation of rural gender orders, it is important to recognize the significance of the frequently repeated argument that the population structure of rural areas does not

Table 14.2 Occupations and gender identities: some illustrations

Occupations	*Gender identities*	*Form of gender relation*	*Case study*
Professional managerial	'Rational man'	Emotional ties minimized. Men divorced as far as possible from emotional ties with child care and domestic work. Women maginalized in workplace.	Pahl and Pahl (1971)
Professional/managerial	'Paternal man'	Management based on authority, which was seen to rely on discretion and autonomy from supervision. Abilities seen to be product of age and a stable home life. Managerial posts therefore often restricted to married men with families.	Savage (1992), Halford and Savage (1995)
Craft worker	'Versatile man'	Prepared to be flexible and to move with the job. Partner must be willing to stay or go as required.	Connell (1987)
Office worker	'The secretary'	Must be able to fuse technical competence with interpersonal skills, attractiveness and compliance	Griffin (1985). Pringle (1989)
Industrial factory worker	'Violently heterosexual man'/'The lads'	Involves a 'cult of masculinity' centred on physical and heterosexual power. Used to deprecate non-manual workers and managers.	Lippert (1977), Willis (1977, 1979)

reflect any specifically rural processes but is largely the outcome of general processes of economic and social change (for example, Day, Rees and Murdoch 1989; Lewis and Maund 1976; Newby 1986; Pahl 1965). While this argument has been most commonly outlined in relation to social class, there may well be an important but hitherto neglected gender dimension to these processes of change. As evidenced in the discussion of the service proletariat, there is clear evidence that some class positions are distinctly gendered. Indeed, a series of studies, summarized in table 14.2, have come to argue that some occupations are 'sex-typed' in that they have particular gender identities and relations associated with them. Much of this work has been concerned with the social practices and relations of the workplace and economy, but they may have implications for understanding the social composition of residential areas. As mentioned at the beginning of the chapter, it is widely recognized that particular classes may colonize particular areas: there are, in effect, 'class colonies'. However, if classes are distinctly gendered then class colonies may also exhibit differential gender identities. The colonization of an area by a professional and managerial service class may, for example, lead to the colonization of the area by 'rational men' or 'paternal men'. Similarly, an area of craft workers may be an area of 'versatile men', whilst the residential areas of industrial workers may be where 'the lads' live. The processes of class colonization in which areas gain a particular class composition may hence be equally a process of composing, and recomposing, the gender order of these areas.

In the context of the present study, it has been suggested that there may be two differentially gendered groups of colonists, namely a heavily feminized service proletariat and a masculinized service class. Boyle and Halfacree (1995) have recently examined the connection between gender and migrational behaviour within the service class. They have suggested that there may be important differences in the migration patterns of male and female service class members, with the former moving either from metropolitan to non-metropolitan regions or between non-metropolitan regions, while service class women's movement was towards metropolitan areas such as London. They note, too, that there are also clear age differences, with female migrants being in general younger than their male counterparts. Drawing these points together, their work implies that many young independent women are attracted to the job opportunities and facilities of the city centre, whereas older service class men generally migrate outwards towards metropolitan areas beyond London and to rural areas.

However, as shown earlier in our study villages (and evident in other rural areas in England and Wales: see Cloke, Phillips and Thrift forthcoming; Phillips 1998b), there appears to be another important group of rural colonizers, a feminized service proletariat. This group may constitute something of a hidden group of migrants in the analysis of Boyle and Halfacree, in that many in this group appear to have moved into the countryside in association with the migration of the male service class. Rural in-migration in the study villages was predominantly undertaken by households: in the sample, 95 per cent of the people interviewed who had moved into the area had migrated as a household of more than one person. It would also appear that the jobs of male partners in the household exerted far more influence on migration decisions than did the jobs of female partners (see table 14.3), particularly where these women were classified as service proletariat. For example, when questioned about their reasons for moving to their current place of residence, a majority of both men and women classified as being in the service class stated that their job was an important influence. However, members of the service proletariat – which as shown already were predominantly women – were more likely to have stated that they moved because of their partner's job than for their own jobs.

The implications of these findings are not only, as noted earlier, that the notion of a service-class-led rural restructuring needs to be used with some caution. In addition, the rural patriarchal gender order might be, at least in part, the outcome of the migration into these areas of a masculinized service class and a feminized service proletariat, often combined within single households in a patriarchal class structure.[3]

The patriarchal gender order as an influence on migration

While migration behaviour may lead to the construction and reconstruction of the gender orders of rural communities on the basis of particular gendered workplace identities, it is also important to recognize that migration behaviour may itself be constituted through gender relations and identities. Indeed, as demonstrated elsewhere in this book, a rising number of studies have come to examine the role that gender can play within the constitution of migration. Recognition of such studies suggests that the migration of households with a patriarchal class structure should not be seen as a random occurrence to be analysed purely in terms of its local impact on rural communities, but that the formation of such households needs itself to be the subject of critical attention.

The presence of gendered class asymmetry within household members and its significance for migration was highlighted a while ago by Jan and Ray Pahl's (1971) study of *Managers and their Wives*. This study focused largely on households with a male managerial husband and a wife with no paid employment. It highlighted how married women often played an important role in the process of 'managerial spiralling', whereby managers rise up their career ladders by moving between localities, either within the same firm or by moving firms.

With rising interest in gender issues and their role within migration, there has been renewed interest in this early study and a series of debates have come to emerge around it, together with some subsequent studies (see Bielby and Bielby 1992; Bonney and Love 1991; Bruegel 1996; Finch 1983). The focus of the Pahls' work was very much on how women's labour in the unofficial economy of the household and the community may act either to encourage –

including to 'take pleasure in meeting new people, seeing new places and having a new house to arrange' (Pahl and Pahl 1971: 62) – or to put a brake on male spiralling in the official economy (see also Crompton 1986). More recent work on gender and migration has tended to focus rather more on how acts of migration are both conditioned by, and have impacts for, participation by women in the official economy. Abercrombie and Urry (1983), for example, argued that involvement of both partners in paid employment might reduce geographical mobility because it can be difficult for them both to find jobs in a new area. A series of studies have documented that migration of such households does occur, but that this migration often has detrimental effects on the participation of the woman partner in the official economy. Thus, the work of Finch (1983), Sandell (1977) and Mincer (1978) has suggested that geographical migration is often highly disruptive for the employment conditions and career prospects of women in households, even though the conditions and prospects for male partners – and household incomes – may increase.

Bonney and Love (1991) have argued that although the conclusions of these studies might apply to households where partners have symmetrical participation in the official economy, in many instances participation is asymmetrical, with women being in more casual, lower-status and less well-paid employment than their male partners. Bonney and Love go on to distinguish between 'dual-income' households, which may be seen to be composed of both asymmetrical and symmetrically classed households, and 'dual-career' households, which are more likely to display class symmetry. Dual-career households may both exhibit a lack of impetus to move frequently and experience declining fortunes in women partners' career prospects and employment. In dual-income households there is much less of a brake on migration and female participation in the official economy, largely because women in these households do not have much of a career to lose.

These distinctions are useful in understanding the role that gender relations can play in migration decision making. The results, however, may be hard to distinguish, in that dual-income households may have class asymmetry prior to migration, while dual-career households may become asymmetrical, at least for a time, after migration.

The rural patriarchal gender order as a product of rural space

In a sense, the preceding discussions of the links between gender and migration are of potentially equal relevance to urban and rural analysis. (Indeed, they come up in many of the more 'urban' chapters in this book.) This raises the question as to whether or not there are any specifically rural connections. Here, three lines of interconnection are signalled.

First, it is possible to argue that the material spatiality of rural areas may encourage the establishment of a patriarchal gender order. One way in which this may be done is through the material spatiality of rural areas fostering more rigid separations between 'official economies' and 'unofficial reproductive economies' (for discussion of these terms, see Phillips 1994). Living in rural areas makes the combining of roles in official and unofficial economies hard: work in the official economy is often located outside the village and may involve considerable travelling time, particularly if car transport is not available (Little 1987, 1994; Little, Ross and Collins 1991). Studies have also demonstrated the impact of poor child-care facilities in rural areas and how this has acted to reduce female participation in the official economy (see Little 1991; Little and Austin 1996; Little, Ross and Collins 1991). Indeed, in one specific example, Hughes (1997) has noted how the temporal lag in gaining access to mains electricity supplies in some rural areas, such as mid-Wales, may have played a part in preventing employment of domestic labour-saving devices such as washing machines.

Second, the material spatiality of rural areas may also strengthen participation in patriarchally gendered social and leisure activities. Studies by Little (1987) and Hughes (1997), for example, have highlighted how institutions with 'emphasized feminine' identities such as the Women's Institute play a particularly important role in the lives of many rural women, in part because of a lack of alternatives. One female respondent in the study certainly saw membership of the Women's Institute being conditioned by material spatiality:

> in a very rural place it attracts everybody, because it is one of the few social outlets in a very rural place. Every woman, pretty well, would join the WI because that's her chance of a night out with the girls and they organize nice trips and interesting speakers and so forth . . . I'm thinking of Yorkshire villages and so on, where everybody, where all the farmers' wives, pretty well turn up to the WI, but here, there's so much going on outside the village.

This respondent differentiates between the rurality of a Yorkshire village and her own, Middle English, village where she sees the constraining influence of space being relatively less strong. Other respondents emphasized how feminine institutions in the villages, such as Mother and Toddler Groups (in all the villages this term was used over the more gender-neutral term Parents and Toddlers), also seemed to benefit from the relatively 'spaced out and less densely populated' (Philo 1992: 202) character of rural space:

Charles: The Toddlers' groups, that must be very much of a social event.
Gillian: It's not just for the children, it's for us too.
Heather: You have a natter and three times the gossip.

One important implication of these remarks about employment opportunities and social activities is that they suggest that one impact of rural space may be to establish patriarchal gender relations and identities *subsequent* to people moving into this space, rather than have such an order as the outcome of the colonization of people with pre-existing gender identities and relations. It is increasingly being recognized that gender is not a fixed and immutable construction but is rather diverse, continually constituted, frequently contested and sometimes transformable; and that these processes are played out in and through spaces and spatialities.

So far in this section, attention has been focused on the significance of rurality as material space. However, a third aspect of rural space points to direct interconnections between the rural patriarchal gender order and migration. It is increasingly being recognized that rural space is not only a material phenomenon but is also very important as a cultural symbol. A series of studies by people such as Brandth (1994, 1995), Davidoff (Davidoff, L'Esperance and Newby 1976; Davidoff and Hall 1987), Hughes (1997), Little (1987, 1997), Nash (1993), Nead (1988) and Rose (1993) have all pointed to the gendering of dominant conceptions of the countryside. Little and Austin (1996: 103), for example, claimed that

> patriarchal gender relations are . . . fundamentally embedded in the creation of the rural idyll. The romantic vision of pastoral England is built on a particular interpretation of masculinity and femininity that sees women representing the innocence of the natural world which 'active masculinity must support, protect and oversee' . . . The image of women encompassed in the rural idyll is one of virtue and morality. The so-called 'lynch pin' of rural community, their actual activities are trivialised except where they are seen to relate directly to the provisioning of men and the sustenance of the male headed household.

In previous work (Agg and Phillips 1998; Phillips 1993) it has been suggested that the gender identities and relations of the patriarchal gender order become bound up in cultural constructions of rurality and act to motivate some people to live in the countryside. In the context of a study of rural gentrification, for example, Phillips (1993) argued that patriarchal gender identities may contribute to the decision to move into the countryside, with many gentrifiers moving 'at a time when they were starting, or had just started, a family' (1993: 137). This argument can be seen to be supported by the results of Little and Austin's (1996) study of women's experiences of rural living and their conceptions of rurality in East Harptree, Avon. They argue, for instance, that there was a clear link between decisions to move to the countryside and patriarchal views about men and women for the majority of women they talked to. Little and Austin suggest that one of the key mediating factors was the family and, in particular, conceptions about the countryside and children. They note, for example, that many households had moved to the village at key stages of their children's life path such as 'on or just before birth . . . or as children reached school age' (1996: 105). They argue that this reflected a positive association being drawn between rural living and the well-being of children – the countryside and village were seen to be healthier, safer and happier environments than towns and cities – and that

> incorporated within the decision to move to the countryside to provide a better environment for 'the family' and in particular for 'bringing up children' were voiced firm views about women's roles as wives and mothers.
> (Little and Austin 1996: 105–6)

Summarizing these voices, one might say that they were stating that if the ideal environment for bringing up children was the countryside then the ideal family set-up was where their mother was able to spend time looking after them.

In the study of five villages in Middle England, evidence was found supporting this argument. For example, the raising of children appeared to be a key factor in many people's decision to move into these villages. As table 14.3 demonstrates, changes in household structure and the well-being of children were almost as likely to be cited as reasons for moving as occupational issues. The well-being of children was particularly likely to be cited by women, with about 20 per cent of those women who cited this influence on their migration behaviour having also stopped working for the same reason. Many more appeared to have taken up part-time or temporary employment, often in jobs within the service proletariat. As mentioned earlier, interviews and group discussion elicited clear expressions of the 'othering' – the making different and devaluing – of non-patriarchal relations. Many of these comments also seemed to infer connections between gender identities and urban and rural space (see also Hughes 1997). Here, for example, are comments about men working in the home and not in the official paid economy:

Bert: . . . there are probably ones on the estate, ones we don't know about . . .
Bert: You get that a lot in towns though, this equal . . .
Anne: Because they probably choose to live there if they're both working in the town and commuters . . . it will be in the modern mode of both working . . . so it will affect their domestic behaviour I suppose.

Indeed, changes in gender relations were sometimes connected with the future of rural life:

Table 14.3 Reasons for moving, by gender

Stated influences on move	*Number of respondents*		
	Men	*Women*	*Total*
Own job	24	12	36
Partner's job	8	42	50
Changing size of household	17	24	41
Well-being of children	15	34	49

Grace: More mothers work, and this makes it more of a dormitory, and if the wife's working, she can't get involved with what goes on in the village.

Olive: As soon as they've got the children off their hands they go to work, most of them, don't they? And the young haven't got time for these [WI] meetings;

Grace: That's very much the thing [as the membership of the WI falls] . . . you lose part of village life, we're losing part of, you know, that sort of thing, we lose a lot of traditional things, the idyllic side of village life, the old ways.

Conclusion

This chapter has sought to detail some interconnections between gender, rurality and migration. It began by highlighting how recognition of gender can transform prevailing notions of rural social change, such as the notion of rural areas being a 'middle class territory' and/or undergoing 'service class' colonization and restructuring. The presence in at least some of the villages of Middle England of two quite different and quite distinctly gendered service classes was highlighted. It was then argued that gender should be seen as being constitutive of more than just class positions and the notion of distinct gender orders was outlined. Drawing on divisions in the performance of domestic labour and in participation in social and leisure activities, it was claimed that, at least in the five Middle England villages, there was something of a 'patriarchal' or 'hegemonically masculine' gender order. Attention then switched to considering how such a gender order might be both an outcome and constituent of migrational processes. It was highlighted how processes of differential class colonization could also be processes by which people with different gender identities could colonize particular areas. Attention was also drawn to how divisions in the performance of official employment and domestic work might impact migration, and also how rural space may materially and symbolically act to constitute a rural patriarchal gender order.

Acknowledgements

Particular thanks go to Jenny Agg, who conducted the group discussions on which this paper draws, and to all those involved in the questionnaire survey. The research project on 'Recent social change in the Leicestershire and Warwickshire countryside' was conducted in collaboration with the Leicestershire and Warwickshire Community Councils and was financed by Coventry University. The support of these organizations is acknowledged, as is the debt owed to all the women and men of the five villages who participated in the research.

Notes

1 The class positions were calculated on the basis of last position of employment. This has the

advantage that it recognizes that groups such as the unemployed, retired and those working in the home often have very different economic, social and cultural assets which are derived from their earlier class positionings.

2 Connell's notion of gender orders has clear similarities with the notion of 'modes of regulation' where these are seen to be largely contingent creations of an often disparate range of institutions, agents and actants.

3 The term patriarchal class structure is used here to refer to a gender asymmetry in class positions defined solely with regard to positions within the official money economy. It is not used to imply anything about the division of domestic labour within households, as it is used, for example, in the work of Sylvia Walby (1986, 1990).

References

Abercrombie, N. and Urry, J. (1983) *Capital, Labour and the Middle Classes*, London: Allen and Unwin.

Agg, J. and Phillips, M. (1998) 'Neglected gender dimensions of rural social restructuring', in P. Boyle and K. Halfacree (eds) *Migration into Rural Areas: Theories and Issues*, London: Wiley, pp. 252–79.

Bielby, W.T. and Bielby, D.D. (1992) 'I will follow him: family ties, gender-role beliefs, and reluctance to relocate for a better job', *American Journal of Sociology* 97: 1241–67.

Bonney, N. and Love, J. (1991) 'Gender and migration: geographical mobility and the wife's sacrifice', *Sociological Review* 39: 335–48.

Boyle, P. and Halfacree, K. (1995) 'Service class migration in England and Wales, 1980–1991: identifying gender-specific mobility patterns', *Regional Studies* 29: 43–57.

Brandth, B. (1994) 'Changing femininity: the social construction of women farmers in Norway', *Sociologia Ruralis* 34: 127–49.

Brandth, B. (1995) 'Rural masculinity in transition: gender images in tractor advertisements', *Journal of Rural Studies* 11: 123–33.

Bruegel, I. (1996) 'The trailing wife: a declining breed? Careers, geographical mobility and household conflict in Britain, 1970–89', in R. Crompton, D. Gallie, and K. Purcell (eds) *Changing Forms of Employment: Organization, Skills and Gender*, London: Routledge, pp. 235–58.

Champion, A. (1989) *Counterurbanization: the Changing Pace and Nature of Population Deconcentration*, London: Edward Arnold.

Chant, S. (ed.) (1992) *Gender and Migration in Developing Countries*, London: Belhaven.

Cloke, P., Phillips, M. and Thrift, N. (1995) 'The new middle classes and the social constructs of rural living', in T. Butler and M. Savage (eds) *Social Change and the Middle Classes*, London: UCL Press, pp. 220–38.

Cloke, P., Phillips, M. and Thrift, N. (1998) 'Class, colonisation and lifestyle strategies in Gower', in P. Boyle and K. Halfacree (eds) *Migration to Rural Areas: Theories and Issues*, London: Wiley, pp. 166–85.

Cloke, P., Phillips, M. and Thrift, N. (forthcoming) *Moving to Rural Idylls*, London: Paul Chapman.

Cloke, P. and Thrift, N. (1987) 'Intra-class conflict in rural areas', *Journal of Rural Studies* 3: 321–33.

Cloke, P. and Thrift, N. (1990) 'Class change and conflict in rural areas', in T. Marsden, P. Lowe, and S. Whatmore (eds) *Rural Restructuring*, London: David Fulton, pp. 165–81.

Connell, R. (1987) *Gender and Power: Society, the Person and Sexual Politics*, Cambridge: Cambridge University Press.

Crompton, R. (1986) 'Women and the "service class"', in R. Crompton and M. Mann (eds) *Gender and Stratification*, Cambridge: Polity, pp. 119–36.

Crompton, R. and Sanderson, K. (1990) *Gendered Jobs and Social Change*, London: Unwin Hyman.

Davidoff, L. and Hall, C. (1987) *Family Fortunes: Men and Women of the English Middle Classes 1780–1850*, London: Hutchinson.

Davidoff, L., L'Esperance, J. and Newby, H. (1976) *Landscape with Figures*, Harmondsworth: Penguin.

Day, G., Rees, G. and Murdoch, J. (1989) 'Social change, rural localities and the state: the restructuring of rural Wales', *Journal of Rural Studies* 5: 227–44.

Esping-Andersen, G. (1993) *Changing Classes: Stratification and Mobility in Post-industrial Societies*, London: Sage.

Finch, J. (1983) *Married to the Job: Wives' Incorporation in Men's Work*, London: Allen and Unwin.

Goldthorpe, J. (1982) 'On the service class, its formation and future', in A. Giddens and G. McKenzie (eds) *Social Class and the Division of Labour*, Cambridge: Cambridge University Press, pp. 162–85.

Goldthorpe, J., Llewellyn, C. and Payne, C. (1980) *Social Mobility and the Class Structure in Modern Britain*, Oxford: Oxford University Press.

Gouldner, A. (1979) *The Future of Intellectuals and the Rise of the New Class*, New York: Continuum.

Griffin, C. (1985) *Typical Girls?*, London: Routledge and Kegan Paul.

Halford, S. and Savage, M. (1995) 'Restructuring organizations, changing people: gender and careers in banking and local government', *Work, Employment and Society* 9: 97–122.

Hoggart, K. (1997) 'The middle classes in rural England 1971–1991', *Journal of Rural Studies* 13: 253–73

Hughes, A. (1997) 'Rurality and "cultures of womanhood"', in P. Cloke and J. Little (eds) *Contested Countryside Cultures: Otherness, Marginalisation and Rurality*, London: Routledge, pp. 123–37.

Lewis, G. and Maund, D.J. (1976) 'The urbanisation of the countryside: a framework for analysis', *Geografiska Annaler* 58B: 17–27.

Lippert, J. (1977) 'Sexuality as consumption', in J. Snograss (ed.) *For Men Against Sexism*, Albion, Cal.: Times Change Press, pp. 207–13.

Little, J. (1986) 'Feminist perspectives in rural geography: an introduction', *Journal of Rural Studies* 2: 1–8.

Little, J. (1987) 'Gender relations in rural areas: the importance of women's domestic role', *Journal of Rural Studies* 3: 335–42.

Little, J. (1991) 'Women in the rural labour market: an evaluation', in T. Champion and C. Watkins (eds) *People in the Countryside: Studies of Social Change in Rural Britain*, London: Paul Chapman, pp. 96–107.

Little, J. (1994) Gender relations and the rural labour process', in S. Whatmore, T. Marsden and P. Lowe (eds) *Gender and Rurality*, London: David Fulton, pp. 11–30.

Little, J. (1997) 'Employment marginality and self-identity', in P. Cloke and J. Little (eds) *Contested Countryside Cultures: Otherness, Marginalisation and Rurality*, London: Routledge, pp. 138–57.

Little, J. and Austin, P. (1996) 'Women and the rural idyll', *Journal of Rural Studies* 12: 101–11.

Little, J., Ross, K. and Collins, I. (1991) *Women and Employment in Rural Areas*, London: Rural Development Commission.

Mincer, J. (1978) 'Family migration decisions', *Journal of Political Economy* 86: 749–73.

Murdoch, J. (1995) 'Middle class territory? Some remarks on the use of class analysis in rural studies', *Environment and Planning A* 27: 1213–30.

Murdoch, J. and Marsden, T. (1994) *Reconstituting Rurality: Class, Community and Power in the Development Process*, London: UCL Press.

Nash, C. (1993) 'Remapping and renaming: new cartographies of identity, gender and landscape in Ireland', *Feminist Review* 44: 39–57.

Nead, L. (1988) *Myths of Sexuality: Representations of Women in Victorian Britain*, Oxford: Basil Blackwell.

Newby, H. (1986) 'Locality and rurality: the restructuring of rural social relations', *Regional Studies* 20: 209–16.

Pahl, J. and Pahl, R. (1971) *Managers and their Wives: a Study of Career and Family Relationships in the Middle Class*, London: Allen Lane.

Pahl, R. (1965) 'Class and community in English commuter villages', *Sociologia Ruralis* 5: 5–23.

Phillips, M. (1993) 'Rural gentrification and the processes of class colonisation', *Journal of Rural Studies* 9: 123–40.

Phillips, M. (1994) 'Habermas, rural studies and critical social theory', in P. Cloke, M. Doel, D. Matless, M. Phillips and N. Thrift *Writing the Rural: Five Cultural Geographies*, London: Paul Chapman, pp. 89–126.

Phillips, M. (1998) 'Investigations of the British middle classes – parts 1 and 2', *Journal of Rural Studies* 14: 411–43.

Philo, C. (1992) 'Neglected rural geographies: a review', *Journal of Rural Studies* 8: 193–207.

Pringle, R. (1989) *Secretaries Talk: Sexuality, Power and Work*, London: Verso.

Rose, G. (1993) *Feminism and Geography: the Limits of Geographical Knowledge*, Cambridge: Polity.

Sandell, S. (1977) 'Women and the economics of family migration', *Review of Economics and Statistics* 59: 406–14.

Savage, M. (1992) 'Women's expertise, men's authority: gendered organization and the contemporary middle classes', in M. Savage and A. Witz (eds) *Gender and Bureaucracy*, Oxford: Blackwell, pp. 124–51.

Savage, M., Barlow, J., Dickens, P. and Fielding, T. (1992) *Property, Bureaucracy and Culture: Middle Class Formation in Contemporary Britain*, London: Routledge.

Thrift, N. (1987) 'Manufacturing rural geography', *Journal of Rural Studies* 3: 77–81.

Walby, S. (1986) *Patriarchy at Work*, Cambridge: Polity.

Walby, S. (1990) *Theorizing Patriarchy*, Oxford: Blackwell.

Whatmore, S. (1991) *Farming Women: Gender, Work and Family Enterprise*, London: Macmillan.

Willis, P. (1977) *Learning to Labour: How Working Class Kids Get Working Class Jobs*, London: Saxon House.

Willis, P. (1979) 'Shop floor culture, masculinity and the wage form', in J. Clarke, C. Critcher and R. Johnson (eds) *Working Class Culture*, London: Hutchinson, pp. 185–98.

Wright, E.O. (1978) *Class Crisis and the State*, London: New Left Books.

Wright, E.O. (1979) *Class Structure and Income Determination*, New York: Academic Press.

15 Residential change

Differences in the movements and living arrangements of divorced men and women

Lynn Hayes and Alaa Al-Hamad

Introduction

The work we report here arises from a project that examines how residential change is linked to family or life-course events. In particular, the project focuses on two life events: residential change associated with the breakdown of relationships and movement associated with the care needs of elderly relatives. The present chapter concentrates on the first of these and uses data from the British 1 per cent Household Sample of Anonymized Records (SAR) to look at differences in the movement and living arrangements of men and women who were coded as divorced in the 1991 Census. Early on in the project we undertook a basic analysis of SAR data to compare the movement patterns of married and divorced people (Hayes, Al-Hamad and Geddes 1995). The paper found significant differences between the two groups, including gender differences. However, we were aware that more subtle differences probably existed *within* the gender groups. In particular, our analysis did not look at the households in which our divorced sample were living: how many of them lived alone, how many were living with a partner or with children – in short, the circumstances that cut across and help to make sense of aggregate movement patterns. This chapter addresses these issues, looking in greater detail at the characteristics and circumstances of the SAR divorced mover sample.

The authors of the chapter are members of a multi-disciplinary team working on a project that combines qualitative and quantitative methods. Lynn Hayes is a qualitative researcher whose background lies in research on family relations and Alaa Al-Hamad's background is in quantitative data analysis. Working together on this and other papers has meant each of us approaching our data in new ways, drawing upon each other's experience and ideas to do so. In this chapter we have dissected the data to a greater degree than is normally the case in quantitative research, and we believe our analysis has greater depth and balance as a result. The chapter is, therefore, the product of our collaborative efforts but is also a product of our different academic backgrounds and the combination of our research experience.

Divorce

Over the latter half of this century various reforms have made divorce in Britain easier. Haskey (1996) calculates that around four in ten marriages would end in divorce if the divorce rate continues at its 1993–4 level. Second marriages are more susceptible to breakdown than first marriages (Clulow 1991; Haskey 1996). This is especially the case when both partners have been married before. In 1991, 158,745 divorces were granted: 17 per cent of these divorcing men had been divorced previously and half of these were divorcing a partner who had also been divorced previously (OPCS 1991: table 4.1c). Figures for divorcing women are similar (OPCS 1991: table 4.1d). Cohabitation frequently precedes second marriages (Clark 1987) and couples who cohabit prior to marriage have higher divorce rates than those who do not cohabit (Haskey 1992). There is little information available as yet on the extent of cohabitation breakdown (Murphy 1990) but it seems reasonable to assume that many relationships of this kind do end. Divorced men and women are more likely to enter into cohabiting relationships than never-married people and some of these relationships will end in second marriage, others will break down, and others will continue as stable cohabiting unions. Hence, within the divorced population we might expect to find a range of situations, some akin to those of single people and others more akin to those of married people. Studies of the divorced population need to take this into account.

Turning to migration, Devis (1983) notes that divorced people have higher rates of mobility than married people and suggests that this is not accounted for simply by the event of divorce itself. As we note above, some movement will be linked directly to the divorce or to the changing economic and personal circumstances that accompany it. However, other moves may be linked to new partnerships whilst the movement of those living in stable 'as married' relationships will be different again.

Analysis of SAR data (Hayes, Al-Hamad and Geddes 1995) shows that divorced people have a higher level of mobility than married people (13 and 7 per cent respectively) and divorced men have higher levels of mobility than divorced women (14 and 11 per cent respectively). The SAR does not include information on timing, so it is impossible to differentiate the recently-divorced from those whose marriages ended some considerable time ago. Clearly, this affects movement patterns and is a problem – whichever way analysis is tackled. However, looking in more detail at the living arrangements of the divorced goes some way to separating out different categories of divorced people. This chapter addresses these issues, looking in greater detail at the characteristics and circumstances of the SAR divorced mover sample and those they live with. The main focus is gender differences in movement behaviour of divorced people, looking specifically at who moves with them and who they join. However, in order to do this we had to engage in some recoding of the data, and the next section discusses how and why we did this.

Use of the Sample of Anonymized Records (SAR)

The SAR represents a welcome and innovative departure from earlier census data in that it allows researchers to access individual-level information, which means that more detailed analyses are possible than was the case previously. However, this raises a number of issues when handling the data. As with any secondary analysis of data, care needs to be taken when interpreting the data and some recoding was necessary. The SAR includes details of all household members and their relationship to the head of household. For some purposes this is fine. However, we were interested in the relationship of our divorced movers to those who had moved with them as well as those to whom they had moved. The way that we approached this is discussed below.

Our sample comprises all those who were coded as divorced at the time of the 1991 Census and who had moved in the twelve months preceding the census, together with all those who moved with them and those to whom they moved. This gives us a total sample of 7,167 individuals, 3,203 of whom were divorced movers. The SAR does not include information on the date or duration of divorce. Hence, some of our sample will be recently divorced whilst others will have been divorced for some time. However, our purpose is not to examine the impact of divorce itself on movement but to demonstrate the variety of living arrangements of divorced movers.

Figure 15.1 gives details of the breakdown of our sample. We began by separating our sample into two groups: households in which *every* member was a mover and households where *some* people (including a divorced person) had moved but others had not. In effect, the first group represents people moving into 'new' homes (either alone or with others) whilst the second represents those joining someone else's household. We wanted to see how many of our divorced sample had actually moved alone and how many had moved with other people; how many appeared to be in stable relationships or family groups. So we further subdivided each of the two main groups to separate out households where the divorced person was the only mover from those where people had moved together and those whose members had moved from *different* addresses. This was possible using the SAR 'distance moved' variable.

The SAR includes a variable of 'wholly moving household' but this was not helpful here since the SAR definition of a wholly moving household is one in which every member is a mover, irrespective of where they came from. We wanted to know whom people moved with and whom they met up with or joined. A further problem with SAR coding arises with households that include infants under the age of one. Infants born since the household moved are coded as non-migrants. Technically this is correct, since the child never lived at a different address. However, we found that in many of these households the child was the only 'non-migrant'. Hence, a cohabiting couple who moved together then had a child at their new address is coded as a 'partially moving household' in the SAR. In practice, some of these infants would be born shortly after the move whilst others may have been conceived after the

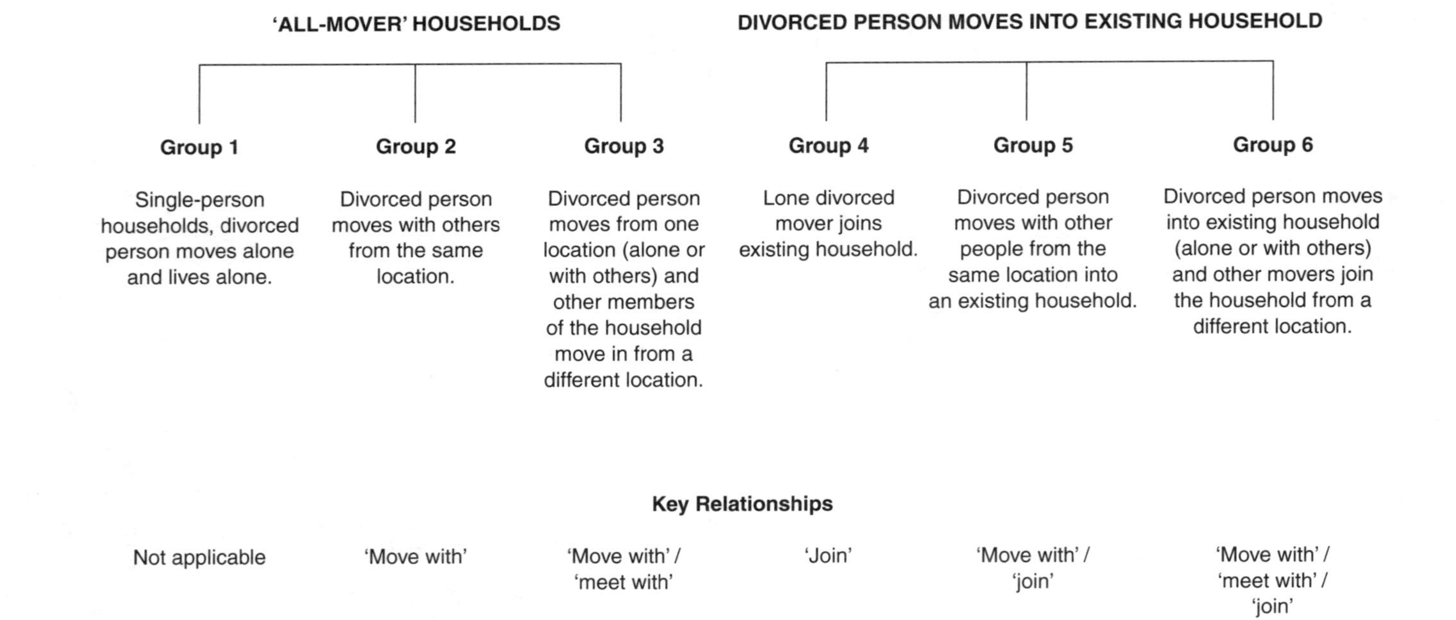

Figure 15.1 Divorced mover types

move, and we therefore regard these households as 'all-mover households' since that was the situation at the time of the move.

Having broken down the sample into our six groups we were then faced with the problem of how to examine the relationships within the households. Figure 15.2 shows the direction of movement and relationships within each group. 'A' is the divorced person. Our first category was straightforward since its members had moved, and were living, alone. The second group included other people who had moved with our divorced person and we wanted to explore the relationship between them. The next group was more complicated because it could have contained people who had moved with the divorced person as well as the person(s) who had moved into the household from a different location. Here we were dealing with two sets of relationships: who moved with our divorced person and whom did our divorced person meet up with. We have used the expression 'meet up with' for these relationships because we do not know the timing of the moves, so cannot say who moved into the household first. However, we use the phrase 'join' in our subsequent groups because those households contain non-movers. In our fourth group the divorced person was the only mover, and he or she moved into an existing household. Here we wanted to know whom our divorced person joined. In the fifth group our divorced person was moving with others from the same location into an existing household, so there were two relationships to explore: who moved with our divorced person and whom did they join. Households in our final group were more complicated still, since they included people moving with our divorced person, movers coming from a different location and people who had not moved at all. Hence this group has three sets of relationships: who does the divorced person 'move with' (if anyone), 'meet up with' and 'join'.

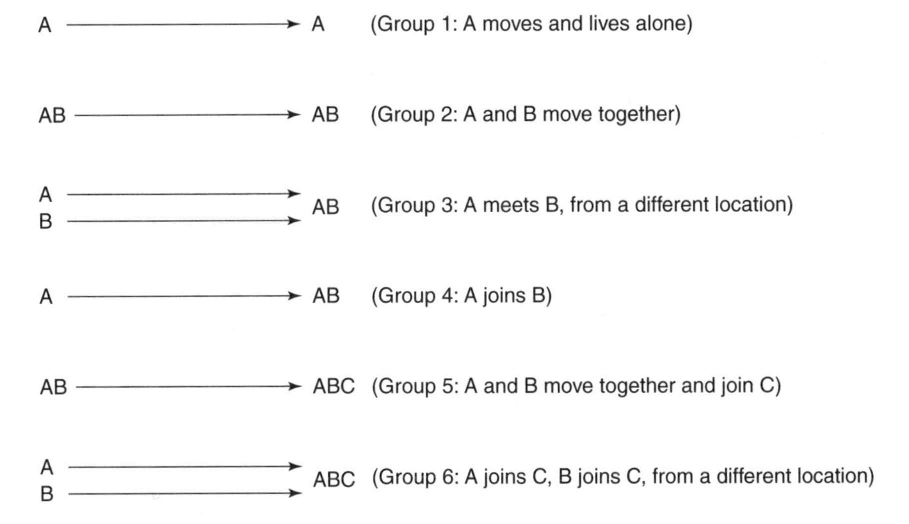

Figure 15.2 Movement and relationship flows

The SAR includes information on relationships within households, coded in terms of their members' relationship to the head of household. However, this was not always our divorced person, as some households contained more than one divorced person, and in any case we were interested in the various sets of relationships in these households. We therefore needed to devise a set of relationship codes that could be applied to our new variables of 'move with', 'meet up with' and 'join'. These are:

- partner;
- partner and children ('family' group, including children of either partner);
- children (but no resident partner);
- parents;
- other relatives;
- non-relatives only;
- complex household.

Complex households include multi-generational households, those including more than one family and those where the coding did not allow us to make reasonable judgements about the relationships in the household.[1]

The SAR variables of 'family number' and 'family type' provided further clues to assist us in our coding. For example, if a woman and her child join an unrelated person who is head of household both are coded as 'unrelated' in the SAR relationship coding. However, the 'family number' code reveals that they are in fact related to each other and the 'family type' code shows the relationship, such as lone parent/dependent child family, thus allowing us to code the 'move with' relationship accurately.

General patterns

We begin with some general comparisons across the six groups, then move on to examine each group in more detail. Finally, we return to an overview of the data, discussing the implications of the relationship patterns noted in our discussion of each group.

Gender

The SAR divorced mover population includes slightly more women than men (47 per cent men, 53 per cent women), reflecting gender divisions in the divorced population as a whole (Haskey 1996). However, a rather different picture emerges when we look at the gender composition of our six groups in table 15.1, with men dominating some groups and women dominating others. Both 'move alone' groups (Groups 1 and 4) contain more men than women, whilst there are higher percentages of women in the 'move with someone' groups (Groups 2 and 5), mainly because women were moving with their children.

Table 15.1 Gender composition of divorced movers in the six groups

Divorced mover group	*Men*	*Women*
1: Move alone	450 (54%)	388 (46%)
2: Move with	465 (35%)	852 (65%)
3: Meet with	191 (54%)	164 (46%)
4: Join	326 (65%)	177 (35%)
5: Move with/join	38 (28%)	99 (72%)
6: Meet with/join	36 (68%)	17 (32%)
Total	1506 (47%)	1697 (53%)

Age

Table 15.2 gives the ages of our divorced movers broken down into three age bands. Group 1 (those who move and live alone) has the greatest variation. In particular, low percentages of men and women in this group were under 30 years of age, with a correspondingly high percentage over 50, especially women. It might be the case that younger people prefer to move in with someone else when their relationships break down, or it may be the case that they have more opportunity to do so. Several factors come into play here. Younger people are more likely to enter new partnerships. In most cases they will have parents alive to return to (McCarthy and Simpson 1991: 116) and may also have unmarried friends who are able to offer accommodation. On the other hand, younger people face a series of constraints to independent living and they may find it more difficult to set up a home of their own on separation. Younger people (and those whose marriages are of shorter duration) may not have the resources to consider buying a home of their own when their marriage ends (McCarthy and Simpson 1991), whilst local authority housing may not be available to those on their own (Cole and Furbey 1994).

The high percentage of over-50-year-olds moving and living alone needs a different explanation. It is the case that divorced men are more likely to remarry than divorced women. They also tend to marry or cohabit with

Table 15.2 Percentage of divorced movers in age bands

	Under 30 years		*20–49 years*		*50 years and over*	
Divorced mover group	*Men*	*Women*	*Men*	*Women*	*Men*	*Women*
1: Move alone	8	10	65	47	27	43
2: Move with	19	26	68	66	13	8
3: Meet with	20	29	69	64	11	7
4: Join	18	26	68	59	14	15
5: Move with / join	16	31	63	63	21	6
6: Meet with / join	19	29	53	59	28	12

women younger than themselves and tend to marry or cohabit with women who have not been married before (Haskey 1995). This means there is a gender imbalance in the divorced population as a whole and, in particular, an excess of older divorced women who are not likely to enter new relationships. Women of this age will not be able to turn to parents, as younger people do. Their children are also likely to be grown up with families of their own (note the low percentages of women over 50 years in our 'move with' and 'meet with' groups). It would seem, then, that women of this age have fewer options and are therefore more likely to end up alone.

Tenure

Owner occupation is the dominant form of tenure in Britain (Saunders 1990). In 1961 around a third of households owned (or were buying) their own homes. By 1989 this had increased to two-thirds (*Social Trends* 1991). The rise in home ownership has been accompanied more recently by a decline in the availability of local authority housing (Cole and Furbey 1994) and a long-term decline in the availability of private rented accommodation (Saunders 1990). For the total SAR 1 per cent Household sample of 531,170 individuals, the figure in owner occupation is 70%, with 23 per cent in local authority or housing association rented accommodation and 7 per cent in the private rented sector. Studies of the housing patterns of divorced people consistently show movement down the housing ladder from owner occupation into rented housing, or downward movement *within* a sector (such as from detached to terraced property) (McCarthy and Simpson 1991; Symon 1990). This is especially the case for women (Austerberry and Watson 1983; Grundy 1985, 1989; Murphy 1990; Wasoff and Dobash 1990).

The SAR includes information on tenure and type of property occupied. Table 15.3 gives a breakdown for our first three groups. Tenure type for the latter three groups refers to the tenure of the household our divorced person joined, so is less helpful in assessing the circumstances of the divorced movers in those groups. Owner occupation within our sample is lower than we would expect to find in the general population. However, it is much lower for some groups than others. In particular, those moving and living alone have very low rates of owner occupation, which is probably linked to their economic situation and their different housing needs. In all our groups, the percentage occupying private rented accommodation is much higher than that of the general SAR population. Finally, there are also gender differences in migration into rented housing, with women showing a greater association with state-funded housing than men and less of an association with privately rented housing.

Analysis by group

The general patterns noted above give an indication of the different circumstances of the divorced people in our six groups, something we can explore further

Table 15.3 Percentage of divorced men and women in each group by tenure type

	Owner-occupier		*Rented (local authority/ housing association)*		*Private rented*	
Divorced mover group	*Men*	*Women*	*Men*	*Women*	*Men*	*Women*
1: Move alone	31	39	32	38	37	23
2: Move with	53	39	22	40	25	21
3: Meet with	49	50	17	21	34	29

by looking at each group in greater detail. They also appear to point towards the disadvantaged housing position of some of our groups in comparison to other groups and the population in general. We explore this further below. We begin by looking at those divorced people who occupy lone-person households.

Group 1: 'Moving alone, living alone'

Our first group comprises divorced people who had moved and were living alone at the time of the census. In total we have 838 individuals, 450 men and 388 women (two of whom had a child after the move). This is our simplest group since no co-resident relationships are involved. The age curve for this group peaks at 35–39 years for men and 40–44 years for women, reflecting the greater tendency for divorced men to enter new relationships (Clark 1987; Leete and Anthony 1979; and see below). In fact, this group is characterized by younger men and older women, with more men in each of the younger five-year age bands and more women in each of the five-year bands over 50 years.

As this group live alone we might expect to find lower rates of owner occupation and a tendency to occupy rented accommodation of various kinds. Access to one wage (if any) is likely to limit housing options. Housing needs are also likely to be different for this group compared with the others. The data support this. There are low levels of owner occupation and relatively high rates of occupancy of private rented flats, rented rooms and bedsits – a pattern which is more pronounced for men (37 per cent of men and 23 per cent of women were living in private rented accommodation: see table 15.3). Nearly three-quarters (73%) of the men in this group were under 50 years of age, and some of them will be non-custodial fathers (though we have no way of knowing how many). McCarthy and Simpson (1991) suggest that men in this position have difficulty obtaining local authority housing and therefore tend to depend on the private rented sector to a higher degree. Our data suggest that women who are on their own also tend to occupy private rented accommodation to a greater degree than women with partners or children.

Group 2: All-mover households, 'moving together'

Our second group is made up of households where the divorced mover and everyone else in the household moved together. These represent the most established households and partnerships in our sample, since the household's composition has not been changed by the move. Within this group there are 3,232 individuals, of which 1,409 were men and 1,823 were women. However, the gender division for the divorced population in the group is much more pronounced: 465 men and 852 women. The reasons for these differences become clear when we look at the composition of the households in table 15.4.

When we examine the relationships between our divorced movers and those who move with them we find very strong and predictable gender differences: 80 per cent of the divorced men in this group had moved with a partner (or partner and children) compared to 35 per cent of the divorced women. Over half the women in the group had moved with their children (and no partner) compared to just 8 per cent of the divorced men. In fact 93 per cent of those moving with children were women and 81 per cent of these women were under 40 years of age. Clearly, this reflects the tendency for women to be granted custody of the children on divorce.[2]

The age curve for this group of divorced movers peaks at 30–35 years for both men and women, in contrast to the lone movers discussed above who were older on the whole. Indeed, 62 per cent of men and 54 per cent of women moving with partners were aged between 25 and 39 years. Those moving with a partner and children also tended to be in the younger age bands: 74 per cent of women and 64 per cent of men were aged between 25 and 39 years.

Gender differences were noted in terms of the tenure occupied. Divorced men in this group had higher rates of owner occupation than women (53 per cent men, 39 per cent women) whilst a higher percentage of women occupied local authority or housing association rented property (22 per cent men, 40 per cent women: see table 15.3). When we break this down further we

Table 15.4 Group 2 (move with households) – who do the divorced move with by gender of divorced movers?

	Men		*Women*	
Relationship	*Count*	*%*	*Count*	*%*
Partner	256	55	165	19
Partner and child	116	25	137	16
Child	38	8	485	57
Parent	6	1	15	2
Other relative	7	2	9	1
Non-relative	35	7	23	3
Complex household	7	2	18	2
Total	465	100	852	100

see that men moving with children were in a much better position than women: 42 per cent of men with children were buying their homes compared with 21 per cent of women with children. However, 63 per cent of divorced women moving with a partner and 52 per cent of women moving with a partner plus children were buying their homes, which lends support to McCarthy and Simpson's (1991: 110) suggestion that divorced women are able to move up the housing ladder through new relationships.

Group 3: All-mover households, 'meeting up'

Our third group comprises households where all members had moved into the household at some point in the twelve months preceding the census but where some of them had come from a different location to the divorced mover. Hence, they are 'new' households whose members have 'met up' at the new location, as opposed to the established households discussed above. In exploring the composition of these households we were interested in two sets of relationships: who did the divorced person move with and whom did they meet up with, plus how many were living in established partner relationships and how many partner relationships were new ones. This group included 820 individuals (413 men, 407 women). Of these, 355 were divorced movers (191 men, 164 women) with an age distribution similar to that of our second group (where all members of the household moved together). Table 15.5 shows that substantial numbers of these divorced people had actually moved alone to meet others in new households: 83 per cent of divorced men and 54 per cent of divorced women. A further 30 per cent of women move with their children only, so the key issue here is who do they meet. Table 15.6 gives details for selected relationships.

Table 15.5 shows that small numbers only had moved with a partner (4 per cent of men and 5 per cent of women). However, from table 15.6, over half the group (57 per cent of men and 50 per cent of women) met up with a partner at

Table 15.5 Group 3 (meet with households) – who do the divorced move with by gender of divorced movers?

	Men		*Women*	
Divorced moves with	*Count*	*%*	*Count*	*%*
Alone	158	83	90	54
Partner	6	3	6	4
Partner and child	2	1	2	1
Child	9	5	50	30
Parent	1	0.5	1	1
Other relative	1	0.5	2	1
Non-relative	12	6	10	6
Complex household	2	1	3	2
Total	187	100	162	100

Table 15.6 Group 3 (meet with households) – who do the divorced meet with by gender of divorced movers?

Relationship	*Men*	*Women*
Move alone, meet partner	80 (42%)	52 (32%)
Move alone, meet partner and child	27 (14%)	2 (1%)
Move alone, meet non-relative	39 (20%)	19 (12%)
Move with child, meet partner	2 (1%)	27 (17%)
Move with child, meet non-relative	1 (.5%)	12 (7%)
Total	191	164

the new household. These partners had moved from different locations to the divorced mover and so these are new cohabiting unions. As we noted earlier, we do not know the timing of the moves so cannot say whether the partners moved into the household together or whether one moved in before the other. However, the point is that ultimately the moves resulted in new cohabiting partnerships or new partner and child 'family' groups. New partnership couples have the highest levels of home ownership of all our divorced movers: 67 per cent of women and 63 per cent of men who had formed new cohabiting relationships were living in owner occupation, a level approaching that of the population in general. In contrast, divorced people in the group who had met up with non-relatives were mainly living in private rented accommodation (59 per cent of men and 43 per cent of women).

Joining existing households

We now move on to examine the three groups of households in our sample which contain non-movers. Each of the three groups is smaller than the corresponding 'all-mover' group, which suggests that divorced people prefer to live in households of their own rather than share someone else's home. Studies that examine the experiences of divorced people have pointed to the difficulties involved in sharing accommodation with friends or relatives after separation

Table 15.7 Group 4 (join households) – who do the lone divorced join by gender of divorced movers?

	Men		*Women*	
Divorced joins	*Count*	*%*	*Count*	*%*
Partner	44	13	58	33
Partner and child	71	22	9	5
Child	6	2	14	8
Parent	88	27	39	22
Other relative	15	5	10	5
Non-relative	85	26	42	24
Complex household	17	5	5	3
Total	326	100	177	100

or divorce (Brailey 1986; McCarthy and Simpson 1991: 74). After living independently it is perhaps understandable that divorced people would want to live in a home of their own rather than depend on relatives or friends, and some of the moves in this group may well be temporary arrangements – a point raised by Sullivan (1986) in her analysis of Labour Force Survey data.

Group 4: Lone divorced movers joining existing households

Our fourth group comprises households where the divorced mover is the only mover in the household. These households contain 1,469 individuals in total (769 men, 700 women), of whom 503 were divorced movers (326 men, 177 women). Table 15.7 shows who these lone divorced movers join, and in common with Group 3, this group includes divorced people entering new cohabiting relationships with predictable gender differences. Women were moving into the home of a partner whilst men were moving into the home of a partner plus her children: 13 per cent of men and 33 per cent of women had joined a partner only, while 22 per cent of men and 5 per cent of women had joined partner and child households. This group also features parents as a destination in substantial numbers, with men slightly more likely to return to parents than women. This finding is in line with an analysis of 1981 Labour Force Survey data, where Sullivan (1986) found that a quarter of divorced and separated men aged 30–34 years were recorded as living with parents. The corresponding figures for women were much lower, though for our data they are similar (men 27%, women 22%). Around a quarter of the divorced men and women in this group moved in with non-relatives, presumably friends. This figure is higher than is the case in our other groups, where non-relatives feature as a destination for the divorced in relatively small numbers only. In both instances it is perhaps easier to return home to parents or to move in with friends if you are on your own. McCarthy and Simpson (1991: 116) note that younger people who divorce will have had

Table 15.8 Group 5 (move with/join households) – who do the divorced move with by gender of divorced movers?

	Men		*Women*	
Divorced moves with	*Count*	*%*	*Count*	*%*
Partner	2	5	2	2
Partner and child	2	5	2	2
Child	12	32	84	85
Parent	2	5	0	0
Other relative	2	5	4	4
Non-relative	18	48	7	7
Total	38	100	99	100

Table 15.9 Group 5 (move with/join households) – who do the divorced meet with by gender of divorced movers?

Relationship flow	*Men*	*Women*
Move with non-relative, join non-relative	13 (34%)	5 (5%)
Move with child, join parent	1 (21%)	33 (33%)
Move with child, join partner	1 (3%)	26 (26%)
Move with child, join non-relative	1 (3%)	14 (14%)
Total	38	99

shorter marriages, and suggest it might be easier for people in that situation to return to the parental home.

Group 5: Moving together to join existing households

Our fifth group comprises households where our divorced mover plus someone else had moved into an existing household. There are 570 people in total in this group (265 men, 305 women), of whom 137 were divorced movers (38 men, 99 women). This is a small group and the data should be treated with caution. Nevertheless, two sets of relationships stand out; almost half of the men were moving with a non-relative compared to only 7 per cent of the women. In contrast, as shown in table 15.8, 32 per cent of men and 85 per cent of women had moved with their children.

Table 15.9 shows who these divorced people join for selected relationships. A third of the divorced men in this group had moved with a non-relative and joined others who were not related to them. However, only a small number of women were in a similar situation (5%). A fifth of men had joined their parents' households, taking their children with them and a third of the women in this group had also returned to parents with their children. This is in contrast to the group of lone movers above, where we had more men returning to parents. Of course, this reflects the fact that women are more likely to have custody of children after divorce. This is also clear when we look at who joined a partner: 3 per cent of men and 26 per cent of women had moved with their children into the home of a partner.

Group 6: Moving with, meeting with and joining

Our final group comprises households where our divorced person has moved from A (either alone or with someone else), another person has moved from B, and they have joined an existing household at C. Potentially, this is our most complicated group since we are dealing with three sets of relationships. However, it is also the smallest group. There are 236 individuals in this group (133 men, 103 women) of whom 53 were divorced movers (36 men, 17 women). All but three of the divorced men in the group moved alone.

Table 15.10 Movement patterns and living arrangements of divorced movers

	Lives with partner		*With partner and child*		*No partner*	
	Men	*Women*	*Men*	*Women*	*Men*	*Women*
Group 1	0	0	0	0	450	388
Group 2	256	165	116	137	93	550
Group 3	89	86	31	6	71	72
Group 4	44	58	71	9	211	110
Group 5	1	1	9	37	28	61
Group 6	3	1	7	2	26	14
Total	393 (26%)	311 (18%)	234 (16%)	191 (12%)	879 (58%)	1195 (70%)

Half of these men were living in households with unrelated people only, and a fifth had met or joined partners and children. A third of the divorced women movers were living with non-relatives only, either moving alone into non-related households or moving with non-relatives to join other unrelated people. Only three of the women in this group had a partner.

Discussion

At the beginning of the chapter we suggested that divorced people are not a homogeneous group but are likely to be living in a variety of situations, some akin to married people, others akin to single people. Our analysis divided the divorced mover population into six groups and examined the movements and living arrangements of each group by looking at who moved with our divorced movers and who they met up with or joined. The members of each of our groups were moving under different circumstances and the relationships within the households in each group cast further light on the situation, highlighting differences *within* each group and differences and similarities *between* them. We now want to bring this information together to give an overview of SAR divorced movers.

Table 15.11 Couple and couple-and-child households – numbers in couple relationships *before* the move and numbers in cohabiting relationships *after* the move

	Men		*Women*	
	Existing couple	*New couple*	*Existing couple*	*New couple*
Group 2	376	0	304	0
Group 3	8	108	8	82
Group 4	0	115	0	67
Group 5	4	6	4	34
Group 6	0	10	0	3
Total	388 (62%)	239 (38%)	316 (63%)	186 (37%)

Haskey (1995) suggests that divorced men enter new relationships more frequently than divorced women do and our data support this. Table 15.10 shows how many divorced movers were living in partner relationships, how many were living in family groups with a partner and children and how many did not have a co-resident partner. The gender differences are clear: 70 per cent of our divorced women had no resident partner compared with 58 per cent of divorced men. In contrast, more men were living with partners or with partners and children.

Table 15.11 looks in more detail at those who were living with a partner (or with a partner plus children) at the time of the census and shows how many of these divorced people were living as a cohabiting couple *prior* to the move (i.e. how many of our divorced people moved with their partner) and how many cohabiting partnerships were formed *by* the move (i.e. how many of our divorced people joined or met up with a partner). The table shows that a surprisingly high number of these cohabiting relationships are new ones. In our sample a third of divorced people who were cohabiting at the time of the census had not been living with their partner twelve months earlier. The rate of new partnerships applies equally to men and women. Relatively little is known at present about the breakdown of cohabiting relationships. However, recent analysis of longitudinal data from the British Household Panel Survey indicates that the separation rate for cohabiting couples is around four times higher than that of married couples (Buck and Scott 1994). The high level of new relationships among our divorced movers would seem to suggest a high turnover of relationships among the divorced population, which in turn goes some way to explaining the high rates of mobility of our divorced sample – accounting for 13 per cent of all divorced movers. A further 22 per cent of the divorced sample were living in established cohabiting couple relationships so in all, 35 per cent of our divorcees had partners living with them.

Table 15.12 gives details of those divorced movers who were not living with a partner at the time of the census. It shows that half of these men and a third of the women were actually living alone, and 43 per cent of women were living with their children only, compared with 6 per cent of men. This means, overall, that 35 per cent of the divorced mover population were liv-

Table 15.12 Living arrangements of divorced movers with no resident partner by gender of the divorced movers (all groups)

Divorced lives with	*Men*	*Women*
Alone	450 (51%)	388 (32%)
Non-relatives only	204 (23%)	103 (9%)
Child only	53 (6%)	518 (43%)
Parent only	94 (11%)	55 (5%)
Other relatives	27 (3%)	27 (2%)
Complex household	51 (6%)	104 (9%)
Total	879	1195

ing in a cohabiting relationship, 26 per cent were living alone, 10 per cent were living with non-relatives only, 17 per cent were living with children only, 5 per cent with parents only, 2 per cent with other relatives and 5 per cent in complex or multi-generational households.

When we began our analysis we expected to find divorced movers living with partners and/or children. We also expected that parents and, to a lesser degree, other relatives would feature in the relationships in the households in our sample. However the percentage of divorcees living with parents was less than 2 per cent for each of our first three groups. In the latter three groups (where the divorced mover was joining an existing household) there is more parental involvement, as illustrated in table 15.13. Those going back to parents were mainly the lone movers in Group 4 (88 men and 39 women). The numbers in Groups 5 and 6 are small but the pattern is clear, nevertheless. Earlier studies have shown that divorced men are more likely to return to the parental home than divorced women (Sullivan 1986) and this was certainly the case for our group of 'lone' movers: 27 per cent of lone mover men and 22 per cent of lone mover women moved into their parents' home. However, in our group where divorced movers move with other people there is a slightly higher percentage of women over men, as women return to the parental home with their children. Other relatives feature as a destination in very small numbers. However, non-relatives do seem to be an important destination for the divorced men in some of the groups (tables 15.6, 15.7 and 15.9) and for men without partners in particular (table 15.12).

Conclusion

Examining the relationships in divorced mover households shows that divorced movers are not all the same and puts into question any analysis that does not take account of the important differences in living arrangements noted here. In quantitative research it is common to compare marital status groups. Frequently, 'married' and 'remarried' people are put together as one group, and 'single', 'widowed' and 'divorced' are put together as another in order to separate those with partners from those without. However, our analysis demonstrates that marital status is not as straightforward as it seems, for a third of our divorced movers actually do have a partner living with them. Who people move with, meet or join influences (and in some instances explains) movement behaviour. Some of the moves we investigate here were

Table 15.13 Returning to the parental home

Divorced mover type	*Men*	*Women*
Alone to join parent	88 (27%)	39 (22%)
With to join parent	12 (32%)	36 (36%)
With, to meet and join parent	3 (8%)	5 (29%)
Total	103	80

related to the formation of new cohabiting relationships. Other moves were made by divorced people who were living in established couples or partner and children 'family' groups, and other moves were made by lone parent families, headed in the main by women.

Moving with someone – be it a partner or one's children – is different to moving alone, and joining an existing household or moving to meet up with a partner is different to moving and living alone. Gender cuts across this, for there are more divorced women in the population, though they have lower rates of mobility than divorced men. Divorced men are more likely than women to be living with a partner and women have custody of children in the majority of cases when marriages break down, so they are more likely to have children living and moving with them. All of these factors need to be taken into account in any study that compares divorced men and women, or their movements.

Acknowledgements

Our chapter arises from work on the Migration, Kinship and Household Change Project funded by the Economic and Social Research Council (reference L315253007). The project is concerned with residential change related to two key life events: movement associated with the breakdown of relationships and movement associated with the care needs of elderly people. The project is directed by Robin Flowerdew and Richard Davies at Lancaster University and Jennifer Mason at Leeds University. Lynn Hayes and Alaa Al-Hamad are research associates on the project. They would like to thank their colleagues on the team for helpful comments on earlier drafts of the chapter. The SAR are crown copyright and are provided through the Census Microdata Unit of the University of Manchester, with the support of the ESRC/JISC/DENI.

Notes

1 For example, one household contained a divorced man and woman who were coded as joint heads of the household. This may have been a cohabiting couple, a brother and sister or two unrelated friends. We have no way of knowing, so placed the household in our 'complex' category.

2 In 1991, 45,590 custody orders were granted to women compared with 4,968 granted to men (OPCS 1991: table 4.10a).

References

Austerberry, H. and Watson, S. (1983) *Women on the Margins: a Study of Single Women's Housing Problems*, Housing Research Group, City University, London.

Brailey, M. (1986) 'Splitting up – and finding somewhere to live', *Critical Social Policy* 17: 61–9.

Buck, N. and Scott, J. (1994) 'Household and family change', in N. Buck, J. Gershuny, D. Rose and J. Scott (eds) *Changing Households: the British Household Panel Survey*, ESRC Centre on Micro-social Change, University of Essex, Colchester, pp. 61–82.

Clark, D. (1987) 'Changing partners: marriage and divorce', in G. Cohen (ed.) *Social Change and the Life Course*, London: Tavistock, pp. 106–33.

Clulow, C. (1991) 'Making, breaking and remaking marriage', in D. Clark (ed.) *Marriage, Domestic Life and Social Change*, London: Routledge, pp. 167–87.

Cole, I. and Furbey, R. (1994) *The Eclipse of Council Housing*, London: Routledge.

Devis, T. (1983) 'People changing address: 1971 and 1981', *Population Trends* 32: 15–20.

Grundy, E. (1985) 'Divorce, widowhood, remarriage and geographic mobility among women', *Journal of Biosocial Science* 17: 415–35.

Grundy, E. (1989) *Women's Migration: Marriage, Fertility and Divorce*, London: Office of Population Censuses and Surveys, Longitudinal Study 4.

Haskey, J. (1992) 'Pre-marital cohabitation and the probability of subsequent divorce: analyses using new data from the General Household Survey', *Population Trends* 68: 10–19.

Haskey, J. (1995) 'Trends in marriage and cohabitation: the decline in marriage and the changing patterns of living in partnerships', *Population Trends* 80: 5–15.

Haskey, J. (1996) 'The proportion of married couples who divorce: past patterns and current prospects', *Population Trends* 83: 25–36.

Hayes, L., Al-Hamad, A. and Geddes, A. (1995) 'Marriage, divorce and residential change: evidence from the Household Sample of Anonymised Records', *Migration, Kinship and Household Change Working Paper* 3, Department of Geography, Lancaster University.

Leete, R. and Anthony, S. (1979) 'Divorce and remarriage: a record linkage study', *Population Trends* 16: 5–11.

McCarthy, P. and Simpson, B. (1991) *Issues in Post-divorce Housing*, Aldershot: Avebury.

Murphy, M. (1990) 'Housing consequences of marital breakdown and remarriage', in P. Symon (ed.) *Housing and Divorce*, Centre for Housing Research, Studies in Housing 4, University of Glasgow, Glasgow, pp. 1–46.

OPCS [Office of Population Censuses and Surveys] (1991) *Marriage and Divorce Statistics*, Series FM2, No. 19, London: OPCS.

Saunders, P. (1990) *A Nation of Homeowners*, London: Unwin Hyman.

Sullivan, O. (1986) 'Housing movements of the divorced and separated', *Housing Studies* 1: 35–48.

Symon, P. (1990) 'Marital breakdown, gender and home-ownership: the owner occupied home in separation and divorce', in P. Symon (ed.) *Housing and Divorce*, Centre for Housing Research, Studies in Housing 4, University of Glasgow, Glasgow, pp. 110–38.

Wasoff, F. and Dobash, R. E. (1990) 'Moving the family: changing housing circumstances after divorce', in P. Symon (ed.) *Housing and Divorce*, Centre for Housing Research, Studies in Housing 4, University of Glasgow, Glasgow, pp. 139–66.

16 Gender, migration and household change in elderly age groups

Emily Grundy and Karen Glaser

Introduction

Migration in elderly age groups

Age is a key variable in migration research and studies from a range of countries show a high degree of regularity in age-related variations in migration rates (Rogers 1988; Serow 1992). Migration rates are typically low in older adult age groups, although in a number of countries there is an identifiable peak around 'retirement age' (increasingly difficult to pinpoint through information on age alone) and a further increase in later old age (Rogers and Watkins 1987; Warnes 1983; Grundy 1987a; Bean et al. 1994; Warnes 1996). In England and Wales, rates of migration among very old people in their nineties are higher than in any other five-year age group in the population aged 55 years and over (Grundy 1987a). This pattern clearly represents the outcome of different types of events and motivations for moving. Younger 'retirement' migrants, no longer constrained locationally by the demands of paid work or the needs of children still at home, are regarded as moving for 'amenity' or 'lifestyle' reasons to increase their supply of what Graves and Linneman (1979) term 'non-traded' goods. These include features such as an attractive environment. A wish to release capital and reduce housing costs may also be an important motivation, particularly in prompting migration from metropolitan areas with high property prices (Clark and Davies 1990; Steinnes and Hogan 1992; Cribier and Kych 1992; Stuart 1987). Such migrants, predominantly couples, tend to be better educated, economically advantaged and in better health than non-migrants of the same age band (Grundy 1987a; Speare and Meyer 1988; Morrison 1990; Rogers, Watkins and Woodward 1990; Bean et al. 1994). The moves made by these migrants are often location-specific, with destinations chosen on the grounds of climate, proximity to the coast or mountains, or other environmental grounds (Warnes and Law 1982; Drysdale 1991).

The gender dimension in this type of migration that, as noted above, predominantly involves couples, is generally not regarded as an important issue and has received little specific attention. Arguably this is an omission of importance now that the implicit assumption often made that only men retire is no longer valid. In Britain in 1991, migration rates amongst older married couples in their late fifties and early sixties are higher when both have retired (Grundy 1987a) and there would seem a need for researchers now to consider the 'dual-retiree' household as well as the 'dual-career' one. This issue, however, is not the main concern of this chapter, which focuses to a larger extent on moves in later old age, among whom women are recognized to predominate.

Moves in later old age are more often viewed as a response to new constraints, rather than releases from former ones. In particular, deteriorating health and widowhood are posited as reasons prompting moves nearer to relatives and, for the more seriously disabled, into institutional care, the homes of relatives or other supported settings (Baglioni 1989; Bradsher et al. 1992; Grundy 1993; Silverstein 1995). Litwak and Longino (1987) are among those who have produced a typology of moves in later life which, in their version, posits a progression from amenity moves close to retirement, through 'kin-orientated' moves nearer relatives – perhaps as a response to widowhood or moderate health limitations – and, finally, 'disability-driven' moves in late old age (Biggar 1980; Wiseman 1980; Meyer and Speare 1985; Speare, Avery and Lawton 1991; Bradsher et al. 1992).

Numerous empirical studies suggest that this typology is a useful one, and that the characteristics and geographic destinations of 'young old' and 'old old' migrants differ substantially (Serow 1996). However, there are some limitations to such life-cycle or age-related typologies. First, use of terms such as 'developmental' to describe different types of migration in later life may be taken to imply that such moves are normative, whereas long-distance moves around retirement are in fact made by only small minorities. 'Kin' and 'disability-driven' moves in later old age may be more common, but they are far from universal. Moreover, recent research indicates that, as might be expected, reasons for migration in later life are mixed; retirement age 'amenity' moves, for example, are sometimes motivated by a wish to move nearer children (Warnes 1986; Ford and Warnes 1993). The migration of parties other than elderly people themselves also need to be considered, as a need for more support may result not in the elderly person moving, but in the move of a child nearer to them (including in some cases a move by a child to an elderly person's household) (Silverstein 1995). The classic differentiation between 'young old' and 'old old' migrants may also obscure other sources of heterogeneity in the elderly population which may be more important (Clark and Davies 1990; Bean et al. 1994).

Gender and migration in later life

The elderly population, and more particularly the very old population, is a predominantly female one. In the United Kingdom women constitute nearly 60 per cent of the population aged 60 years and over; two-thirds of the population aged 75 and over and three-quarters of the very old population aged 85 years or more (Grundy 1996). This necessarily implies that most 'older old' migrants are women. Moreover, the gender differential in mortality that underlies the unbalanced sex ratio in older age groups means that far more women experience widowhood than men do and that the duration of widowhood is longer. Moreover widowhood is more likely to have economic consequences for women (Disney, Grundy and Johnson 1996). Additionally, the prevalence of disability in older age groups is higher among women than men (Grundy 1997). As widowhood, disability and reductions in income are posited as important factors prompting migration in later old age, this would suggest that such migration is likely to be more common among women than men. Possibly there are other factors which may also differentiate male and female propensity to move in later old age. Women, for example, may have stronger attachments to their residential locations than men, or stronger links with children. However, virtually nothing is known about this, and only a few studies of gender differences in motivations for migration have been undertaken. In this chapter we use data from the Office of National Statistics (ONS) Longitudinal Study (LS) to examine gender differentials in the migration of the population aged 65 years and over during 1971–81 and 1981–91. Our main aims are to quantify differences in the migration behaviour of men and women in these age groups and to see how observed differences were related to gender variations in the experience of household change and poor health. The approach taken is largely descriptive, as this represents a necessary first step towards developing an understanding of the importance of gender in later-life migration patterns and specifying appropriate, testable hypotheses.

Data and definitions

The LS is a record linkage study based on a 1% sample of the population enumerated in the 1971 Census of England and Wales. Information on vital events, such as deaths, has been added to the records on sample members, together with data collected in the 1981 and 1991 censuses. The sample has been maintained through the addition of 1% of new births and immigrants. The strengths of the LS lie in the large sample size and the ability to track the circumstances of surviving sample members through three censuses. The data collected in these three censuses on migration are, of course, available within the LS. However, the relatively small proportions moving in the year before census (only a one-year migration question was included in the 1981 and 1991 censuses) restrict the utility of this indicator for studying the migration patterns of population subgroups, such as the very old. A further problem with one-year migration data is that yearly fluctuations in migration reflecting economic up- and down-turns are quite considerable.[1]

Fortunately, the linked nature of the individual data in the LS means that it is possible to examine ten-year migration rates based on the proportions who changed address between one census and the next, and it is these data that we use here to examine gender variations in the patterns of migration in elderly age groups. In order to look both at differences between time periods and at changes as sample members aged, we have adopted a cross-sequential approach and compare moves in the period 1971–81 and 1981–91 among the population aged 65 years and over at the start of the relevant decade. This means that the populations considered are separate but overlapping, as those in the 75 years and over age group considered in the second decade comprise all sample survivors who were aged 65 years and over in the 1971–81 period (plus any immigrants of the appropriate age subsequently added to the sample).

The ten-year migration indicators were derived in slightly different ways for the 1971–81 and 1981–91 decades. In both decades, comparisons of district, county and region codes on sample members' records allow identification of those moving between these administrative areas.[2] Additionally, following the 1981 linkage of LS members' reoccurs to 1971 information, a manual comparison of addresses in 1971 and 1981 was undertaken by ONS. This allowed generation of an 'any mover' variable which, together with the coded locality data, can be used to identify those who moved within districts. This costly exercise was not repeated after the linkage of 1991 Census records. Instead, ONS used a computerized system to allocate 1981 and 1991 addresses to enumeration district (ED) centroids and allocated a mover code to anyone whose ED centroid in 1991 differed from that in 1981. Enumeration districts typically include only some 200–300 households, so this approximation is sufficient to capture all movers except those moving very locally. A further limitation of the 1981–91 mover code derivation was that it was only applied to those living in private (non-institutional) households in 1991, while the 1971–81 exercise was carried out for everybody, whether resident in a private or a non-private household (Gleave 1997). In the analyses presented here we have assumed that all those who moved from a private household in 1981 to a non-private household in 1991 also changed address. This assumption should be valid except for a tiny minority living in households that became non-private between 1981 and 1991. This might apply, for example, to people living in sheltered housing schemes in 1981 where residents catered for themselves, if subsequently meals started being provided which would mean that by 1991 residents were classified as living in non-private households.[3]

A limitation of using ten-year indicators of change is that of course these can only be derived for surviving sample members. Although LS attrition rates for reasons other than death are very low (Hattersley and Creeser 1995), in elderly age groups attrition from death is of course high. Our data are also only sufficient to identify movers, not moves, so those moving several times in a decade cannot be distinguished from those moving only once. Moreover, we have no information about the timing of moves. Against these limitations must be set the strengths of large sample size referred to above and the ability to examine migration in relation to household change, which among the older old is hypothesized to be a key variable associated with moving. Other census variables are also available for analysis in relation to migration. In 1991 these included an indicator of health status based on responses to a question on long-standing illness that limited activities.

Results

Gender differences in migration 1971–81 and 1981–91

Table 16.1 shows migration rates 1971–81 and 1981–91 among men and women aged 65 years and over at the start of the interval. Movers within and between counties are distinguished. The table shows that women accounted for some two-thirds of the movers in the group initially aged 65–74 years and nearly 80% of movers aged 75 years or more at the first point of observation. To a large extent this simply reflects the increasing predominance of women in older age groups, a result of gender differentials in mortality. However, some rates of migration were also higher among women. Among those aged 65–74 years at the start of the relevant decade, the proportion of local within-county movers was some 10% higher among women than men, although there was no gender difference in rates of migration between counties. In the 75 years and over age group, both within- and between-county migration rates were higher among women than men and the extent of this female 'excess' was

Table 16.1 Migration status 1971–81 and 1981–91 by gender and age at start of interval

	Age and migration status							
	64–74 years				*75+ years*			
Period	*Within-county migrant*	*Between county migrant*	*All migrants*	*Non-migrant*	*Within-county migrant*	*Between county migrant*	*All migrants*	*Non-migrant*
1971–81								
% men	26.9	7.4	34.3	69.7	28.5	7.1	35.6	64.4
% women	30.3	7.4	37.7	62.3	35.6	9.12	44.7	55.3
Ratio w:m (x100)	113	100	110	89	125	128	126	86
Women as % of total	67.9	65.4	67.4	64.0	79.6	79.8	79.7	72.8
1981–91								
% men	25.6	8.5	34.1	65.9	30.5	8.6	39.1	60.9
% women	27.9	8.3	36.2	63.8	36.1	10.4	46.5	53.5
Ratio w:m (x100)	109	98	106	97	118	121	119	88
Women as % of total	65.1	62.6	64.5	62.3	78.2	78.4	78.2	72.6

greater. In both age groups female migration rates were slightly higher relative to those of men in the 1971–81 decade than in 1981–91. Previous empirical and theoretical research would suggest that these gender and period differences are likely to reflect different 'exposures' to events that may trigger migration, including widowhood, other changes in household circumstances and deteriorating health. Data within the LS allow some examination of these issues as it is possible to analyse migration in relation to household change and, for the 1981–91 period, in relation to long-standing illness.

Family/household type, household change and migration

Table 16.2 shows the proportions of men and women who moved between 1971 and 1981, or 1981 and 1991, by family/household type at the start of the relevant decade. It includes all surviving LS members regardless of household type at the end of the relevant decade (including those who by then were in non-private households) but excludes those already in non-private households at the start of the decade. The classification used has been derived from the census information on family membership[4] and position in the household. The first two categories shown in table 16.2 are self-explanatory. The category 'married couple+' comprises those living in a couple, one of whom is the head of household, together with never-married child(ren) and/or other relatives or friends. Those living with other people who are not members of their family unit we describe as living in 'complex' households; in the age groups considered here most of these are living with ever-married children. The very small proportion of people living in two-family households has also been assigned to this category. This group would thus include a widow living with a daughter and son-in-law, or, less usually, a widow living with a married sister and brother-in-law. However, an LS member who was married and lived with her spouse and a widowed sister would be assigned to the 'married couple+' category.

Table 16.2 shows that in the younger of the two age groups, and among older men, migration rates tended to be lower for those initially living in families than for those living alone or in complex households. Among women aged 75 years and over, rates of migration were high for those initially living just with a spouse, considerably higher than for men in the same type of household. This reflects the fact that far fewer women than men who lived with a spouse at the start of each reference period were still in that type of household at the end of the period. As shown in table 16.3, fewer than a quarter of women aged 75 years and over in 1981 and living in a married couple household were still in that type of household ten years later; nearly half by then living alone and nearly a fifth resident in an institution. Among the equivalent group of men, by contrast, 56% were still living with a spouse, just over a quarter were living alone and 11% were in an institution. Differences in the younger age group considered were similarly marked,

Table 16.2 Migrants (%) 1971–81 and 1981–91 among men and women aged 65 years and over by age and family/household type at start of interval

	Period, age and gender							
	1971–81				*1981–91*			
	65–74 years		*75+ years*		*65–74 years*		*75+ years*	
Family/household type	Men	Women	Men	Women	Men	Women	Men	Women
Solitary	44.0	40.7	37.4	47.7	35.5	37.0	44.2	47.9
Married couple alone	33.6	37.5	36.7	50.2	34.1	36.5	36.1	47.4
Married couple+	28.6	27.4	19.1	33.3	30.3	30.9	35.3	27.3
Lone parent	28.3	35.5	35.6	31.2	30.6	29.8	39.5	34.1
Complex	40.6	38.5	41.4	41.5	40.0	39.5	50.9	48.9
All	34.3	37.7	35.6	44.7	34.1	36.2	39.1	46.5

although the proportions of both men and women still in married couples were understandably much higher.

Consolidating this analysis, figures 16.1 and 2 show migration rates for men and women initially in a married couple household by type of household at the end of the relevant decade. The proportion of migrants was much higher for those who had made a transition to living in a complex household and, in this group, slightly higher for those in the older age group. However, differences between men and women in these rates of migration (and between the younger and older age group) were generally slight. This indicates that the migration rates of women initially living with a spouse were higher than those of men, not because widows are more likely to move than widowers, but because more women experience the loss of a spouse.

Of course the end of a marriage represents only one type of household

Table 16.3 Family/household type in 1981/91 of men and women who ten years earlier lived in married couple households[1] and were then aged 65 years and over

		Age group ten years earlier			
		65–75 years		*75+ years*	
Family/household type		*Women*	*Men*	*Women*	*Men*
1981	Married couple	75.1	46.6	55.3	25.7
	Solitary	16.1	38.4	24.2	38.9
	Lone-parent	2.1	4.8	3.7	5.2
	Complex	4.8	7.2	9.9	13.9
	Non-private	1.8	3.0	6.9	15.4
1991	Married couple	77.1	48.8	55.6	23.5
	Solitary	16.5	39.3	26.7	45.9
	Lone-parent	1.5	3.4	2.1	3.8
	Complex	2.5	3.9	4.6	7.1
	Non-private	2.4	4.6	11.1	19.7

Note

1 Those in married couple only and married couple+ households.

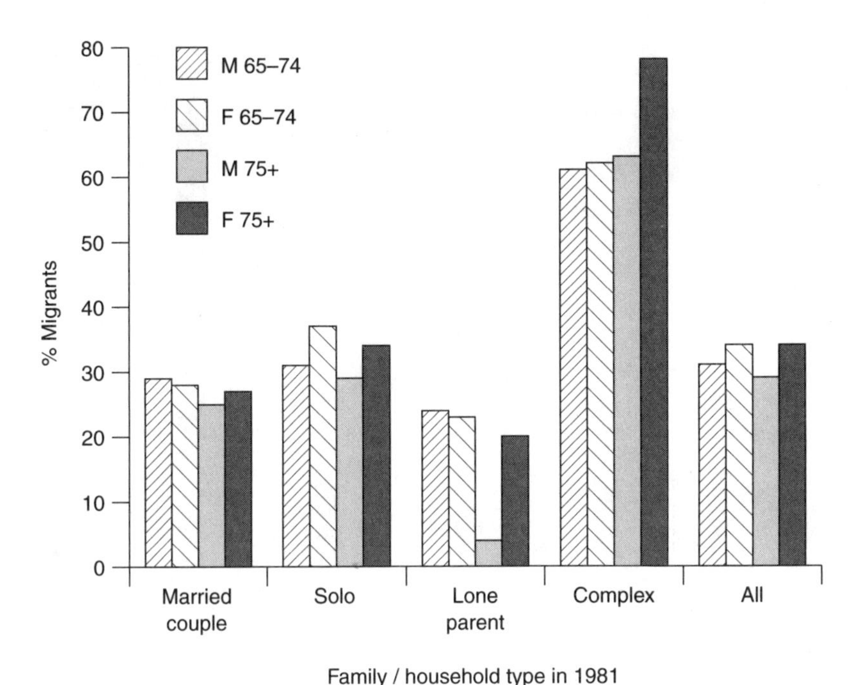

Figure 16.1 Migrants 1971–81 in married couples in 1971 and by household type in 1981

change. The departure, or death, of other co-residents (children, siblings, and so on) will also result in changes in household composition and decreases in household size. Changes in household composition may also be associated with increases in household size. Such changes are particularly relevant in the context of analysing the migration patterns of elderly people as in nearly all circumstances they will involve the move of at least one person. An elderly person living alone in one census but with others in the next must have either moved himself or herself or have been joined by someone who moved. Not surprisingly, numerous studies show strong associations between household change and migration (Grundy 1987b; Bartiaux 1988; Speare and McNally 1992). Table 16.4 reiterates this point, showing that migration rates were much higher among those experiencing a change in family/household type than among those in the same broad family/household type in two successive censuses. This table shows the proportion of men and women who experienced a change in the type of household they lived in,[5] the proportion in the same type of household and the proportions of these categories that moved.

Despite the higher incidence of widowhood among women, there was very little difference in the younger age group in the proportions of men and women who made a transition from one type of private household to another. For those aged 75 years and over this proportion was lower among women than men. Among those who changed from one type of private household to anoth-

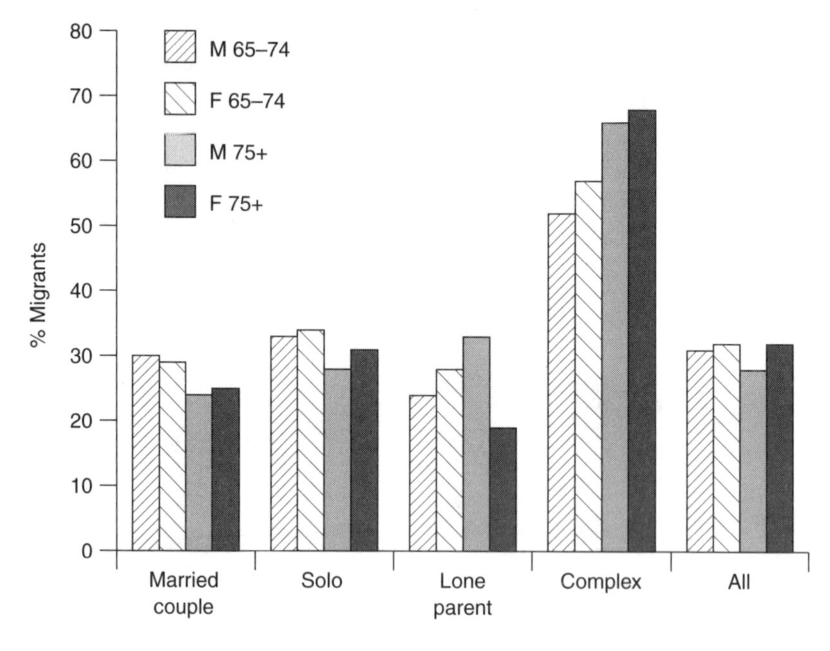

Figure 16.2 Migrants 1981–91 in married couples in 1981 and by household type in 1991

er, rates of migration were, however, higher for women than for men in the 1971–81 decade, particularly among those aged 75 years and over, but much less so in the 1981–91 period. This difference between decades may be because of the higher rate of transitions to institutions in 1981–91 (Grundy and Glaser 1997). By definition, all those moving to non-private households were migrants. Among women aged 75 years and over, transition rates to such households were high and much higher than those of men. Indeed in the second decade this type of transition was at the same level for very elderly women as the rate of transition between different types of private household. In short, frail elderly women in need of support were more likely to move to an institution, rather than to a different type of private household, in the second decade, partly reflecting policy changes which made institutional care more available in the 1980s (Laing 1993; Grundy and Glaser 1997).

Household change, health and migration

As noted earlier, the 1991 census included a question of limiting long-standing illness. Indicators of health status have been shown to be associated with differentials in the migration rates of elderly people in a wide range of studies (Grundy 1987a; Baglioni 1989; Speare, Avery and Lawton 1991; Al-Hamad, Flowerdew and Hayes 1997). Migrants who move long distances

Table 16.4 Household change, and migration among those living in a different type of household at the end of the interval, by age at start of the interval 1971–81 and 1981–91

Period and household change		*Age* 65–74 *years* *Percentage*	*Percentage of these migrants*	75+ *years* *Percentage*	*Percentage of these migrants*
1971–81					
Different type of private household	Men	32.3	38.0	32.7	36.1
	Women	35.3	41.5	25.6	49.0
Non-private household type	Men	2.7	100.0	9.6	100.0
	Women	4.7	100.0	16.0	100.0
All in changed household type	Men	35.4	42.6	42.2	50.5
	Women	40.0	48.4	41.6	68.6
Same household type	Men	64.6	29.6	57.8	24.9
	Women	60.0	30.7	58.4	27.8
1981–91					
Different type of private household	Men	29.9	36.7	30.1	36.6
	Women	34.0	37.7	23.5	39.3
Non-private household	Men	3.3	100.0	14.7	100.0
	Women	6.1	100.0	24.3	100.0
All in changed household type	Men	33.1	42.8	44.8	57.4
	Women	40.1	46.7	47.7	70.2
Same household type	Men	66.9	29.8	55.2	24.3
	Women	59.9	29.2	52.3	24.9

around the age of retirement, for example, appear to be healthier than non-migrants, suggesting that, as would be expected, health limitations restrict this type of migration (Grundy 1987a). However, in late old age increased frailty may provide a major impetus for migration for support reasons (Speare and Meyer 1988; Longino et al. 1991; Speare, Avery and Lawton 1991; Grundy 1993). Table 16.5 shows rates of migration 1981–91 for men and women aged 65 years and over in 1981 by their family/household type then, by change in family/household type by 1991, and by whether or not in 1991 they had a long-standing illness which limited their activities.

For those in the same family/household type in 1981 and 1991, migration rates for men and women, and for those with and without a long-standing illness were remarkably similar. Some differences by age were apparent; 65–74-year-olds with long-standing illness and living alone or with a spouse had slightly higher migration rates than their counterparts aged 75 years and over in 1981. For those without long-standing illness, a similar age differential was evident for those in married couples in 1981. Migration rates among those

Table 16.5 Percentage migrants 1981–91 among men and women aged 65 years and over in 1981 by household/family type in 1981, whether in a different household/family type in 1991 and health status in 1991

Age and family/household type in 1981	*Household/family type and health status in 1991*							
	Same household/family type				*Changed household/family type*			
	With illness[1]		*No illness*		*With illness*		*No illness*	
	Men	*Women*	*Men*	*Women*	*Men*	*Women*	*Men*	*Women*
65–74 years								
Solitary	27.7	31.5	24.1	27.7	79.7	86.8	66.7	58.9
Married couple	30.4	30.1	30.9	29.3	44.6	44.6	33.7	35.5
Lone parent/complex	30.0	28.5	24.7	25.3	56.2	45.9	42.4	39.8
75+ years								
Solitary	19.2	25.1	24.1	22.5	90.9	94.1	92.9[3]	79.2
Married couple	23.3	25.5	25.3	22.5	53.4	58.0	40.8	35.6
Lone parent/complex	25.0	26.5	43.2[2]	29.3	75.4	72.9	27.8[3]	45.1

Notes
1 Illness = limiting long-standing illness.
2 Denominator < 50.
3 Demoninator < 20.

whose family/household type had changed were much higher and, in both age groups and for both men and women, were highest among those with a long-standing illness. Indeed, migration rates for those with both long-standing illness and changed family/household type were very high, particularly for those who had been living alone in 1981. This reflects to a large extent the high rate of institutionalization in this group (Grundy and Glaser 1997). As with those in the same household type, gender differences within categories were slight. It is likely that those with a long-standing illness whose family/household type had changed had more severe health limitations than those with an illness whose family/household remained the same; unfortunately the health status variable within the data set gives no indication of the severity of health related functional limitations. In short, family/household change may well often have been a response to disabilities requiring more support than available in sample members' initial household circumstances.

Conclusion

These results show that most migrants in the population aged 65 years and over are women and that women contribute nearly four-fifths of migrants in the 75 years and over age group. Most of this 'excess' is demographically determined and simply reflects the preponderance of women in these age groups. Rates of migration, however, were also slightly higher among women than men in both 1971–81 and 1981–91. This seems to reflect gender differences in household circumstances, household change – including transitions to institutions – and health, rather than any more specific difference by gender in propensity to move. Family/household change is a particularly important factor in this age group and rates of migration among those experiencing such changes were much higher than among those remaining in the same family/household type. In the second decade considered here, rates of transition to institutions were much higher, and rates of transition to complex households much lower, than in 1971–81 and, as a result, the relative balance of moves to institutions and moves to other types of private household changed. However, the overall volume of migration in the two decades considered was very similar. Finally, the analyses presented here are necessarily constrained by the data, which include no directly gathered information on motivations for moving. We are unable to say whether women and men have similar or different frameworks in which they reach decisions about moving. Research on this question would seem an important topic for the future.

Notes

1 Migration rates in the early 1980s, for example, were atypically low as a result of recession and an associated downturn in the housing market and so the one-year migration data collected in the 1981 Census were also atypically low (Stillwell, Rees and Boden 1992).
2 There was a major reorganization of administrative units in 1974, but ONS recoded 1971 addresses to post-1974 boundaries.
3 It is possible to assess the extent of this potential bias by seeing what proportion of the 1971–81 sample changed from living in a private to a non-private household without changing address. Among those aged 65 and over in 1971, this proportion was only 0.8 per cent, strongly indicating that the assumption made for 1981–91 that all making this type of household change were also movers is valid.
4 ONS, in common with most other census offices, defines families in strictly nuclear terms to include those living with a spouse (or cohabitee), a never-married child (of any age) or, for those who themselves are never married, a parent. Grandparents living with never-married grandchildren are also counted as families if the intervening generation is missing. People living alone or with friends and relatives other than spouses or never-married children are classified as living outside a family, although the household they live in may include a family unit.
5 The indicator of change in household type is based on a comparison of family/household type at the start and end of each time period and so does not capture all transitions. In some cases there will be missed changes of circumstances in the inter-censal period, in others people may be in the same broad household type at both points of measurement, but not with the same people (for example, those who were widowed and then remarried or those living with a divorced daughter at one census but with a sibling in the other).

Acknowledgements

The Economic and Social Research Council as part of its Population and Household Change Programme (reference L31525301) supported the work reported here. We thank the Office of National Statistics for access to Longitudinal Study data and Kevin Lynch and colleagues in the LS Support Unit at SSRU (City University) for considerable help with data extraction. The LS Support Unit is itself supported by the ESRC.

References

Al-Hamad, A., Flowerdew, R. and Hayes, L. (1997) 'Migration of elderly people to join existing households: some evidence from the 1991 household Sample of Anonymised Records', *Environment and Planning A* 29: 1243–55.

Baglioni, A.J. (1989) 'Residential relocation and health of the elderly', in K.S. Markides and C.L. Cooper (eds) *Aging, Stress and Health*, New York: John Wiley and Sons, pp. 119–37.

Bartiaux, F. (1988) 'A household dynamics approach to the analysis of elderly migration in the United States', *University of Colorado Population Program Working Paper* 86–1, Boulder, Colorado.

Bean, F.D., Myers, G.L., Angel, J.L. and Galle, O.R. (1994) 'Geographic concentration, migration and population redistribution among the elderly', in L.G. Martin and S.H. Preston (eds) *Demography of Aging*, Washington DC: National Academy Press, pp. 319–51.

Biggar, J.C. (1980) 'Who moved among the elderly 1965–70: a comparison of types of older movers', *Research on Aging* 2: 73–91.

Bradsher, J.E., Longino, C.F., Jackson, D.J. and Zimmerman, R.S. (1992) 'Health and geographic mobility among the recently widowed', *Journal of Gerontology* 47: S261–S268.

Clark, W.A. and Davies, S. (1990) 'Elderly mobility and mobility outcomes: households in the later stages of the life course', *Research on Aging* 12: 430–462.

Cribier, F. and Kych, A. (1992) 'La migration de retraite des Parisiens: une analyse de la propension au départ', *Population* 3: 677–718.

Disney, R., Grundy, E. and Johnson, P. (eds) (1996) 'Retirement and retirement plans: analysis of the follow-up survey', Report submitted to the Department of Social Security, United Kingdom.

Drysdale, R. (1991) 'Aged migration to coastal and inland centres in NSW', *Australian Geographical Studies* 29: 268–84.

Ford, R. and Warnes, A. (1993) 'The process of mobility decision making in later old age: early findings from an original survey of elderly people in South East England', *Espace, Populations, Sociétés* 3: 523–32.

Gleave, S. (1997) 'Introduction to migration data in the LS', in S. Gleave and L. Hattersley (eds) *Migration Analysis Using the ONS Longitudinal Study*, London: HMSO.

Graves, P.E. and Linneman, P.D. (1979) 'Household migration: theoretical and empirical results', *Journal of Urban Economics* 7: 383–404.

Grundy, E. (1987a) 'Retirement migration and its consequences in England and Wales', *Ageing and Society* 7: 57–82.

Grundy, E. (1987b) 'Household change and migration among the elderly in England and Wales', *Espace, Populations, Sociétés* 1: 109–23.

Grundy, E. (1993) 'Moves into supported private households among elderly people in England and Wales', *Environment and Planning A* 25: 1467–79.

Grundy, E. (1996) 'Population review: the population aged 60 and over', *Population Trends* 84: 14–20.

Grundy, E. (1997) 'The epidemiology of aging', in J.C. Brocklehurst, R. Tallis and H. Fillit (eds) *Textbook of Geriatric Medicine and Gerontology*, 5th edition, Edinburgh: Churchill Livingstone, pp. 3–20.

Grundy, E. and Glaser, K. (1997) 'Trends in, and transitions to, institutional residence among older people in England and Wales, 1971–91', *Journal of Epidemiology and Community Health* 51: 531–40.

Hattersley, L. and Creeser, R. (1995) *Longitudinal Study 1971–1991: History, Organization and Quality of Data*, London: Office of Population Censuses and Surveys, Longitudinal Study 7.

Laing, W. (1993) *Financing Long-term Care: the Crucial Debate*, London: Age Concern.

Litwak, E. and Longino, C.F. (1987) 'Migration patterns among the elderly: a developmental perspective', *The Gerontologist* 27: 266–72.

Longino, C.F., Jackson, D.J., Zimmerman, R.S. and Bradsher, S.E. (1991) 'The second move: health and geographic mobility', *Journal of Gerontology* 46: S218–S224.

Meyer, J.W. and Speare, A. (1985) 'Distinctly elderly mobility: types and determinants', *Economic Geography* 61: 79–88.

Morrison, P.A. (1990) 'Demographic factors reshaping ties to family and place', *Research on Aging* 12: 399–408.

Rogers, A. (1988) 'Age patterns of elderly migration: an international comparison', *Demography* 25: 355–70.

Rogers, A. and Watkins, J.F. (1987) 'General versus elderly interstate migration and population redistribution in the United States', *Research on Aging* 9: 483–529.

Rogers, A., Watkins, J.F. and Woodward, J.A. (1990) 'Interregional elderly migration and population redistribution in four industrialized countries: a comparative analysis', *Research on Aging* 12: 251–93.

Serow, W.J. (1992) 'Unanswered questions and new directions in research on elderly migration: economic and demographic perspectives', *Journal of Ageing and Social Policy* 4: 73–89.

Serow, W.J. (1996) 'Demographic and socioeconomic aspects of elderly migration in the 1980s', *Journal of Ageing and Social Policy* 8: 19–38.

Silverstein, M. (1995) 'Stability and change in temporal distance between the elderly and their children', *Demography* 32: 129–46.

Speare, A., Avery, R. and Lawton, L. (1991) 'Disability, residential mobility and changes in living arrangements', *Journal of Gerontology* 46: S133–S142.

Speare, A. and McNally, J. (1992) 'The relation of migration and household change among elderly persons', in A. Rogers (ed.) *Elderly Migration and Population Redistribution*, London: Belhaven Press, pp. 61–76.

Speare, A. and Meyer, J.W. (1988) 'Types of elderly residential mobility and their determinants', *Journal of Gerontology* 43: 74–81.

Steinnes, D.N. and Hogan, T.M. (1992) 'Take the money and run: elderly migration as a consequence of gains in unaffordable housing markets', *Journal of Gerontology* 47: S197–S203.

Stillwell, J., Rees, P. and Boden, P. (1992) 'Internal migration trends: an overview', in J. Stillwell, P. Rees and P. Boden (eds) *Migration Processes and Patterns. Volume 2. Population Redistribution in the United Kingdom,* London: Belhaven Press, pp. 28–55.

Stuart, A. (1987) 'Migration and population turnover in a London Borough: the incidence and implications of retirement out-migration', *Espace, Populations, Sociétés* 1: 137–51.

Warnes, A.M. (1983) 'Migration in late working age and early retirement', *Socio-Economic Planning Science* 17: 291–302.

Warnes, A.M. (1986) 'The residential histories of parents and children, and relationships to present proximity and social integration', *Environment and Planning A* 16: 1581–94.

Warnes, A.M. (1996) 'Migrations among older people', *Reviews in Clinical Gerontology* 6: 101–14.

Warnes, A.M. and Law, C.M. (1982) 'The destination decision in retirement migration', in A.M. Warnes (ed.) *Geographical Perspectives on the Elderly*, Chichester: Wiley, pp. 53–81.

Wiseman, R. (1980) 'Why older people move: theoretical issues', *Research on Aging* 2: 141–54.

17 Differential migrations through later life

Anthony Warnes

Introduction

This chapter examines the contrasting migration experience of older men and older women in contemporary Britain, with occasional references both to the past and to other affluent nations. It has two main sections; the first reviews aggregate migration differentials in the British population with special reference to men and women at different stages of later life and in different marital statuses and housing situations; and the second considers gender differentials in the migration process, specifically in the motivations for moves, the constraints and opportunities to move, the immediate consequences of moves, and their long-term implications. The focus is on migrations into general or community housing and not into institutions.

While the large field of migration studies has been concerned predominantly with migrations by working-age people and families, interest in the moves made by older people has grown strongly over the last twenty years (see Bean et al. 1994; Longino 1996; Rogers et al. 1992; Warnes 1996). As with other changes in old-age lives, academic studies in recent years have been stimulated and partly guided by the rising concern of governments with the cost of elderly people's support. The age group receives a large share of public expenditure, primarily on pensions and the health services, but also on various forms of specialized and 'supported' accommodation and, in areas of rising retirement populations, on the physical infrastructure. Well-conceived and timely migrations make a contribution to the prolongation and improved quality of independent living among older people. There is a practical return from increasing our understanding of the housing and locational requirements of older people in various household, social network and health states.

Contemporary later life is extended. The last three decades have seen accelerating improvements in later life survival in most affluent countries (Kendrick and Warnes 1997; United Nations Organization 1993; Warnes 1998). In England and Wales by the early 1990s, the mean life expectancy of women aged 65 years had attained 18 years, an increase of 25 per cent in four decades. This is shown in table 17.1. Mean life expectancy at 80 years increased even faster during 1951–91, for women by 68 per cent. Until the early 1970s, there were widening sex differentials in survival after 65 years, but the trend has since reversed. Increased survival is, however, but one of several fundamental changes in old-age lives: two others of recent decades are massive reductions in old-age poverty (although it is far from eliminated even in the richest countries) and substantial reductions in later-life economic activity rates. Older people live longer, they have considerably more resources than previous generations, and fewer continue to work.

These changes have stimulated a pervasive reconceptualization and reconstruction of the nature, roles and activities of old age; and they have a hand in the increasing differentiation of older people by age and income. Commentators not surprisingly speculate about the principal divisions or stages of contemporary old age (Erikson, Erikson and Kivnick 1989). Peter Laslett (1989) has articulated a thesis of a 'third age', which follows work and child-raising and precedes frailty, incapacity and (particularly among single and widowed women) poverty. He argues that the third age will increasingly become the most fulfilling stage of people's lives, when there is time for the enjoyment of intimate and social relationships, and for the pursuit of intellectual and creative interests. Few American commentators have adopted the concept, but in both proselytizing and analytical literature many do subscribe to the associated ideas of 'successful' and 'positive' ageing (Baltes and Baltes 1990; Rowe and Kahn 1987).

The inferences most relevant to the themes of this book are that the residential requirements in later life are dynamic and at 65 years of age they differ substantially from those at 95 years. The two ages are distinguished by con-

Table 17.1 Increase of mean life expectancy at various base ages, England and Wales

Base age		*Mean life expectancy (years)*				*Annual rate of increase (%)*		
		1891–00	*1950–2*	*1970–2*	*1990–2*	*1896-&**	*1951–71*	*1971–91*
Males	E^0	44.1	66.4	69.0	73.2	0.53	0.19	0.30
	E^{65}	10.3	11.7	12.2	14.2	0.34	0.21	0.76
	E^{80}	4.2[1]	4.7	5.7[2]	6.4	0.44	0.88	0.65[3]
Females	E^0	47.8	71.5	75.2	78.7	0.52	0.25	0.22
	E^{65}	11.3	14.3	16.1	17.9	0.48	0.60	0.53
	E^{80}	4.6[1]	5.0	7.3[2]	8.4	0.64	1.74	0.78[3]
Female: male ratio	E^{65}	1.1	1.2	1.3	1.3	1.41	2.86	0.70
	E^{80}	1.1	1.1	1.3	1.3	1.45	1.98	1.20

Source: OPCS 1994, table 15.

Notes

E^x is the conventional notation for representing men remaining expectancy of life at age x (in years) for a particular year or short period (indicated iun the column headings).

1 Estimated from schedule of age-specific death rates.

2 1972–4.

3 1973–91.

trasting income, health and housing characteristics, and these underlie the different frequencies, motivations, distances and destinations of the migrations undertaken by young and by old elderly people. Litwak and Longino's (1987) developmental model proposed three successive phases of residential requirements, each of which *may* stimulate a migration. The first phase, early retirement, is when good health, income and wealth permit a positive outlook on life and developmental changes. Among the moves made by older people, environmental, amenity and 'lifestyle' considerations are most evident at relatively young ages and mark the life-course transition to retirement. Second-phase moves mark the onset of restrictions of lower income, frailty or ill health, and give more emphasis to a location accessible to services and support. Defensive migrations become more common at this stage. Third-phase migrations take place when a person is unable to live independently and must move either to live with or close to informal supporters or carers, or into a 'supported living environment'. The three phases are not tied to particular chronological ages, nor does every older person progress neatly through the sequence. But the stages do represent the changing composition of migration types with increasing age. They emphasize that migrations in old age are markedly heterogeneous. The interpretation of aggregate patterns and of their variations by age must therefore proceed carefully.

Gender, age and marital-status migration differentials in later life

The following differentials among the migrations undertaken by older people have been demonstrated in many studies:

- The migration rate among people aged 60 years or more is generally low relative to working-age adults.
- The average distance of migrations by older men is generally greater than those by older women.
- Around the modal retirement age, men display a modest and brief peak in their migration rates, which is more sharply defined than a similar peak at a slightly younger age for women.
- There is an exponentially rising rate of migration with increased age after the mid-seventies, and the few who survive to their late eighties and beyond have unusually high rates.
- In advanced old age (75 years and above), women's age-specific rates of migration are higher than men's.
- In advanced old age, women tend to have a higher rate of migration into institutional accommodation than men, which is generally ascribed to the larger proportion that become widowed and their higher rates of disability.

The United Kingdom decennial censuses collect the current address and the address one year before the census night of every enumerated person

(Rhind 1983). This comprehensive record of the population's residential mobility enables detailed profiles and analyses of migrants and migrations to be produced. Table 17.2 presents the 1990–91 migration rates for the resident population of Great Britain by sex and marital status and for selected older age groups. The rates for the both-sex population are given, and the ratios of the sex-specific to the overall rates. During the year, 3.9 per cent of the 50+ years population moved, just 30 per cent of the rate for younger people. Older men were slightly less likely to have migrated than older women, but the differential was mainly the result of the age and marital-status compositions of the male and female populations. A higher proportion of men are married (with a low migration rate), and a higher proportion of women are widowed (with a high rate). As table 17.2 shows, up to 79 years of age, men in all marital-status groups had higher age-specific migration rates than women. It is above these ages that women are more likely to move. An approximation to a period 'migration expectancy', or the overall likelihood of undertaking migrations over a period, is estimated by summing the rates

Table 17.2 Migration rates by age, sex and marital status, persons aged 50+ years and at the statutory retirement ages, Great Britain, 1990–91[1] (percentages)

		Marital status				
Age group	*All*	*Never married*	*First marriage*	*Remarried*	*Widowed*	*Divorced*
Rates per 100 of the both sex population						
50–64	3.7	4.5	2.8	5.8	4.4	7.3
65–79	3.4	4.1	2.6	4.0	4.3	5.3
80+	6.2	7.0	4.0	4.1	7.0	6.0
50+	3.9	4.8	2.8	5.1	5.2	6.8
1–49	12.7	14.0	10.4	11.6	8.1	16.4
50+/1–49	0.31	0.34	0.27	0.44	0.64	0.41
1+	9.9	13.4	7.1	8.6	5.3	13.0
Ratio of male to both-sex rate						
50–64	1.03	1.04	1.03	1.03	1.00	1.14
65–79	0.96	1.11	1.02	1.03	1.02	1.13
80+	0.84	1.04	0.94	0.96	0.96	1.15
50+	0.96	1.01	1.02	1.03	0.98	1.15
Ratio of female to both-sex rate						
50–64	0.97	0.94	0.97	0.96	1.00	0.88
65–79	1.03	0.92	0.98	0.96	1.00	0.91
80+	1.07	0.99	1.09	1.07	1.01	0.94
50+	1.03	0.99	0.98	0.97	1.00	0.88

Source: OPCS/GRO (S) 1994, table 4.

Note
1 At a different address within or outside Great Britain one year before census night in 1991.

for each single year of age. For those who live from 50 to 80 years of age (a little over the mean life expectancy), the average number of moves made by men and women in the 1990–91 schedule was identical at 1.06. Among the few who survived to 95 years, however, a sex differential emerges, with men accumulating 2.1 moves and women 2.3.

Turning to marital status, older people in first marriages had the lowest annual migration rate (2.8 per cent), and the lowest mobility relative to younger adults. The most migratory group of older people was the divorced, for their 6.8 per cent annual rate was 74 per cent above the general figure. The other three marital-status groups (never-married, remarried and widowed) shared an intermediate migration rate of around 5 per cent per year, but with contrasting age associations through later life. The rates for first-married, single and widowed persons climbed steeply through the oldest ages, whereas those for the remarried and divorced were highest before 65 years of age. The greatest gender differentials were among the divorced, with the male rate being around 14 per cent above the both-sex figure throughout later life, and the female rate 6–12 per cent below.

Further interesting differentials by sex, age and marital status are shown by the five-year age group rates in figure 17.1, although all are dominated by the steep rise in the migration rate after 75 years of age. This exponential rise, dubbed the 'late age slope', is common to many western countries and was first established by Rogers and Castro (1986). It has only recently been confirmed for the United Kingdom, largely because of poor age breakdowns in the pre-1991 census migration tabulations (Rees 1992; Warnes 1992). During 1990–91, men and women in first marriages had a very low annual migration rate throughout their fifties and sixties, but among those in their late seventies and older a steep exponential rise and a clear sex difference appears. The progressively widening sex differential in advanced old age is partly explained by the generally lower age of wives than husbands. In advanced old age, moves by married couples (or partners), either to more convenient dwellings or locations or into supported or sheltered accommodation, raise the age-specific migration rate of women who are younger than men.

There are interesting contrasts between people in first marriages and second (and subsequent) marriages. Remarried people in their fifties have a relatively high annual migration rate – many of the moves are probably associated with the formation of the new relationship. On the other hand, remarried people in their eighties and over had a low migration rate. The average difference between husband's and wife's ages is greater in second than first marriages, and this greater variation reduces the average age of a couple registered by its oldest member. This in turn reduces the likelihood of either partner having an incapacitating illness or very low income and lowers the incidence of impelled migrations. Various selection effects for health and 'willingness to provide care' can also be postulated in second marriages, particularly those contracted very late in life.

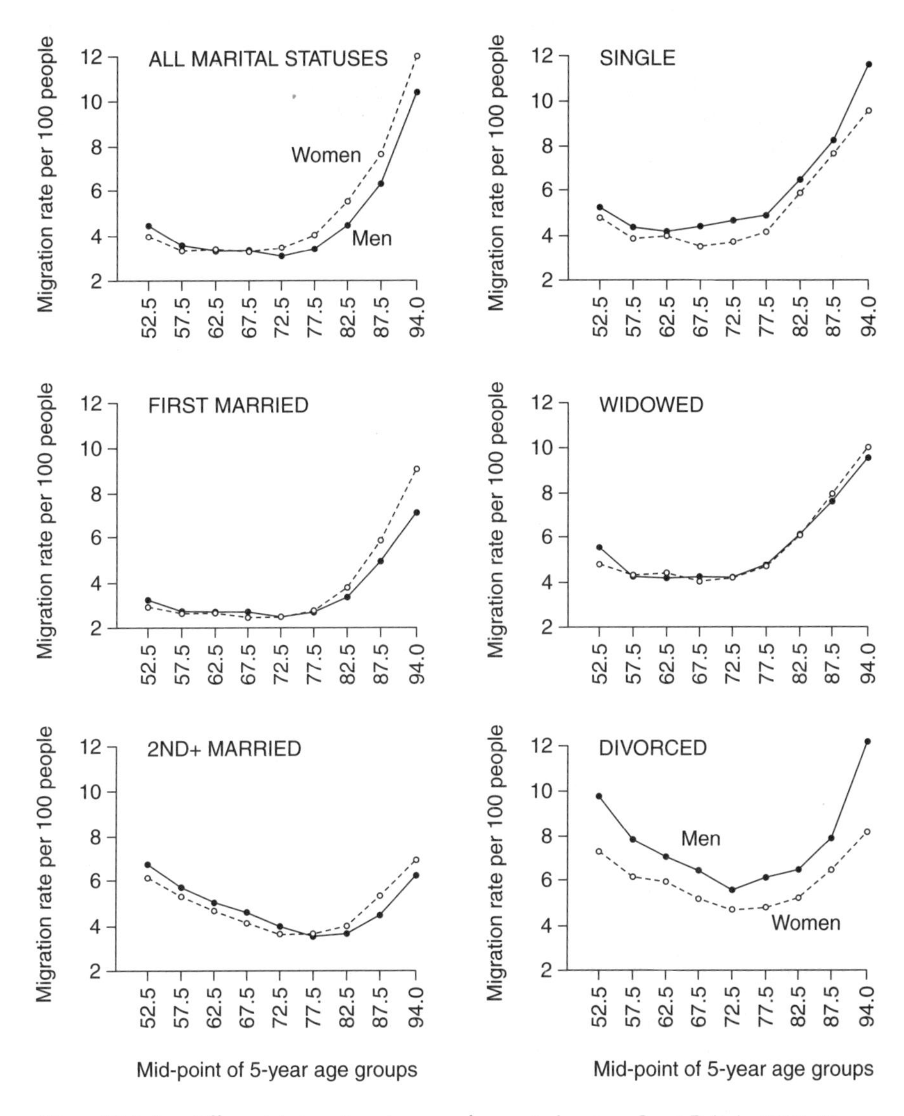

Figure 17.1 Sex differentials in migration rates by marital status, Great Britain 1990–91

It has been noted that divorced people have unusually high migration rates: they also show the most distinctive age relationship, for a clear 'U-shaped' function is shown with the lowest rates among people in their mid-seventies. Among single (never-married) older people, men were more likely to have moved than women, and some unusual if minor differentials are found. The most interesting is that the lowest male rate of 4.3 per cent is at 63 years of age, after which each additional year of age to the late seventies brought a moderately higher rate, while in contrast the lowest female rates (3.4–3.6 per cent) occurred from 66 to 72 years of age. It is as though the male retirement peak extends to older ages by delayed retirements and then merges with the exponential rise of the late age slope.

The five-year age-group rates mask a short peak of migrations around the modal retirement age, which are seen in the single years series for men at 64–65 years and for women at 59–62 years (Stillwell, Rees and Duke-Williams 1996: 295). The 'retirement peak' is most commonly found among migrations that are long-distance, undertaken by the more affluent, and which originate in the largest metropolitan areas, and it was more pronounced in Great Britain at the previous two censuses than in 1990–91 (Warnes 1983). The actual migration rates in extreme old age are likely to be higher than the census estimates for all marital status groups. At these advanced ages, mortality rates are very high: they may be higher among migrants than non-migrants. Many people at these ages, and more women than men, move into residential and nursing homes and hospitals for their last weeks and months of life (Harrop and Grundy 1991). Some of these moves will be actual changes of address, others will be temporary absences from the normal residence. As the average duration of these last migrations is short, the census 'one-year' migration questions will record but a fraction – perhaps a minority.

While generalizations about age, sex and marital-status differentials in migration are empirically well founded they are slight in explanatory content. Only the most apparent proximate causes are indicated, and only the most obvious consequences revealed. A sociological understanding of gender differences in migration requires study of moving house as a process, with attention to the ways in which decisions are made, implemented and frustrated, to the motivations and aspirations of the migrants, and to the consequences of the move in terms of the participants' activities, convenience of living, social relationships, standard of living and morale.

Gender and the motivations of moves in later life

Information on the motivations and decision processes involved in migration comes mainly from surveys, most of which have samples of fewer than 500 people. Most surveys fail to detect or diminish the extent of gender differences, either because the questions assume that the motivations of all members of a migrating household are shared, or from a tendency for respondents when not directed otherwise to report consensus motivations rather than their individual assessment. It is rare for individual members of a household to be interviewed separately.

In a 1976 study of 201 retired migrants into North Wales and Dorset, and of 100 matched non-movers in Greater Manchester and the London Metropolitan Area, some evidence of a gender differentiation in the enthusiasm to move was found (Law and Warnes 1982: 66). There were 144 married couple (plus) households in the mover sample and 86 in the non-mover sample. Of the latter, 44 had actively considered moving for retirement. When asked who initiated the move, 'our question . . . often prompted . . . knowing smiles' (1982: 66) and it was believed that the 40 per cent who claimed both husband and wife exaggerated 'the prevalence of equal enthusiasm' (ibid.). Among the movers, the husband initiated the idea in 35 per cent and the wife in 26 per cent of the cases, while among the non-movers, 18 wives (41 per cent), double the number of husbands, proposed that a move should be made. From the pooled samples, it turns out that 88 per cent of the husbands' proposals resulted in a move compared to 67 per cent of the wives'. The overall finding was that, 'for a retirement migration to occur among married couples, a period of discussion and consideration leading to the agreement of both parties normally occurs. Husbands' views were more likely to have prevailed and wives' more likely to have unsuccessfully recommended a move' (ibid.).

Subsequently, a study of 'Residential Mobility in Later Life' (MILL) in 1991–93 investigated housing satisfactions and stresses in later life, including the triggers of and motivations for moves, the life events and decision-making process associated with the most recent move, and future residential intentions (Warnes and Ford 1995a, 1995b). Its final survey was a postal inquiry of 813 men and 1,092 women aged 60+ years in South East England. 'Primary' respondents were randomly selected in the survey areas, and where there were two or more people in a household, the two principal members were asked to complete questionnaires independently. Some 631 replies were received from 'second' household members, 565 being spouses of primary respondents, and both husband and wife completed questionnaires for 97.0 per cent of the surveyed married couples. The data enable a detailed examination of gender differences in housing stress and migration motivations.

Focus groups and pilot in-depth interviews identified the 14 most prevalent 'life events' and 'main reasons' associated with moving during the previous five years. The postal survey respondents were asked which of the 14 events had occurred during the previous five years, and which had been the 'principal reason' and which 'an important event' in causing them to move. The relative frequencies, shown in table 17.3, indicate that the most commonly reported event by men was a significant decline in income (22 per cent), and by women a decline in own health (21 per cent). Widowhood was more common among women (11 per cent) than men (6 per cent), among whom a decline in a co-resident's health and a burglary were more frequently reported. Other relatively frequent events were bereavement and the decline in health of non co-residents.

As for the 'main reasons' for moving given in table 17.3, the most cited (excluding encouragement from others) by both men and women was a decline in one's own health but for only a low percentage of the event's occurrences (8 per cent among males, 15 per cent among females). Widowhood was the second most frequent principal reason among women, with nearly twice the incidence of the third, a decline in income. For men, widowhood and a decline

Table 17.3 Events in last five years and reasons for moving: men and women aged 60+ years in general housing, South East England 1993

	Prevalence of events				*Main reasons for move*				
	Women	*Men*	*Men*	*Ratio*	*Women*	*Men*	*Women*	*Men*	*Ratio*
	No.	*No.*	*%*	*W:M*	*No.*	*No.*	*%*	*%*	*W:M*
Decline in own health	213	125	16.0	1.3	31	10	33.7	20.4	1.7
Decline in income	160	175	22.4	0.7	8	4	8.7	8.2	1.1
Widowhood	111	45	5.8	1.9	15	5	16.3	10.2	1.6
Decline of co-resident's health	78	74	9.5	0.8	5	5	5.4	10.2	0.5
Bereavement of non co-resident	61	35	4.5	1.4	0	0	0.0	0.0	
Burglary or break-in at home	55	59	7.5	0.7	4	6	4.3	12.2	0.4
Improvement in own health	28	26	3.3	0.8	5	6	5.4	12.2	0.4
Decline of non co-resident's health	25	15	1.9	1.3	0	0	0.0	0.0	
Increase in income	24	23	2.9	0.8	2	0	2.2	0.0	
Bereavement of co-resident	23	10	1.3	1.8	3	1	3.3	2.0	1.6
Improvement of co-resident's health	6	7	0.9	0.7	1	1	1.1	2.0	0.5
Another person encouraged move	87	56	7.2	1.2	18	11	19.6	22.4	0.9
Number of respondents	1,002	782			92	49			

Note
1 The percentages for women may be read by dividing the frequencies by 10.

in a co-resident's health were equally likely to be a principal reason for moving, but rather less likely than either an improvement in health or a burglary or similar crime event. It is pleasing that positive changes, particularly an improvement in either one's own or a co-resident's health, were the most effective of all recent events in being linked to a change of address (although this may have resulted from selective reporting). Around one-fifth of both men and women who reported improved health said that it was the main reason for a recent move. On the other hand, income decline was rarely identified as a principal reason for moving. Despite the relatively low number of 'main reasons', the overall picture is one of marked gender differentials in the *ex post facto* attribution of main reasons for having moved.

Turning to a wider range of housing and environmental evaluations, women more often reported the upkeep and suitability of the house and garden and the residential environment as main reasons for a move, while men reported locational reasons more frequently. However, none of the gendered

Table 17.4 Main reasons for move[1] during last five years and dissatisfaction with current dwelling

Attribute	*Main reasons for move*			*Unsatisfactory at dwelling*			*Gendered housing stress*[2]
	Women	*Men*	*Ratio*	*Women*	*Men*	*Ratio*	
	A	*B*	*C*	*D*	*E*	*F*	*G*
Location (instrumental)	17.4	25.1	0.7	6.5	6.0	1.1	1.6
(95% confidence interval)	(11.9–22.9)	(18.6–31.6)		(5.5–7.5)	(4.9–7.1)		
Convenience of location	15.2	23.4	0.6	4.6	4.2	1.1	1.7
Handiness for outdoor activities	0.4	0.6	0.8	1.0	1.2	0.8	1.1
Distance from carers	0.0	1.2	0.0	0.6	0.3	1.7	0.0
Distance from cared-for	1.8	0.0	—	0.3	0.2	1.5	
Location (social)	13.4	14.6	0.9	9.8	7.5	1.3	1.3
(95% confidence interval)	(8.4–18.4)	(9.3–19.9)		(8.6–11.0)	(6.3–8.8)		
Distance from relatives	12.5	12.9	1.0	7.3	5.8	1.2	1.3
Distance from friends[3]	0.9	1.8	0.5	2.5	1.7	1.5	3.0
Security[4]	2.2	2.3	1.0	11.1	10.9	1.0	1.1
(95% confidence interval)	(0.1–4.4)	(0.1–4.6)		(9.8–12.3)	(9.5–12.3)		
Upkeep and suitability	26.8	21.1	1.3	28.1	26.5	1.1	0.9
(95% confidence interval)	(20.3–33.2)	(14.9–27.2)		(26.3–30.0)	(24.5–28.6)		
Workload of house	5.8	2.9	2.0	8.1	7.2	1.1	0.6
Workload of garden	7.6	4.7	1.6	13.0	12.5	1.0	0.6
Size or form of dwelling	13.4	13.5	1.0	7.0	6.8	1.0	1.0
Costs	11.2	9.9	1.1	16.9	21.5	0.8	0.8
(95% confidence interval)	(6.6–15.7)	(5.5–14.4)		(15.3–18.4[5])	(19.6–23.4[5])		
Running costs	7.1	7.6	0.9	13.0	15.5	0.8	0.9
Capital tied up in house	4.0	2.3	1.7	3.8	6.0	0.6	0.4
Environmental	19.6	17.0	1.2	8.4	10.6	0.8	0.7
(95% confidence interval)	(13.9–25.4)	(11.3–22.6)		(7.3–9.5)	(9.1–12.0)		
Pleasantness of surroundings	12.9	14.6	0.9	3.4	3.7	0.9	1.0
Noisiness of the neighbourhood	6.7	2.3	2.9	5.0	6.9	0.7	0.3
Other reasons	9.4	9.9	0.9	3.9	3.8	1.0	1.1
Nothing can be improved	—	—	—	15.3	13.3	1.2	—
Number of main reasons/complaints	181	171		2314	1780		

Notes
1 Compulsory moves are excluded
2 Women:male ratio of attribute unsatisfactory, divided by women:male ratio of main reasons for moving, i.e. an index calculated from columns (D/E)/(B/A)
3 Distance from friends and social activities
4 Risk of burglary or intruders
5 Significantly different at $p < 0.05$.

differences for either these broad categories or their component reasons, given in table 17.4, were significant ($p = 0.05$). Expressed dissatisfactions with the dwelling were inevitably more numerous. Their relative frequencies and gender differentials contrasted with those for the main reasons for mov-

Table 17.5 Events in last five years and their association with a change of address

	Occurrence during last five years				*Citations of event as a main reason for moving*			
	Males		*Females*		*Males*		*Females*	
Age group (years)	*Number*	*%*	*Number*	*%*	*Number*	*%*	*Number*	*%*
Serious decline in own health								
60–69	53	15	63	15	4	8	3	5
70–79	43	14	73	20	4	9	8	11
80+	29	22	79	34	2	7	20	25
Total	125	16	215	21	10	8	31	14
Widowhood								
60–69	8	2	32	8	2	25	2	7
70–79	20	7	45	12	3	15	5	11
80+	17	13	34	16	0	0	8	24
Total	45	6	111	11	5	11	15	14
Income decline								
60–69	117	33	112	29	4	4	4	2
70–79	44	15	33	9	0	0	3	9
80+	14	11	16	7	0	0	2	13
Total	175	22	161	16	4	2	9	6
Encouragement from others to move								
60–69	22	6	34	8	4	18	7	21
70–79	22	7	29	8	2	9	5	17
80+	12	9	24	11	5	42	6	25
Total	56	7	87	9	11	20	18	21

ing. Upkeep and suitability complaints were again the most prevalent, but cost complaints supplanted instrumental aspects of location in second place. Men were significantly more likely than women to mention costs as a problem, while women more frequently mentioned problems of poor access to their relatives, friends and social activities (the difference in means being significant at $p = 0.0375$).

The incidence of the various reasons for moving show contrasting gender differences by age. Table 17.5 shows that while 19 per cent of the entire sample reported a serious health decline in the previous five years, both the sex differential and the age gradient among males were moderate. At ages 60–69 years, females were no more likely to report a health decline than males, but at 80+ years the relative frequency from women was 50 per cent higher. Around 8 per cent of males stated that their health decline was a main reason for moving and there was no significant association with age. For women, health declines altered from a rare reason for moving among those aged

60–69 years (5 per cent) to a common reason among those aged 80+ years (25 per cent).

Turning to recent spouse bereavement, 9 per cent had experienced this loss and of those 13 per cent said that it was the principal reason for moving. Some 192 moves were reported in the last five years, and in approximately one-tenth bereavement of a spouse was explicitly implicated. Disaggregation by age and sex reveals the expected concentration of bereavement at older ages but also an interesting contrast between men and women: for the former there is a negative age relationship between spouse bereavement and moving, while for women the age association is positive. While noting the small number of cases, one-quarter of the recent widowers aged 60–69 years gave their bereavement as the main reason for changing address, but none aged 80+ years reported this association.[1] Among recent widows, however, the relationship of moving to age was reversed, with only 7 per cent of those aged 60–69 years but 35 per cent of those aged 85+ years stating the association.[2] The gender inversion in the age association of marital *status* and moving is therefore repeated in the *event* data.

Recent income decline was reported more frequently than spouse bereavement but with approximately the same incidence as a serious decline in own health, *viz.* 336 cases (19 per cent). Its age and sex distribution was distinctive, with a clear concentration into early retirement and more frequent reporting (by about one-third) among men than women. It was much less influential as a reason for moving, there being only 13 cases. The relatively high incidence of income decline after the normal retirement ages is consistent with the life-course patterning of individual and household income, and the positive association among women between age and migration is consistent with the distribution of old-age poverty.

Finally in this sequence of reasons for moving, encouragement from others has a different character. It is not endogenous to the person, as with a decline in health, and while it may occur independently of the previously-considered life events, it is probably often a reaction. Table 17.5 shows that its expression occurred with a similar frequency to widowhood and that men reported it a little more often. In one-fifth of the cases it was associated with a recent move, in every case as the 'principal reason', so was more 'effective' in producing residential change than any of the endogenous events. Its operation is the converse of the ineffectuality of an income decline in causing moves. There is a suggestion that the effectiveness of encouragement from others in producing moves is relatively invariant by age for women, but most marked for men at the oldest ages.

Gender differences in the impact of the various housing stresses have been summarized in an index of 'gendered housing stress'. This expresses the female to male ratio of 'complaints' to 'reasons for moving'. The index scores high when an attribute produces for women more than men relatively high dissatisfaction but few moves, and scores low when a source of complaint among women results in a relatively high frequency of moves (table 17.4, column G). The strongest gender differential is in the inconvenience experienced in the

Table 17.6 Probability of events being a principal reason for moving in general housing, South East England 1993

Events of last five years	*Both sexes (60+ years)*		*Males (65–79 years)*		*Females (80+ years)*	
	%	*Cases*	*%*	*Cases*	*%*	*Cases*
Spouse bereavement	13	20/156	22	5/23	24	8/34
Serious health decline	12	41/338	10	7/68	25	20/79
Significant income decline	4	12/335	1	1/95	13	2/16
Encouragement by others	20	29/143	13	4/31	25	6/24
Spouse bereavement and health decline	21	8/39	29	2/7	42	5/12
Spouse bereavement and income decline	14	6/43	11	1/9	33	1/3
Spouse bereavement and encouragement	33	5/15	50	1/2	50	3/6
Health decline and income decline	1	1/86	3	1/29	0	0/7
Health decline and encouragement	29	15/51	8	01/12	47	7/15
Income decline and encouragement	21	8/38	22	2/9	50	1/2
Spouse bereavement, health decline and encouragement	40	2/5	0	0/1	67	2/3

home's location, particularly for practical (as opposed to social) purposes. The index should be considered alongside the raw percentages, which show that men achieve moves more frequently for locational convenience reasons, but women more frequently find the home's location unsatisfactory. Women also experienced slightly more 'stress' from the fear of crime. Relatively fewer women complain of the costs of a dwelling than men, although a higher proportion actually move for expressed financial reasons, so the 'cost stress' index is lower for women than men. The issue is complex and more subtle comparisons would require disaggregation by household size, marital status and age, and with consideration of the role of tenure, the proportion of income expended on housing, and entitlement to state housing benefits.

Concurrent life events and their association with moves

One further step can be taken in the examination of gender differentials in the relation between various life events and subsequent moves. The concurrence of adverse events in the previous five years generally produced a higher likelihood of moving than the events occurring independently, but clear variations are found according to the combination and the age-sex group. Two groups, males aged 65–79 years, and females aged 80+ years, provide sufficient cases for comparative examination, as shown in table 17.6.

Beginning with spouse bereavement, among both young elderly men and 'old old' women, the loss had a higher association with a move than in the general sample. Both serious declines of health and spouse bereavement during the previous five years were reported by 11 men and 28 women aged 60+

years. In eight (21 per cent) of these cases, one or both adverse life events were given as a main reason for changing address. The association of a bereavement with encouragement to move from others also more than doubled the likelihood of one of the events being cited as the main reason for moving, while an association with an income decline generally had much less effect and among young elderly men halved the association.

The association of a decline in own health with moving house was stronger among women in advanced old age than young elderly men or all older people. Coincident spouse bereavement, and more variably the coincident encouragement to move, raised the probability of moving. Young elderly men were exceptional in this respect, for concurrent encouragement to move slightly depressed the association of the events with a move. A more surprising finding is that a conjoint income decline virtually eliminated moves in all age-sex categories (only one person in 86 citing either event as the main reason for moving).

There was virtually a null association of an income decline with moving in the general sample and among young elderly men, and only 13 per cent of the women in advanced old age who reported the change gave it as the main reason for moving. Nor did the conjoint occurrence of a health decline change this lack of association: indeed among the seven women in advanced old age who experienced both events, none reported either as a principal reason for moving. On the other hand, the additive affect of 'encouragement' was massive, for it resulted in one-fifth of the aggregate sample who experienced both events reporting one of them as the main reason for moving. The additive effect of spouse bereavement was generally only one-half as great. The effect of an income decline will differ according to the level of income at which it occurs. Decreases from formerly 'comfortable' incomes are more likely (and able) to produce a housing adjustment than decreases from already low levels. There are also likely to be pronounced housing tenure influences on people's reaction to income decline.

Conclusions

The reports from this sample of older people contribute fascinating insights into the diverse housing and migration experience and behaviour of older men and older women. They also emphasize the contrasts between the effects of various adverse life events at different ages. The clearest age-gender differentials are amongst widowed persons. When young elderly men are widowed, they are much more likely to move than their older counterparts, but the opposite age relationship is found among widows. This contrast is probably the complex outcome of male–female differences in early and late old age in: income levels and sources and the ability to finance a move; health and functioning and the responses of both statutory services and relatives as informal carers to the person living alone; the likelihood of remarriage; and the individual's psychological and practical preparedness for 'coping' and living alone. The overall conclusion from the analysis is that an unusually complex nexus of influences conditions the propensity to move in later life.

Notes

1 Table 17.4 presents data for three aggregated age groups but the data have been examined by five-year groups from 60–64 years to 90+ years.

2 The numbers are relatively small but in each of the five-year age groups of females from 70–74 to 80–84 years, 11–12 per cent of widows give the bereavement as a main reason for moving (N = 17, 28 and 17). The value rose to 33 per cent among the 15 recent widows aged 85–89 years.

References

Baltes, P., and Baltes, M. (eds) (1990) *Successful Ageing: Perspectives from the Behavioural Sciences,* Cambridge: Cambridge University Press.

Bean, F.D., Myers, G.C., Angel, J.L. and Galle, O.R. (1994) 'Geographic concentration, migration and population redistribution among the elderly', in L.G. Martin and S.H. Preston (eds) *Demography of Aging,* Washington DC: Committee on Population, United States National Research Council, National Academy Press, pp. 279–318.

Erikson, E.H., Erikson, J.M. and Kivnick, H.Q. (1989) *Vital Involvement in Old Age*, New York: Norton.

Harrop, A. and Grundy, E. (1991) 'Geographic variations in moves into institutions among the elderly in England and Wales', *Urban Studies* 28: 65–86.

Kendrick, A. and Warnes, A.M. (1997) 'The demography and mental health of elderly people', in I.J. Norman and S.J. Redfern (eds) *Mental Health Care for Elderly People*, Edinburgh: Churchill Livingstone, pp. 3–20.

Laslett, P. (1989) *A Fresh Map of Life: the Emergence of the Third Age,* London: Weidenfeld and Nicolson.

Law, C.M. and Warnes, A.M. (1982) 'The destination decision in retirement migration', in A.M. Warnes (ed.) *Geographical Perspectives on the Elderly*, Chichester: Wiley, pp. 53–81.

Litwak, E. and Longino, C.F. (1987) 'Migration patterns among the elderly: a developmental perspective', *The Gerontologist* 27: 266–72.

Longino, C.F. (1996) 'Migration', in J.E. Birren (ed.) *Encyclopaedia of Gerontology, Volume 2,* San Diego: Academic Press, pp. 145–50.

OPCS [Office of Population Censuses and Surveys] (1994) *Mortality Statistics 1992: England and Wales, General*, London: HMSO.

OPCS/GRO(S) [General Register Office (Scotland)] (1994) *1991 Census: Migration, Great Britain, Part 1 (100% Tables), Volume 1*, London: HMSO.

Rees, P. (1992) 'Elderly migration and population redistribution in the United Kingdom', in A. Rogers, W.H. Frey, P.H. Rees, A. Speare and A.M. Warnes (eds) *Elderly Migration and Population Redistribution: a Comparative Study*, London: Belhaven, pp. 203–25.

Rhind, D. (ed.) (1983) *A Census Users' Handbook*, London: Methuen.

Rogers, A. and Castro, L.J. (1986) 'Migration', in A. Rogers and F. Willekens (eds) *Migration and Settlement*, Dordrecht: Reidel, pp. 157–209.

Rogers, A., Frey, W.H., Rees, P.H., Speare, A. and Warnes, A.M. (eds) (1992) *Elderly Migration and Population Redistribution: a Comparative Study*, London: Belhaven.

Rowe, J.W. and Kahn, R.L. (1987) 'Human aging: usual and successful', *Science* 237: 143–9.

Stillwell, J., Rees, P.R. and Duke-Williams, O. (1996) 'Migration between NUTS Level 2 regions in the United Kingdom', in P.R. Rees, J. Stillwell, A. Convey and M. Kupiszewski (eds) *Population Migration in the European Union,* Chichester: Wiley, pp. 275–307.

United Nations Organization [UNO] (1993) *Demographic Yearbook 1993, Special Issue: Population Aging and the Situation of Elderly Persons*, New York: UNO.

Warnes, A.M. (1983) 'Migration in late working age and early retirement', *Socio-Economic Planning Sciences* 17: 291–302.

Warnes, A.M. (1992) 'Temporal and spatial patterns of elderly migration', in J. Stillwell, P. Rees and P. Boden (eds) *Migration Processes and Patterns, Volume 2: Population Redistribution In the United Kingdom*, London: Belhaven, pp. 248–70.

Warnes, A.M. (1996) 'Migrations among older people', *Reviews in Clinical Gerontology* 6: 101–14.

Warnes, A.M. (1998) 'Population ageing over the next few decades', in R. Tallis (ed.) *Increasing Longevity: Medical, Social and Political Implications*, London: Royal College of Physicians, pp. 1–15.

Warnes, A.M. and Ford, R. (1995a) 'Migration and family care', in I. Allen and E. Perkins (eds) *The Future of Family Care*, London: HMSO, pp. 65–92.

Warnes, A.M. and Ford, R. (1995b) 'Housing aspirations and migration in later life: developments during the 1980s', *Papers in Regional Science* 74: 361–87.

18 Inside and outside the Pale

Diaspora experiences of Irish women

Bronwen Walter

Introduction

Diaspora offers a conceptual framework within which Irish women's migration to Britain may be re-evaluated. It allows the specificity of emigrant experience in Britain to be highlighted by comparing and contrasting women's lives according to their destinations, whilst simultaneously viewing different destinations as part of the same process. In this chapter the comparison is made with the very large outflow of Irish women to the United States over the last two centuries. The aim therefore is to explore both differences and connections, providing new perspectives on migration flows and subsequent settlement.

I am using 'the Pale' as a boundary of the colonial relationship with the British centre, so that Irish women migrating to Britain can be seen as 'inside', and those emigrating to North America, particularly the United States, as 'outside'. In fact, of course, the Pale has been a very permeable boundary and British attitudes towards the Irish have leaked heavily across the Atlantic. Moreover, Irish women in Britain may also be 'beyond the Pale'. Such ambiguity is integral to the concept of diaspora which challenges 'old localizing strategies', such as centre and periphery, replacing them with single transnational communities (Clifford 1994).

Comparison and linkage of the experiences of several million women over two centuries is a very ambitious project, and there are many practical problems. Particularly important is the periodization of Irish emigration by destination. By far the largest numbers and proportion emigrated to America in the nineteenth century. The conventional estimate is 80 per cent of the total flow (Kennedy 1973), though Cormac O'Grada (1973) has suggested a serious undercounting of the numbers of Irish emigrants settling in Britain during the period, possibly as many as one million, which would reduce though not remove this over-weighting. Since the 1920s Britain has been by far the most important destination for Irish emigrants, accounting for approximately 80 per cent of the total by 1939. However, in the late 1980s, the choice of destinations had once again shifted somewhat. Between 1987 and 1994, the distribution of out-migrants showed 59 per cent going to the United Kingdom, 25 per cent to the United States, 8 per cent to the rest of the European Union and 13 per cent to the rest of the world (CSO 1994). A major consequence of the sharp overall change in the direction of flow is the different balance of generations in America and Britain. Most Irish-Americans are in the fifth or sixth generation, with far fewer Irish-born than in Britain. In 1970, 250,000 Irish-born people were recorded in the United States, of whom 40 per cent were born before 1925, whereas 950,000 were recorded in the 1971 British census. Another important issue is the danger, but also the necessity, of ignoring intra-national differences when using such a broad-brush approach. I am aware that there are many regional and local differences in the experiences, but will not address these here (cf. Walter 1984).

So why attempt the project at all? A central concern is to theorize and contextualize Irish experiences in Britain and in particular to throw light on the invisibility of the Irish as a racialized minority ethnic group in Britain (see Walter 1998). A recent report for the Commission for Racial Equality (CRE) documents systematically for the first time evidence of discrimination against the Irish community (Hickman and Walter 1997). It highlights the situation whereby the Irish are constructed as 'different' and treated less favourably because of their origins, yet there is a refusal to accept that they experience racial discrimination because they are 'white'.

One way of destabilizing entrenched denials of anti-Irish racism is to juxtapose Irish experience in Britain and America. Irish people in Britain readily do this. As part of the CRE survey, Irish people in London and Birmingham were asked how they felt the Irish were viewed in America and Britain. Only 1 per cent said they felt the Irish were viewed negatively in the United States, compared with 22 per cent who gave this opinion about Britain. These crude figures require complex elaboration, but they serve as a starting point for the investigation.

The inclusive framework provided by the notion of diaspora enables us to explore these differences. This is a notion that is now 'loose in the world' according to James Clifford (1994: 306). There are two reasons why its use has expanded recently. First, it offers a reconceptualization of the settlement of dispersed groups, replacing commonly accepted ideas about the binary relations of minority communities within majority societies, which have characterized migration studies and led to an emphasis on normative processes such as assimilation. The hegemonic political agenda underlying the preoccupation with assimilation as a positive outcome of migrant ethnic interactions over time is rarely acknowledged in academic analyses in the social sciences and history, though the question has been raised (Jackson 1987; Clifford 1994). Both by the scale of analysis and the theoretical frameworks used to examine its consequences, migration studies thus reinforce the ideology of the primacy of the nation-state and its concerns. Second, diaspora discourse recognizes changing transnational patterns whereby dispersed peoples are no longer cut off from their places of origin but are able to retain very close links.

Both these facets of diaspora are highly relevant to the Irish situation. The Irish have not fitted comfortably into minority/majority frameworks in either America or Britain. In the United States, the long-established Irish clearly shifted into the 'mainstream' at some point in the early twentieth century (Greeley 1973), although new arrivals in the 1980s for example did not necessarily partake of this status (Corcoran 1993). In Britain, the assimilationist assumptions of historians of Irish settlement in the nineteenth century have been strongly questioned (Hickman and Walter 1995). Moreover, the increased fluidity of transnational connections clearly characterizes the 1980s migrants, often known as the 'Ryanair generation' (after the Irish airline).

Relevance of the diaspora concept

The notion of diaspora challenges traditional understandings of migration at the global scale. Instead of focusing on linear paths travelled through space and over time, it emphasizes interconnections between individuals and places that may create a 'world wide web'. This analogy with electronic communication is not simply metaphorical. In the contemporary world diasporic communities are held together ever more firmly by instantaneous linkages, digital technologies and voice telephone, both at a private, social level and at the macro-scale as constituent parts of a global labour force. They thus represent key aspects of contemporary society; in Khachig Tololyan's (1991: 5) words, 'diasporas are the exemplary communities of the transnational moment'.

One of the strengths of this concept, as a framework for the study of migrant communities, is its inclusive character over both time and space. Whereas migration is usually recorded as an event affecting mobile individuals and groups, which is completed when they resettle, diasporic experience continues into subsequent generations for as long as the flow has any impact on the lives of descendants and the societies in which they live. The salience of this past may differ for individuals within the group, and change over time, both strengthening and weakening at particular moments. A far greater proportion of the population is thus affected through this intergenerational history of displacement.

Inclusion over space is also fundamental to the concept. The web of connections between people who share a common tie to the homeland from which they are separated is readily comprehended. Avtah Brah defines this as the space 'inhabited' not only by diasporic subjects but equally by those who are constructed and represented as 'indigenous'. As such, the concept of diaspora space foregrounds the entanglement of genealogies of dispersion with those of 'staying put'. In other words, those whose identity is defined by migration are necessarily placed relationally with those who define themselves as fixed. The 'difference' of diasporic people is constructed through this juxtaposition. Adoption of the concept of 'diaspora space' shifts the focus dramatically by foregrounding shared locations, which are also contested. All inhabitants are involved, through the myriad ties which connect them and through the shared experiences of 'staying put', since all occupy the same space. This calls into question the simple binary polarities that underlie conventional thinking about migration and its consequences. Instead, Brah (1996: 16) suggests that diaspora can be envisaged as 'a cartography of the politics of intersectionality'.

Finally, diaspora experiences provide reminders of the global political contexts in which dispersion has occurred, on scales that are lost when migration is measured by the crossing of particular state boundaries. These often-violent origins are hidden by nation-states, which prefer to represent their histories as linear, progressive narratives. None the less, memories of terror and loss originating in these 'constitutive outsides' persist and re-emerge. For example, the continuing currency of such feelings was illustrated in 1997 by the ceremonies of remembrance on the one hundred and fiftieth anniversary of the 'Black '47' Famine year in Ireland, which included the acknowledgement by the British Prime Minister of British responsibility for some of the suffering (*Guardian* 1997).

Gender and diaspora

Gender is a key facet of the intersectionality that defines the concept of diaspora, as power relations of class and gender cross-cut those of ethnicity and 'race'. Migration studies privilege differences attributed to origin, but the shared character of 'diaspora space', inhabited by both diasporic and 'indigenous' peoples, allows the web of interconnections of a much wider range of identities to be traced.

The spatial worlds of both women and men are highlighted more fully in diaspora discourses. In migration studies, the move itself is centred, a public activity which is often implicitly male. However, the notion of diaspora includes both movement and settlement as part of a single process, so that displacement and placement co-exist in parallel rather than operate in sequence. This ongoing intermeshing is what distinguishes diaspora from migration. Brah (1996) argues that diaspora peoples feel 'at home' and anchored in the area of settlement, even though their 'homeland' may be elsewhere. Clifford (1994: 308) describes this state as 'dwelling-in-displacement' and '"not-here" to stay'. He argues that

> Diaspora discourse articulates, or bends together, both roots and routes to construct what Gilroy describes as alternate public spheres (1987), forms of community consciousness and solidarity that maintain identifications outside the national time/space in order to live inside, with a difference.

Displacement, a concept analogous to Gilroy's 'routes', involves far more than the physical transplantation between areas of origin and geographical destinations. All subsequent contacts between migrants and their descendants that juxtapose the locations are consequences, and reinforcements, of the ongoing 'presence' of another place. These contacts range from return visits, to family reunions, letters between relatives, personal memories and the retelling of family and national histories. Women play distinctive roles in maintaining these ties through their management of personal relationships within the family and close connection with childrearing. For example, they have been more reliable remitters of money than men, both because of notions of femininity involving family obligations and because their work enabled them to save. They have also been managers of emotional connections, in letter-writing, telephone conversations and as transmitters of family narratives.

The notion of placement, closely paralleling Gilroy's 'roots', relates to day-to-day living in which the private world of the home is crucial, as well as public spheres such as paid work and political participation. Clifford (1994) argues that women's experiences are particularly revealing as gender subordination may be reinforced or loosened by the change in location, whilst the task of mediating between discrepant worlds on behalf of families also falls more frequently on women.

In fact, displacement and placement are always complexly interlinked, and gender roles in each process are distinctive and pivotal. As yet, however, the gendering of diaspora cultures remains underexamined, as critiques of Gilroy's (1993) *The Black Atlantic* point out (Clifford 1994).

The Irish diaspora

There is no ideal type of diaspora against which to measure Irish experience. Indeed, one of the key features of diaspora is the unique set of experiences, memories and myths, which bind particular people together. Thus, the specificity rather than the commonality of diaspora experiences needs to be explored. However, attempts have been made to identify shared features, without insisting that all are relevant at all times to the same extent.

One of the most systematic attempts to distinguish the key characteristics of what he called 'expatriate minority communities' was made by William Safran (1991: 83–4). He argued that diasporic communities shared several of the following features:

- a history of dispersal from an original 'centre' to at least two 'peripheral' places;
- myths/memories of the homeland;
- alienation in the host country;
- desire for eventual return;
- ongoing support of the homeland;
- collective memory importantly defined by this relationship.

Clifford (1994) has suggested modifications to this list, including the notion that homeland need not be so central in articulation of transnational connections. He believes that decentred, lateral connections and a shared, ongoing history of displacement, suffering, adaptation or resistance may be equally important. Robin Cohen (1997: 23–5) has added three additional features to Safran's list. One is the possibility of 'aggressive or voluntarist' dispersal, in addition to the 'victim' origin implied by Safran. Cohen also believes that it is necessary for time to pass before a group can be defined as diasporic. Finally, he stresses the positive qualities of diaspora identities, in addition to the continuing social exclusion, arguing that: 'The tension between an ethnic, a national and a transnational identity is often a creative, enriching one' (1997: 24).

Without exploring the evidence in detail, it should be clear that arguments could be made for recognizing Irish emigration experiences in both America and Britain having all these characteristics to a greater or lesser extent at different times. For example, the history of dispersal extends back over centuries but reached strikingly high levels after the 1820s. Three strong 'waves' have been identified, the first peaking in the 1860s, the second in the 1950s and the third in the 1980s. It is now estimated that the Irish community worldwide numbers 70 million, compared with the 5 million inhabiting the island of Ireland. Myths and memories of the homeland are attested by the Irish cultural organizations in areas of settlement around the world. A positive aspect is also apparent in the disproportionate contribution over the years of British-based Irish authors to the literary canon, and the currently acclaimed role of Irish music and dance in British popular culture.

The use of the term *diaspora* to describe Irish displacement has, none the less, been questioned. Gerard Chaliand and Jean-Pierre Rageau (1995: xiv) feel it necessary to ask 'Is there an Irish diaspora?' The main reason for doubt in their view is the very large number of emigrants involved relative to the size of the home population. However, they choose to include the Irish in their *Atlas of Diasporas*, and affirm that, in the United States, 'the Irish community's sense of cohesion and solidarity remains very much alive' (1995: 161).

David Lloyd (1994) mounts a much stronger challenge on several grounds. In contrast to Chaliand and Rageau, he argues that Irish-Americans are now 'a fully integrated element of white and mainstream American society' (1994: 3). Moreover, he claims that the idea of return is now a 'mostly sentimental and fetishising desire to establish their genealogy in the homeland' (1994: 4), representing an attempt to jump on the multicultural 'bandwagon' and reap the benefits of cultural distinctiveness. Perhaps most importantly, he believes that the stress on cultural forms depoliticizes the reality of the continuing massive outflow of skilled and unskilled labour from Ireland. However, there are problems with these objections to the application of the term *diaspora* to Irish dispersal. Lloyd assumes that the American experience is the only significant one, although by far the largest proportion of emigrants currently enters Britain, where it cannot be said that all are part of mainstream white society.

The CRE report chronicled a wealth of evidence that anti-Irish discrimination continues to be endemic in British society (Hickman and Walter 1997). Moreover, recent discovery of significant health differentials in second-generation Irish people in Britain provides strong empirical support for continuing disadvantage (Harding and Balarajan 1996). Lloyd also treats diaspora as unmarked by gender, which implicitly privileges a male viewpoint. He stresses economic causes of emigration, which he fears may be sidelined by accepting a cultural frame of reference, whereas for women social reasons for leaving Ireland may be as strong as economic ones (Kelly and Nic Giolla Choille 1995; O'Carroll 1995).

Far from depoliticizing the reality of emigration as Lloyd fears, the notion of diaspora is a highly political concept from the point of view of dispersed populations. Gilroy (1987) argues that it provides a 'third space', an alternate public sphere where global alliances may be made as well as links between members of different diasporic communities. In other words, diaspora brings together accommodation with, and resistance to, host societies, instead of seeing these as opposing tendencies. There may be continuing tension between these processes over many generations.

Using the diaspora framework, it is therefore possible to see the Irish in Britain both as settled and established, and as retaining a separate identity and continuing to experience discrimination. Such a view challenges hegemonic understanding that a long-established, 'white' group cannot be subject to racism and discrimination. The equation of 'whiteness' with sameness remains to be scrutinized.

Gendered constructions of the Irish diaspora

In this section, I begin to explore the specificity of Irish women's diaspora experiences in the United States and Britain. Overall, there is a striking shortage of information about emigrant Irish women's lives. In his wide-ranging survey, entitled *The Irish Diaspora: a Primer*, Donald Akenson (1996) entitles one chapter 'Women and the Irish Diaspora: the great unknown', claiming:

> The single most severe limitation on our knowledge of the Irish diaspora is this: we know surprisingly little about Irish women in the nineteenth and early twentieth centuries, either in the homeland or in their New Worlds. With any ethnic group this sort of deficit would be a serious problem, but with the Irish it is especially debilitating, because females were half the diaspora.
>
> (1996: 157)

Nevertheless, contrasts between the experiences at the two destinations can be highlighted in several ways.

Written material

Data sources are far more readily available in the United States, making Irish women in America highly visible within written sources. However, this does not mean that a great deal of analysis has been carried out, although two comprehensive monographs on Irish women's lives were published in the 1980s (Diner 1983; Nolan 1989). Indeed, Hasia Diner (1983: 155) stresses this historical visibility in her discussion of her data sources:

> The major problem . . . I encountered in putting together this study of immigrant Irish women was that the mountains of material from government, charity and church sources, particularly at the local level, seemed almost insurmountable.

Irish women have also featured strongly in texts without a specifically ethnic focus. For example, Thomas Dublin's (1979) *Women at Work: the Transformation of Work and Community in Lowell, Massachusetts, 1826–1860* provides a wealth of detail about Irish cotton-mill workers, because they took over from locally-born women as the principal source of mill labour in the New England town. Similarly, Faye Dudden (1983) devotes a large amount of space to Irish 'Biddies' in her study of domestic servants, *Serving Women: Household Service in Nineteenth-century America*.

Another fruitful source of data on nineteenth-century Irish women in America is emigrant letters, which survive in a number of collections (Miller 1985). None the less, it is important to point out that material is not available at the national level, as nineteenth-century censuses did not disaggregate birthplace data by gender.

By contrast, it is extremely difficult to write a history of Irish women in nineteenth-century Britain because of lack of information. One of the few published attempts, by Lynda Letford and Colin Pooley (1995), is based on detailed analysis of enumerators' returns from the 1851 census in Liverpool. However, the interpretation of the findings is severely hampered by lack of additional qualitative material and remains very tentative. Letters from emigrants to Britain do not appear to have survived, despite the high level of remittances to Ireland.

Representations of Irish women

A contrast is again strongly apparent in visual representations of Irish women in the United States and Britain. In America, there have been clear stereotypes of Irish women, who were collectively named as 'Bridgets', 'Norahs' and 'Marys' – a much richer variety than the ubiquitous 'Pat' for men. Whereas the attributes of 'Pat' were almost universally negative – drunken, feckless, violent – those of 'Bridget' and her sisters were much more mixed. She was portrayed as lovable, cheerful, innocent and good-hearted, as well as stupid, clumsy and unreliable (Diner 1983). In a cartoon from *Harper's New Monthly Magazine* of 14 February 1856, sixteen social stereotypes were portrayed opening their 'Valentine's Day' cards. The Irish were represented by 'Bridget Maloney', a domestic servant in the kitchen opening a valentine from 'Pat'. She is rosy-cheeked, if coarse-looking, and smiles naively at her rose-coloured view of 'Pat' as a cupid. Only five of the sixteen cartoon stereotypes displayed are women; men represent all other racialized groups (Walter 1997).

By the turn of the twentieth century, Irish women in America had a more strongly positive stereotype. They were being credited with the upward mobility of their families in the classic move 'from shantytown to lace curtain', and described approvingly as 'civilizers' of their large families (Diner 1983). Using evidence from Irish women's letters to family and friends in Ireland, Kerby Miller and his colleagues (1995: 55) argue that domestic service in middle class American households 'certainly increased and refined, if they did not create, bourgeois aspirations both material and socio-cultural'. With the open approval of the Catholic hierarchy, they aspired to create 'the "right" kind of home . . . which then would make them the "right" kind of people' (McDannell 1986: 73).

An important way in which they achieved this was through their experiences as paid domestic workers, which brought them into contact with the majority society at its core and exposed them to the attitudes, behaviour and material goods of the middle classes. They also earned money, over and above their payment in board and lodging, which could be used to provide consumer goods and an education for their children. In other words, the placement of Irish women within homes played a crucial part in the establishment of the Irish within the white middle class hegemony.

Again, this recognition was in stark contrast to Britain, where the invisibility of Irish women is striking (Walter 1995). Stereotypes of the Irish were, and remain, strongly masculine. Collectively, the Irish are 'Paddies' or 'Micks', often regardless of gender. Cartoons depict only men (Curtis 1971). Where women were clearly holding together Irish families in very similar ways to those in America, this was turned on its head and used to reinforce the racialization of the Irish through Irish men. Melanie Tebbutt (1983) showed how Manchester music-hall jokes portrayed Irish men as 'hen-pecked', using implied racial inferiority to account for their inability to control women in their households. The women in question were invisible: there was no public celebration of their strength, indeed very little mention of them at all.

Recognition of Irish identities

The contrast in both visibility and positive evaluation only becomes observable when the two parts of the diaspora are compared. In the United States, an Irish identity is now widely admired and claimed. The 1990 Census gave respondents the opportunity to name whatever parts of their ancestry they chose. In total, 16 per cent claimed Irish origins, second only to the number identifying their German extraction (US Bureau of the Census 1990). In Britain, by contrast, there has been no opportunity in the Census even for those with two Irish-born parents to identify themselves as other than 'White' and concerted requests to amend the form of the 'ethnic question' in 2001 to collect this information have been strongly resisted (*Irish Post* 1997; Walter 1998). Moreover, within the Irish community itself, the jibe 'plastic Paddy' colludes with this denial by ridiculing such claims (Ullah 1985).

Irish invisibility in Britain thus reflects a paradox. On the one hand, Irish people are strongly identified as different and inferior but, on the other hand, they are too much 'the same' for their separate identity to be recognized. Racialization of the Irish, the attribution of characteristics to inheritance, has a long history traceable to at least the eleventh century (Hickman 1995). In the nineteenth century, overtly racist opinions were most strongly expressed and the Irish were portrayed as ape-like and subhuman. Elements of these attitudes have continued into the twentieth century. In the 1950s and 1960s signs proclaiming 'No blacks, no Irish, no dogs' were displayed in windows of lodgings. The CRE survey (Hickman and Walter 1997), carried out in 1995, showed that anti-Irish remarks, drawing on nineteenth-century stereotypes of stupidity and proneness to mindless violence and drunkenness, are still commonplace. Anti-Irish comments at work were reported by 79 per cent of respondents. These took the forms of 'jokes', negative stereotyping, name-calling and the ridiculing of accents. Their intensity increased at times of IRA activity, especially when this was located in Britain, but anti-Irish responses linked with the conflict in Northern Ireland continued a long-established pattern (Hickman and Walter 1997).

Some gender differences could be discerned amongst respondents, although much of the hostility was directed in very similar ways at both women and men. Women were somewhat less likely to report hearing anti-Irish comments, possibly reflecting the masculinity of 'banter' interactions in the workplace. Jerry Palmer (1994) suggests that important gender differences in the joking behaviour of women and men can be identified. Women are less orientated towards jokes and isolated pieces of humour and do not engage in the most aggressive and competitive forms of humour. 'Irish jokes' fall into both these masculine categories of joking interaction. Thus, fewer Irish women than men in the survey saw anti-Irish jokes as 'simply a bit of fun', though they did not necessarily feel able to challenge the perpetrators. Men's apparent acceptance of 'joking' may also mask the threat to their self-esteem posed by this form of harassment.

Irish responses to racist treatment have contributed to its invisibility. One common coping strategy has been to 'keep a low profile' and to downplay an Irish identity. Nearly one-fifth of the respondents in the CRE survey (Hickman and Walter 1997) said that they had done this at some time. Women respondents were much more likely than men to admit to hiding their accents. Linguistic analyses often comment on women's greater use of prestige forms of English, which would highlight Irish dialects unfavourably amongst women (Montgomery 1986). Moreover, women's voices are more often heard in juxtaposition with English accents. Mothers' roles as household intermediaries place them in direct contact with employees of state institutions involving education, health and housing (Lennon, McAdam and O'Brien 1988). Women's workplaces are less segregated than those of many Irish men, and more strongly clustered in service and professional areas where Received Pronunciation is common (Coates and Cameron 1988). Women have also been socialized to be particularly sensitive about the presentation of self, and speech is a key area of social evaluation in Britain (Osmond 1988). Silence ensures that attention is not drawn to one of the prime identifiers of Irishness; accent. Consequently, escalation of harassment is less likely.

Migrants who arrived in the 'second wave' of postwar labour migration were more likely than later arrivals to police themselves in this way (Kells 1995). Yvonne Hayes, a second-generation Irish woman, described life in London in the 1970s for her parents:

> For Irish people, there's nothing that distinguishes them from being English, as long as they keep their mouths shut. And if you are trying to bring up a family and build a home, you just try and fit in with the establishment and don't put yourself out on a limb too much, so you don't get into trouble.
>
> (Lennon, McAdam and O'Brien 1988: 219)

The most extreme response is for Irish people to 'pass' as English by changing their speech patterns, sometimes going as far as taking elocution lessons to make a permanent change. In some cases, this mimicry takes considerable effort and is only partially successful. Philip Ullah (1985) interviewed second-generation children in Birmingham and London in 1980–81, many of whom described people in their parents' generation who adopted an English identity to avoid the stigma attached to being Irish. Over the last twenty-five years, fear of a backlash after IRA bombing incidents in Britain has reinforced the decision to 'keep their heads down'. However, new migrants are less likely to respond by attempting to 'pass'. None of Mary Kells's (1995) sample of young middle class Irish women in London saw this as an option.

An important corollary of these experiences is the ambivalence expressed by Irish people in Britain about open recognition of their distinctive ethnic identity. In the CRE survey (Hickman and Walter 1997), respondents were asked whether they thought the Irish should be recognized as an ethnic group in Britain. A majority (59 per cent) supported the idea, giving four main reasons. These included the similarity they perceived between their own position and that of other, recognized groups – black and Asian – and the entitlement they felt to equivalent benefits. There was also a widely held view that the contribution of Irish labour migrants to the British economy was overlooked and the extent of Irish cultural difference unrecognized. In contrast, a further 13 per cent expressed ambivalent feelings, unwilling to be labelled in ways that had stigmatized 'visible' minorities. Finally, 28 per cent preferred official invisibility, believing that there was, or should be, no difference between the British and the Irish.

Conclusions

The concept of diaspora relocates migrants within a broader space/time framework. It therefore helps to account for the intrusion of the past into the present that underlies ongoing anti-Irish hostility in Britain, experienced by many Irish people as an everyday reality. Whereas a migration focus attempts to contain movement within national boundaries and limited time periods, diaspora acknowledges global processes. These 'constitutive outsides' are ignored or denied by indigenous nationalisms, vividly illustrated by the aphorism: 'The Irish can never forget their history, the English can never remember it.'

By comparing different experiences by destination I have begun to highlight the specificity of the racialization of the Irish in Britain. The invisibility of the Irish cannot be taken for granted and needs to be explained. It is clearly profoundly different from the perceived location of the Irish in the United States. In the CRE survey (Hickman and Walter 1997), 78 per cent of respondents believed that Irish people are viewed positively in America compared with only 31 per cent who believed this about Britain. There is a clear awareness of this difference in the Irish community in Britain through the myriad of family connections between the two destinations.

Over one-fifth of the respondents to the CRE survey stated firmly that the British view the Irish negatively. A higher proportion (47 per cent) had ambivalent feelings. This illustrates the complexity of relationships between the inhabitants of this 'diaspora space'. For example, the acceptance by many Irish people of these attitudes as 'normal', because they are widespread, longstanding and unacknowledged by the majority society, needs to be taken into account. Many respondents replied 'just the usual' when asked whether they had experienced attitudes towards Irish people that they had found objectionable. A large proportion, nearly four-fifths, could give instances of anti-Irish comments that had been directed at them or they had encountered in the media. Despite assertions by social scientists that the trajectory of Irish settlement in Britain is one of movement towards assimilation (for example, Rose 1969; O'Tuathaigh 1985; Ryan 1990), the persistence of these attitudes undermines such confidence. There is already evidence that children of Irish-born parents retain significant elements of difference, although fuller investigation is needed (Ullah 1985, 1990; Lennon, McAdam and O'Brien 1988; Hickman 1990; Harding and Balarajan 1996; Hickman and Walter 1997).

Focus on women's experiences draws out their part in managing the tension between placement and displacement. They have played key roles in maintaining and transforming the 'roots' of the community through paid and unpaid work. At the same time, they have been involved in structuring the 'routes' or patterns of movement between different parts of the diaspora. This combination of the local and the global defines diasporic identities and challenges the discourse of minorities, which privileges national belonging and exclusion. Gender is central to these identities but its widespread absence leaves the discourse of diaspora unmarked and therefore implicitly masculine.

References

Akenson, D. (1996) *The Irish Diaspora: a Primer*, Toronto: P.D. Meany.

Brah, A. (1996) *Cartographies of Diaspora*, London: Routledge.

Chaliand, G. and Rageau, J. P. (1995) *The Penguin Atlas of Diasporas*, London: Penguin.

Clifford, J. (1994) 'Diasporas', *Cultural Anthropology* 9: 302–38.

Coates, J. and Cameron, D. (1988) *Women in their Speech Communities*, Harlow: Longman.

Cohen, R. (1997) *Global Diasporas*, London: UCL Press.

Corcoran, M. (1993) *Irish Illegals: Transients Between Two Societies*, Westport, CT: Greenwood Press.

CSO [Central Statistics Office] (1994) *Annual Population and Migration Estimates 1987–1994*, Dublin: Government of Ireland.

Curtis, L.P. (1971) *Apes and Angels: the Irishman in Victorian Caricature*, Washington DC: Smithsonian Institution Press.

Diner, H. (1983) *Erin's Daughters in America*, Baltimore: Johns Hopkins University Press.

Dublin, T. (1979) *Women at Work: the Transformation of Work and Community in Lowell, Massachusetts, 1826–1860*, New York: Columbia University Press.

Dudden, F. (1983) *Serving Women: Household Service in Nineteenth-century America*, Middleton, CT: Wesleyan University Press.

Gilroy, P. (1987) *There Ain't No Black in the Union Jack,* London: Hutchinson.

Gilroy, P. (1993) *The Black Atlantic: Modernity and Double Consciousness*, London: Verso.

Greeley, A. (1973) *That Most Distressful Nation*, Chicago: Quadrangle.

Guardian (1997) 'Blair says sorry for Britain's "failure" in the Irish famine', 2 June.

Harding, S. and Balarajan, R. (1996) 'Patterns of mortality in second generation Irish living in England and Wales: longitudinal study', *British Medical Journal* 312: 1389–92.

Harper's New Monthly Magazine (1856) 'Valentines delivered in our street', 14 February.

Hickman, M. (1990) 'A study of the incorporation of the Irish in Britain with special reference to Catholic education: involving a comparison of the attitudes of pupils and teachers in selected secondary schools in London and Liverpool', PhD thesis, Institute of Education, University of London.

Hickman, M. (1995) *Religion, Class and Identity: the State, the Catholic Church and the Education of the Irish in Britain*, Aldershot: Avebury.

Hickman, M. and Walter, B. (1995) 'Deconstructing whiteness: Irish women in Britain', *Feminist Review* 50: 5–19.

Hickman, M. and Walter, B. (1997) *Discrimination and the Irish community in Britain*, London: Commission for Racial Equality.

Irish Post (1997) 'Concern at census reply', 22 March.

Jackson, P. (1987) 'The idea of "race" and the geography of racism', in P. Jackson (ed.) *Race and Racism*, London: Allen and Unwin, pp. 3–21.

Kells, M. (1995) '"I'm myself and nobody else": gender and ethnicity among young middle class Irish women in London', in P. O'Sullivan (ed.) *Irish Women and Irish Migration, Volume 4: the Irish World Wide*, Leicester: Leicester University Press, pp. 201–34.

Kelly, K. and Nic Giolla Choille, T. (1995) 'Listening and learning: experiences in an emigrant advice agency', in P. O'Sullivan (ed.) *Irish Women and Irish Migration, Volume 4: the Irish World Wide*, Leicester: Leicester University Press, pp. 168–91.

Kennedy, R. (1973) *The Irish: Emigration, Marriage, and Fertility*, Berkeley: University of California Press.

Lennon, M., McAdam M. and O'Brien, J. (1988) *Across the Water: Irish Women's Lives in Britain*, London: Virago.

Letford, L. and Pooley, C. (1995) 'Geographies of migration and religion: Irish women in mid-nineteenth century Liverpool', in P. O'Sullivan (ed.) *Irish Women and Irish Migration, Volume 4: the Irish World Wide*, Leicester: Leicester University Press, pp. 89–112.

Lloyd, D. (1994) 'Making sense of the dispersal', *Irish Reporter* 13: 3–4.

McDannell, C. (1986) *The Christian Home in Victorian America 1840–1900*, Bloomington: Indiana University Press.

Miller, K. (1985) *Emigrants and Exiles: Ireland and the Irish Exodus to North America*, Oxford: Oxford University Press.

Miller, K., with Doyle, D. and Kelleher, P. (1995) '"For love and liberty": Irish women, migration and domesticity in Ireland and America, 1815–1920', in P. O'Sullivan (ed.) *Irish Women and Irish Migration, Volume 4: the Irish World Wide*, Leicester: Leicester University Press. pp. 41–65.

Montgomery, M. (1986) *An Introduction to Language and Society*, London: Routledge.

Nolan, J. (1989) *Ourselves Alone: Women's Emigration from Ireland 1885–1920*, Lexington: University Press of Kentucky.

O'Carroll, I. (1995) 'Breaking the silence from a distance: Irish women speak on sexual abuse', in P. O'Sullivan (ed.) *Irish Women and Irish Migration, Volume 4: the Irish World Wide*, Leicester: Leicester University Press, pp. 192–200.

O'Grada, C. (1973) 'A note on nineteenth-century Irish emigration statistics', *Population Studies* 29: 143–9.

O'Tuathaigh, M. (1985) 'The Irish in nineteenth-century Britain: problems of integration', in R. Swift and S. Gilley (eds) *The Irish in the Victorian City*, London: Croom Helm, pp. 13–36.

Osmond, J. (1988) *The Divided Kingdom*, London: Constable.

Palmer, J. (1994) *Taking Humour Seriously*, London: Routledge.

Rose, E. (with others) (1969) *Colour and Citizenship: a Report on British Race Relations*, London: Oxford University Press.

Ryan, L. (1990) 'Irish emigration to Britain since World War II', in R. Kearney (ed.) *Migrations: the Irish at Home and Abroad*, Dublin: Wolfhound Press.

Safran, W. (1991) 'Diasporas in modern societies: myths of homeland and return', *Diaspora* 1: 83–99.

Tebbutt, M. (1983) 'The evolution of ethnic stereotypes: an examination of stereotyping, with particular reference to the Irish (and to a lesser extent the Scots) in Manchester during the late nineteenth and early twentieth centuries', MPhil thesis, University of Manchester, Manchester.

Tololyan, K. (1991) 'The nation state and its others: in lieu of a preface', *Diaspora* 1: 3–7.

Ullah, P. (1985) 'Second-generation Irish youth: identity and ethnicity', *New Community* 12: 310–20.

Ullah, P. (1990) 'Rhetoric and ideology in social identification: the case of second generation Irish youth', *Discourse and Society* 1: 167–88.

US Bureau of the Census (1990) *1990 Census of Population, Supplementary Reports, Detailed Ancestry Groups for States*, Washington DC: United States Bureau of the Census.

Walter, B. (1984) 'Tradition and ethnic interaction: second wave Irish settlement in Luton and Bolton', in C. Clarke, D. Ley and C. Peach (eds) *Geography and Ethnic Pluralism*, London: George Allen and Unwin, pp. 258–83.

Walter, B. (1995) 'Irishness, gender and place', *Environment and Planning D: Society and Space* 13: 35–50.

Walter, B. (1997) 'Gender, "race" and diaspora: racialised identities of emigrant Irish women', in J. Jones, H. Nast and S. Roberts (eds) *Thresholds in Feminist Geography: Difference, Methodology, Representation*, Lanham, MD: Rowman and Littlefield, pp. 339–60.

Walter, B. (1998) 'Challenging the black/white binary: the need for an Irish category in the 2001 Census', *Patterns of Prejudice* 32.2: 73–86.

Index